Novel Approaches for Bioremediation of Organic Pollution

Edited by

Raffi Fass
Yehuda Flashner and
Shaul Reuveny
Israel Institute for Biological Research
Ness-Ziona, Israel

Kluwer Academic / Plenum Publishers
New York, Boston, Dordrecht, London, Moscow

Proceedings of the 42nd OHOLO Conference on Novel Approaches for Bioremediation of Organic Pollution, held May 3 – 7, 1998, in Eilat, Israel

ISBN 0-306-46102-1

233 Spring Street, New York, N.Y. 10013

10 9 8 7 6 5 4 3 2 1

A C.I.P. record for this book is available from the Library of Congress.

Printed in the United States of America

PREFACE

Pollution of the environment with organic compounds, mostly due to industrial activities, has become a major environmental issue. The situation is getting worse due to the continued release of large amounts of chemicals and the broadening of the spectrum of xenobiotics that to which the biosphere is exposed to.

The field of biodegradation and bioremediation has experienced a dynamic and intense period of development during the last few years, encompassing all its different disciplines: Isolation and characterization of new microorganisms with novel catabolic activities, understanding the specific enzymatic mechanisms underlying biodegradation, rational design of genetically engineered degrading microorganisms, and studying the interaction of the degrading organism with biotic and abiotic components within the environment to allow development of efficient bioremediation processes.

This book, devoted to the proceedings of the 42nd OHOLO Conference on "Novel Approaches for Bioremediation of Organic Pollution" which was held in Eilat, Israel, in May 1998, obviously does not cover the full range of bioremediation research. However, an attempt has been made to address the major issues in bioremediation of organic pollution. The four sessions of the conference covered a wide spectrum of topics, from the lab research— characterization of enzymes and pathways, genetic design of degradative microorganisms, to the field application issues—bioaccessibility, bioavailability, process development and field studies.

The contributions aimed at understanding of enzymes and pathways are those on the transformation of halogenated and sulfur containing compounds, the elucidation of different pathways for degradation of nitroaromatic compounds, and the necessity of considering streochemistry in biodegradation reactions.

The genetic approaches for the design of engineered degradative microorganisms are exemplified by presentations on the generation and use of biocatalysis database on the World Wide Web, the manipulation of the catalytic performance of dehalogenases and esterases either by the well established strategy of rational site-directed mutagenesis or by the recently developed "irrational" approach of directed evolution. Other genetic approaches presented are using controlled *gfp* expression as a reporter for sensing bioavailability and the construction of a contained engineered degrading strain by the application of regulated suicide systems.

Several strategies for *in situ* bioremediation application are discussed, including specialized techniques of bioaugmentation and bioavailability enhancement using biosurfactant-producing strains, a combined active/passive biotreatment strategy, monitoring biological degradation of organic pollution by determination of $^{13}C/^{12}C$ isotopic fractionation, and use of engineered *lux* gene fusion based biosensor strains for both bioremediation and process monitoring. Recent developments in *ex situ* bioremediation processes, like a new membrane reactor concept, are presented as well.

Although these proceedings can not cover all the recent advances in the field of biodegradation and bioremediation of organic pollution, we believe that the various strategies presented here reflect the major trends of research and development for the coming years.

We would like to take this opportunity to thank again all the contributors to this book, and to the members of the scientific advisory committee: Kenneth Timmis, Gary Sayler, Eugene Rosenberg, Yigal Henis, Gabi Amitai, Baruch Velan, and Avigdor Shafferman.

ACKNOWLEDGMENTS

The Organizing Committee gratefully acknolwedges the generous financial contribution of the following organizations:

Ministry of the Environment, Jerusalem, Israel
Ministry of Science, Jerusalem, Israel
Ministry of Tourism, Jerusalem, Israel
Yad Avi Ha-Yishuv, Jerusalem, Israel

CONTENTS

CHEMICAL STUDIES OF CARBON TETRACHLORIDE TRANSFORMATION BY *Pseudomonas stutzeri* STRAIN KC

Thomas A. Lewis* and Ronald L. Crawford

Institute for Molecular and Agricultural Genetic Engineering
University of Idaho
Moscow, Idaho 83844-1052

1. INTRODUCTION

Carbon tetrachloride (CT) is a toxic, carcinogenic compound which was widely used in the past as a solvent, degreaser, fumigant, and fire extinguisher. As a result of earlier disposal practices, CT is now found in groundwater. Probably the most extensive example of groundwater contamination can be found at the Hanford Nuclear Reservation operated by the U.S. Department of Energy in south-central Washington state (Illman, 1993). An estimated 1000 metric tons of CT was disposed of into unlined cribs over a 20-year period. CT has percolated through the vadose zone, contaminating groundwater over 250 feet below the surface. The plume of contamination extends over approximately 7 square miles and is migrating toward the Columbia River, only a few miles away.

CT is inherently stable in aqueous solution with a slow rate of spontaneous hydrolysis (Jeffers *et al.,* 1996). Biological destruction of this chemical offers the potential of a low-cost technology to eliminate its risk if a suitably effective transformation activity can be identified and promoted *in situ*. Since the carbon atom of CT is at its highest formal oxidation state, the molecule offers no potential nutritional advantage to microbes as a source of energy. It can serve as an oxidant of suitable biological redox carriers, and several studies have demonstrated reductive dehalogenation using biochemical cofactors (Gantzer and Wackett, 1991; Krone *et al.,* 1989a,b; Wackett *et al.,* 1992). These transformations have primarily involved hydrogenolysis, yielding the net replacement of chlorine atoms with hydrogen to form chloroform (CF) and the lesser chlorinated chloromethanes.

* To whom correspondence should be addressed. Telephone 208-885-6580. Fax 208-885-5741. E-mail: tlewis @uidaho.edu

Novel Approaches for Bioremediation of Organic Pollution, edited by Fass *et al.* Kluwer Academic / Plenum Publishers, New York, 1999.

These methods have the potential for creating new hazards with longer half-lives under the reducing conditions necessary for their formation from CT. Bacterial cultures have been screened for their ability to transform CT co-metabolically, that is, while using other compounds supplied as sources of carbon and energy and with no known selective advantage for carrying out CT transformation. These examples of cometabolism have involved enrichments or pure cultures grown under anaerobic conditions to promote reduction of the CT without the competition assumed to result from molecular oxygen (Bouwer and McCarty, 1983; Criddle et al, 1990a,b; Egli *et al.,* 1988; Mikesell and Boyd, 1990). Some of these studies have identified pure cultures of microorganisms capable of carrying out a net hydrolysis of CT to give CO_2 as a major product without the accumulation of CF or other chloromethanes (Criddle *et al.,* 1990a; Egli et al., 1988). One such culture which we found particularly attractive for our study was a strain of the denitrifying bacterium *Pseudomonas stutzeri* designated KC (Criddle *et al.,* 1990a). As a facultative anaerobe isolated from an aquifer, this organism had characteristics which seemed optimal for a site such as the Hanford Reservation.

In order to assess whether the CT transformation activity of strain KC is suitable for the design of remediation technology, and to allow the rational design of such a technology, a detailed knowledge of the process is necessary. Previous physiological characterizations had shown that the activity was present only in cultures grown under conditions of iron deprivation (Criddle *et al.,* 1990a; Lewis and Crawford, 1993; Tatara *et al.*, 1993) that there was a requirement for trace levels of copper (Tatara *et al.,* 1993), that growing cells were required (Dybas *et al.*, 1995), and that the process was inhibited by oxygen but was still observed under aerobic conditions (Lewis and Crawford, 1993). More recently it has been shown that the factor or factors elaborated by strain KC that allow CT transformation were extracellular and of low molecular weight (<500 daltons; Dybas *et al.*, 1995), and were active with several types of microbial cultures, but not as extracellular preparations alone (Tatara *et al.,* 1995). Chemical characterization of the activity has included gas chromatographic analyses of CT transformation products, which have shown CF to be a minor product (Lewis and Crawford, 1993), and tracer studies using ^{14}CT, which identified CO_2 (approximately 55%) and an uncharacterized nonvolatile fraction (approx. 40%) as the major products (Criddle *et al.,* 1990a).

Two pathways to explain the net hydrolysis of CT have been proposed, both of them including an initial reduction step. The first pathway proposed to explain CO_2 production from CT involved a two-electron reduction to give dichlorocarbene, which would hydrolyze to give formate and/or CO (Figure 1; Criddle *et al.,* 1990a). CO_2 could then be generated from these compounds by oxidation with an appropriate dehydrogenase, provided that one was present and had sufficiently low K_m values to efficiently oxidize the low concentrations of substrate formed from the micromolar quantities of CT used in these studies.

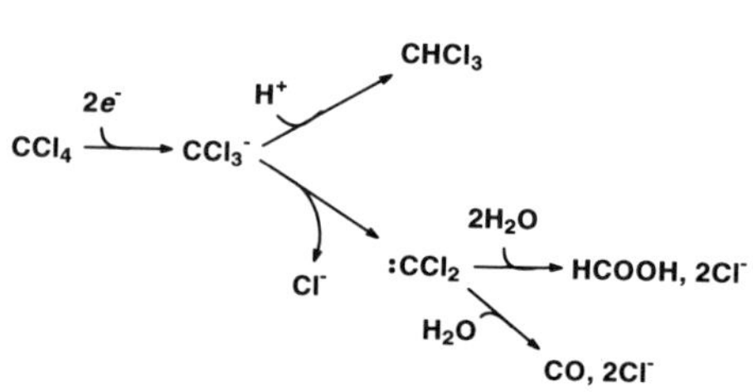

Figure 1. Two-electron reduction of CT.

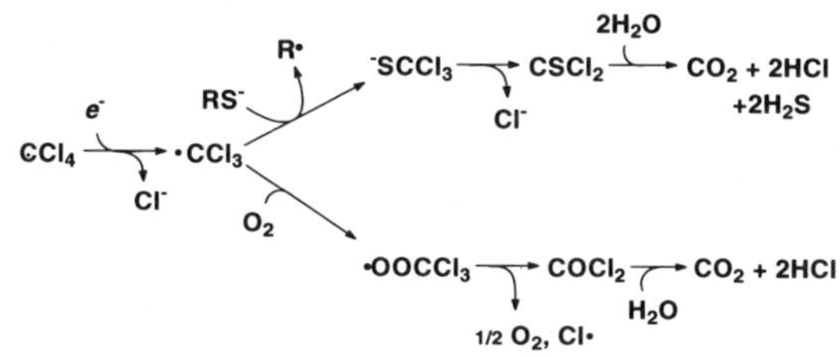

Figure 2. One electron reduction/radical substitution pathway of CT transformation.

The second pathway involves a one-electron reduction to give trichloromethyl radical (Figure 2). Trichloromethyl radical is known to react with molecular oxygen at near the diffusion-limited rate (Slater, 1980) and to give phosgene (Asmus *et al.,* 1985), a reactive electrophile which hydrolyzes to CO_2. Another fate of trichloromethyl radical has been proposed to explain the formation of carbon disulfide and CO_2 from CT in mineral systems under anaerobic conditions (Kriegman-King and Reinhard, 1992). This reaction involves combination of the radical with a reactive sulfur species to give thiophosgene, which can undergo further substitution with sulfur or hydrolyze to give CO_2.

Both of these pathways include reactive intermediates which would not accumulate to measurable quantities in our experiments due to their brief half-lives in aqueous media. We have tried to resolve the relative contribution of each of these pathways to the spectrum of products seen in cultures of strain KC using trapping agents designed to give stable products diagnostic of the respective intermediate, and quantitative product analyses.

2. MATERIALS AND METHODS

2.1. Culture Conditions

Cultures were grown on Medium D (Criddle *et al.,* 1990a), or a medium based on it, but which obviated the need for precipitation of metals. This medium, designated DRM, consisted of 6g K_2HPO_4, 2g sodium acetate, 1g NH_4Cl, and 0.5g $NaNO_3$ per liter, adjusted to pH 7.7–7.9 and autoclaved. $MgSO_4$ (1 mM), $Ca(NO_3)_2$ (0.1 mM), and $CuCl_2$ (5 nM) were added after autoclaving from autoclaved stock solutions. Cultures used to study CT transformation were inoculated from aerobically grown overnight cultures in the respective medium. CT transformation was conducted under anaerobic conditions by transferring 5-ml cultures in 10-ml serum vials to an anaerobic chamber (Forma Scientific, atmosphere: 90%N_2, 10%H_2) before adding CT and sealing with sterile, Teflon-faced butyl rubber stoppers (the West Co., Phoenixville, PA). CT was added from a 10 mg/ml methanolic stock solution. Cultures were incubated at 25°C for 4 days in an inverted position. When O_2 was included it was provided by flushing the headspace with breathing grade O_2 which was passed through a sterile, 0.45-μ (pore size) filter before closing with a stopper. CT was then added by quickly removing and replacing the stopper.

2.2. Analytical Methods

CT and CF were quantitated as described by Lewis and Crawford (1993). Carbon disulfide was quantitated using a Hewlett Packard 5890 gas chromatograph and G1909 purge and trap concentrator with a 5-ml sample vessel, and interfaced with a 5989 mass spectrometer operated in single ion mode (m/z = 76 a.m.u., Hewlett Packard, Avondale, PA). An 11-min purge was used to transfer organic analytes to a trap composed of Tenax, silica gel, and activated charcoal (Supelco, Inc., Bellefonte, PA). Helium was used as the carrier gas. The column used was a 25 m × 0.32 mm PoraPLOT Q (Chrompack, Middelburg, The Netherlands) held at 40°C for 4 min, and then increased to 220°C at 30°/min and held for 5 min. Calibration was performed using a set of external standards spanning the range of concentrations encountered in culture samples.

Fractionation of radioactivity was performed by purging acidified cultures through a trapping train using N_2 at a flow rate of 100 ml/min. The purged vapors passed through Teflon and stainless steel tubing and into three vials of 10-ml Ready Organic scintillation cocktail (Beckman Instruments, Fullerton, CA) sealed with Viton gaskets, and one vial with 10 ml of Carbosorb CO_2-trapping solution (Packard Instruments, Meriden, CT) and a butyl rubber gasket. Vials of Ready Organic were counted directly and 1 ml of the Carbosorb was mixed with 10 ml EcoLite scintillation cocktail (Research Products International, Mount Prospect, IL) before counting in a Packard 2100TR liquid scintillation analyzer. Radioactivity trapped in Ready Organic was designated as "volatile" products; that trapped in Carbosorb as "CO_2"; and that remaining in the medium as "nonvolatile" products. HPLC separation of nonvolatile products was performed using a Hewlett Packard 1090 HPLC system with a diode array detector monitoring at 210 nm. Conditions for separation of products from trapping experiments included a 4.6 × 250 mm Microsorb MV C_{18} column (Rainin Instruments Inc, Emeryville, CA), and acetonitrile (A) and 15 mM H_3PO_4 0.1% triethylamine (B) as eluting solvents at a flow rate of 1 ml/min. The solvent gradient used was 99% B, 0–3 min; 50% B, 16–21 min; 30% B, 23–35 min; 99% B, 37 min. [^{14}C]Formate was analyzed using the same HPLC system but with a Rezex ROA column (Phenomenex, Torrance, CA, 7.8 × 300 mm with 50-mm guard column) operated at 40°C with 5 mM H_2SO_4, 0.5 ml/min, as eluent. Radioactivity in HPLC effluent was quantitated using a β-RAM liquid scintillation detector (IN/US Systems, Inc., Tampa, FL) and IN-Flow scintillation cocktail at 3 ml/min. A calibration curve relating peak area to disintegrations per minute was made using [^{14}C]formate.

2.3. Chemicals

CT and CF of Omnisolv grade were obtained from E. Merck and Co. (Cherryhill, NJ); [^{14}C]CT (4.3 mCi/mmol) from DuPont NEN (Boston, MA); [^{14}C]CS_2 (57 mCi/mmol) from Amersham International (Buckinghamshire, UK); 2,2,6,6-tetramethylpiperidinyl-1-oxyl (TEMPO), hydroxyurea, and sodium [^{14}C]formate from Sigma (St. Louis, MO); and [^{13}C]CT from Cambridge Isotope Laboratories (Andover, MA).

3. CHEMICAL STUDIES OF REACTIVE INTERMEDIATES OF CT TRANSFORMATION BY STRAIN KC

Previous studies have shown that products indicative of thiophosgene and phosgene could be identified using nucleophilic trapping agents such as N,N′-dimethylethylenediamine (DMED; Lewis and Crawford, 1995). Quantitative analyses of this type of experiment and one including the dichlorocarbene-trapping agent 2,3-dimethylbutene (DMB) are shown in Figure 3. DMED substantially diminished production of CO_2 from CT, whereas DMB had no significant effect on this product fraction. The nonvolatile frac-

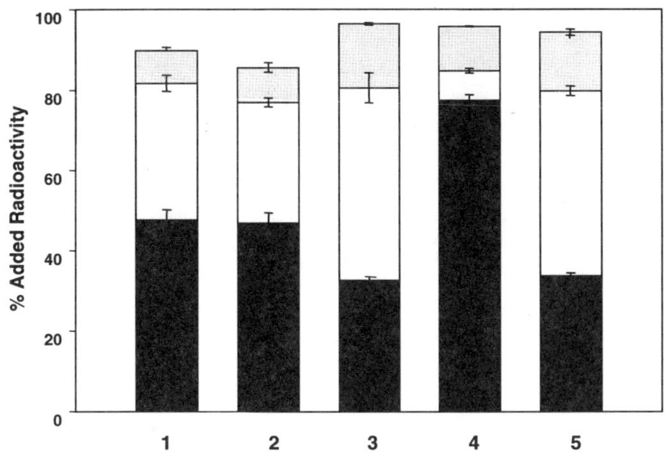

Figure 3. Product profiles of ^{14}CT transformation by strain KC in the presence of trapping agents. Bottom (black) bars represent non-volatile products, middle (white) bars represent CO_2, top (gray) bars represent volatile material trapped in organic scintillation cocktail. 1, strain KC control (no trapping agents); 2, 5 mM DMB; 3, 5 mM formate; 4, 5 mM DMED; 5, strain KC control. Experiments 1 and 2 were done in parallel with Medium D; experiments 3, 4, and 5 were done in parallel with DRM. Results are means of triplicate experiments with error bars representing ± one standard deviation.

tion from experiments including DMED were analyzed to determine if the products seen were those expected of thiophosgene and phosgene trapping. Reversed phase HPLC and liquid scintillation detection showed that the major products eluted were the sulfur- and oxygen-substituted imidazole products seen previously, 2-thioxo-1,3-dimethylimidazole (TDMI) and 1,3-dimethylimidazolidinone (DMIO), respectively. Those products accounted for approximately 66% of the eluted radioactivity (>90% recovery of injected radioactivity). The remainder of the radioactivity eluted in an unknown peak or rather broadly over the radiochromatogram in an anaerobic experiment (Figure 4A). When oxygen was included in the headspace, a profound increase in the oxygen-substituted product was seen, along with a complementary decrease in the amount of the sulfur-substituted product and the unknown peak (Figure 4B).

Though sulfur and oxygen-substituted products clearly could account for the majority of CT transformation products in the trapping experiments mentioned above, it was possible that carbon disulfide was included in the pathway leading to these products. CS_2 can hydrolyze to give CO_2 and can react directly with DMED to give one of the observed products (TDMI). In addition, an intermediate of CS_2 hydrolysis, carbonyl sulfide, can react with DMED to give both of the observed products. CS_2 could be detected by purge and trap GC/MS in cultures of strain KC which had transformed CT. Using ^{13}CT as a mass tracer confirmed that this was a product of CT transformation and not of normal metabolism in the medium used (Figure 5).

Quantitation of CS_2 by this method indicated that it accounted for only 1.8 ± 0.6% of the total CT removal. It was possible that CS_2 accumulated to low steady-state levels due to its further metabolism by the culture. To assess the reactivity of CS_2 in this system, $^{14}CS_2$ was added in an amount similar to CT additions used previously (235 nmol/10-ml vial). In these experiments no apparent CS_2 hydrolysis activity was demonstrated by strain KC when compared with sterile controls. The amount of radioactivity found in base traps used to collect CO_2 from cultures or controls was less than 1.5% of the total. The nonvola-

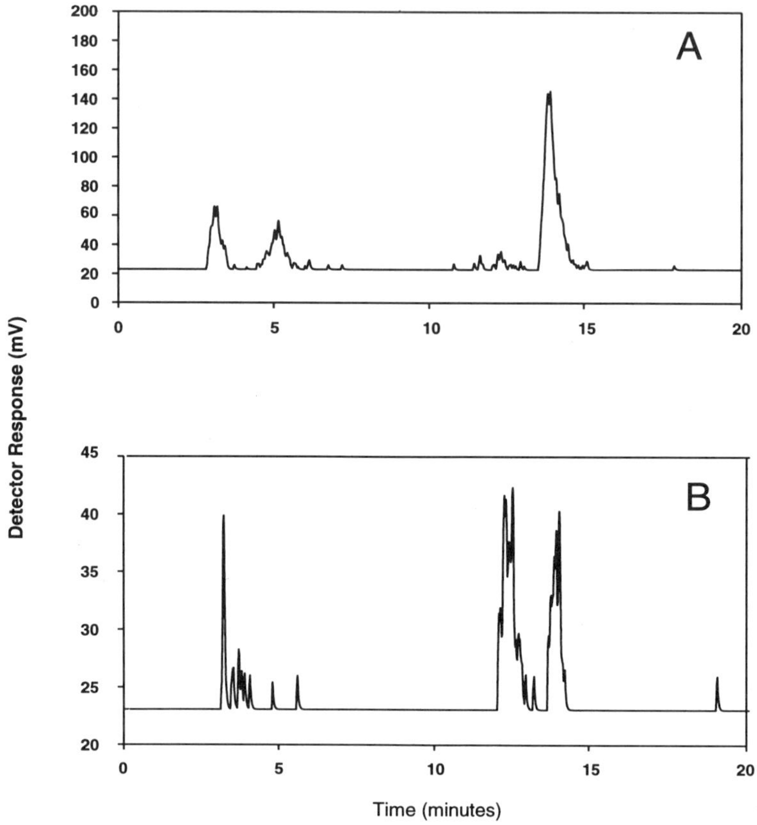

Figure 4. HPLC analysis of DMED-trapped, nonvolatile CT transformation products. A, data from sample of anaerobic culture containing 5mM DMED. B, data from sample of culture incubated with an O_2 headspace and 5mM DMED.

tile products formed in the presence of DMED, TDMI and an unidentified polar fraction, were the same in sterile controls and KC-inoculated cultures, though their relative quantities were reversed (mostly polar material with KC). No DMIO was detected in these experiments.

Other products of CT transformation were examined since their routes of formation and quantities would contribute to a more complete description of the CT-transformation process elicited by strain KC. Chloroform may result from trichloromethyl radical abstracting a hydrogen atom from an organic molecule in the surrounding medium. As mentioned, CF accumulation accounts for a small portion of the total mass balance for CT; however, some phenomena became apparent during our studies which may allow further insight into participating reactive intermediates. We had observed CF accumulations of up to 3% of the CT transformation when a medium buffered with 50 mM N-2-hydroxyethylpiperazine-N′-ethanesulfonic acid (HEPES) was used. This CF production was lower when oxygen was present in the culture vessels (Lewis and Crawford, 1993). Previously, Criddle et al. (1990a) had observed no CF when a phosphate-buffered medium was used. In further experiments with a phosphate-buffered medium, we were able to observe detectable levels of CF which were much lower than with the HEPES-buffered medium. This suggested that the organic buffer had participated in the radical-scavenging

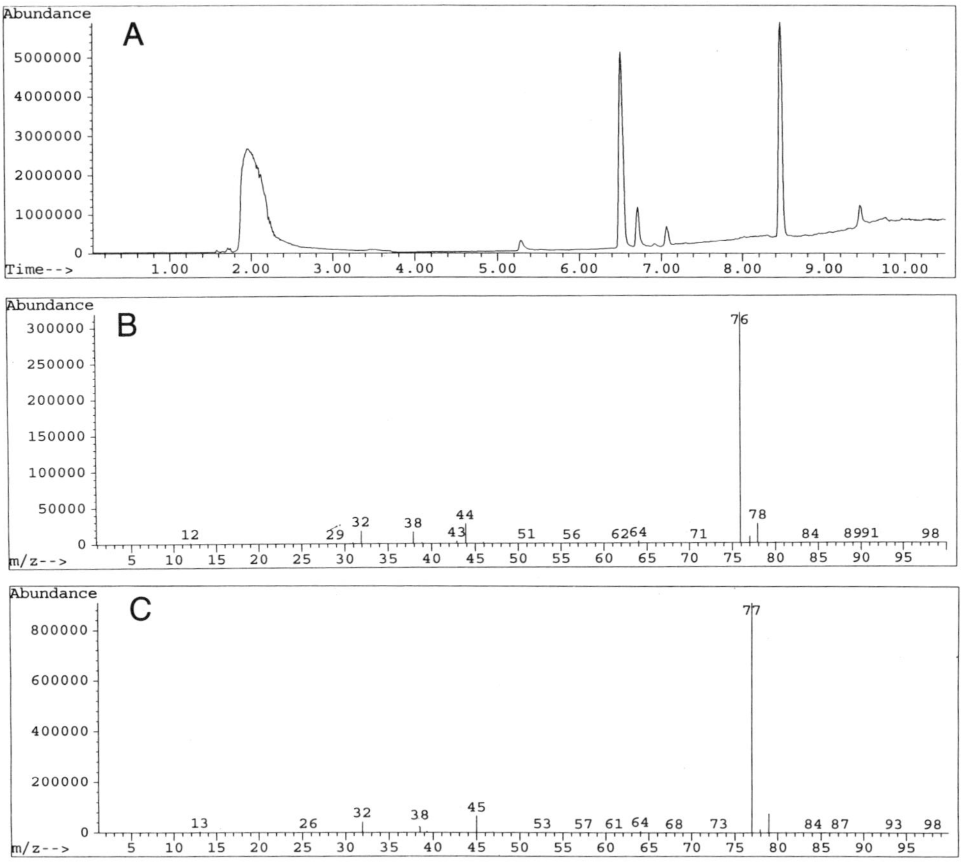

Figure 5. GC/MS identification of carbon disulfide as a product of CT transformation by strain KC. A. GC/MS chromatogram (ions m/z 21–300 a.m.u.) of volatiles trapped by purge and trap concentration from 5 ml of strain KC culture exposed to CT. B. Mass spectrum of CS_2 peak eluting at 7.07 min in above chromatogram. C. Mass spectrum of CS_2 peak from culture grown with ^{13}CT. Peaks eluting earlier than CS_2 include CO_2, 2.0 min; ethanol, 5.27 min; acetone, 6.52 min; and isopropanol, 6.71 min. These peaks were not labeled by ^{13}C. Peaks eluting later than CS_2 include CF, 8.45 min; and CT, 9.42 min. These were present initially.

reaction leading to CF. We conducted another series of experiments to test whether a known radical scavenger, hydroxyurea, and a stable free radical, TEMPO, would influence the accumulation of CF in a manner consistent with their predicted reactivity with trichloromethyl radical. Hydroxyurea is an effective inhibitor of ribonucleotide reductase of some organisms by virtue of its radical-quenching effect (Lammers and Follman, 1983). However, *P. stutzeri* has been shown to possess a corrinoid-type of ribonucleotide reductase (Gleason and Hogenkamp, 1972), and strain KC is resistant to high concentrations of this compound (our unpublished results). Addition of 50 mM hydroxyurea resulted in a ninefold increase in CF accumulation (Table 1). The addition TEMPO resulted in a lower mean CF accumulation in this and other experiments ($P \geq 0.1$).

Formate has been identified as a product of CT transformation by strain KC (Dybas *et al.,* 1995). In our system, when [^{14}C]formate was quantitated from tracer experiments using organic acid HPLC and liquid scintillation detection, radioactivity eluting in the re-

Table 1. Effect of radical scavengers on CF formation

Addition	Mole % CF [a]
None	0.33 ± 0.03
50 mM hydroxyurea	2.73 ± 0.27
1.3 mM TEMPO	0.20 ± 0.05

[a] means ± standard deviation of triplicate cultures analyzed in duplicate

gion corresponding to formate accounted for approximately 4.5% ± 0.6% of the total CT addition. To determine if this amount was the net result of formate production and formate oxidation, 5 mM formate was added in an experiment to examine the relative contribution of formate and formate dehydrogenase to $^{14}CO_2$ production. This addition of formate at more than 300-fold molar excess to the CT addition had no significant effect on CO_2 production (Figure 3, 48 ± 3.7% with formate versus 46 ± 1.2% without formate, ≥96% recovery of ^{14}C), as had been noted previously (Dybas *et al.*, 1995).

4. DEVELOPMENT OF A CELL-FREE ASSAY OF CT TRANSFORMATION ACTIVITY

Studies of cometabolic transformation of xenobiotic chemicals using bacterial cultures are complicated by the fact that they represent the process being studied as well as metabolic functions which support that activity. Consequently, in studies designed to determine the type(s) of factors elicited by strain KC, the effects of inhibitors on the process under study cannot be resolved from their possible effects on cell metabolism. In addition, determining the stoichiometry of electrons consumed per CT molecule transformed would be very difficult with growing cultures. For these purposes, a highly defined cell-free system is necessary. Tatara and co-workers (Dybas *et al.*, 1995; Tatara *et al.*, 1995) have described a system whereby CT transformation activity could be observed in strain KC culture supernatants. Those experiments used cells of several types of bacterial cultures and a yeast, presumably as a source of reductant; the supernatant alone had little or no activity. They also determined that this activity would pass through a low-molecular-weight-cutoff membrane. We have studied this activity first by testing chemical reductants rather than whole cells. In these experiments, a crude cell-free supernatant was prepared by centrifugation and filtration through a 0.2-μ (pore size) membrane. Without added reducing agent, no CT removal was seen; however, with either Ti^{III} citrate or dithiothreitol significant CT removal was seen in overnight incubations. As a test of whether the addition of these potent reductants gave an artifactual CT removal not related to the activity seen with whole cells, these reactions were conducted with supernatants prepared from cultures known to be inactive for CT transformation, and with substrates toward which strain KC has shown no activity. Some of these results are shown in Figure 6.

Only iron-limited KC culture supernatant showed CT removal in this system. In addition, no activity was seen toward CF or fluorotrichloromethane ($CFCl_3$), consistent with the activity described for whole cultures (Criddle, 1990b; Lewis and Crawford, 1993) and not with that of reduced cofactors such as vitamin B_{12} (Gantzer and Wackett, 1991; Krone et al, 1989b). The addition of iron or iron-supplemented culture supernatant did not inhibit the activity of the iron-limited culture supernatant, indicating that this system is not sensitive to any "exacerbating factors" observed in studies with whole cells (Dybas et al., 1995). This activity is not sensitive to 2 mM sodium azide, which inhibits the growth of

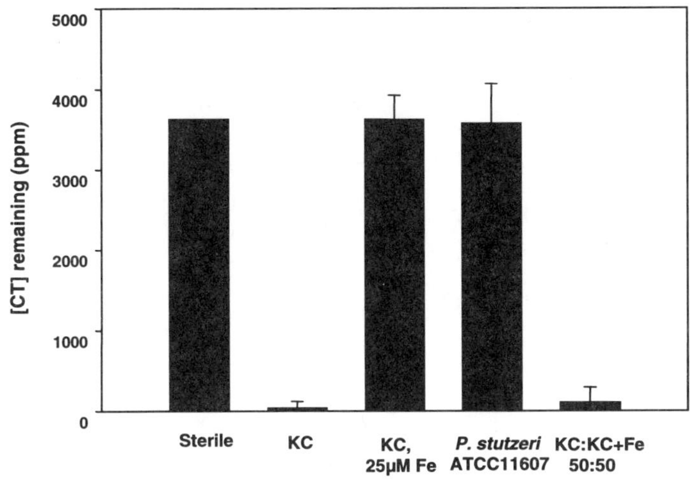

Figure 6. CT removal by cell-free supernatant preparations with Ti^{III} citrate.

strain KC. It is rapidly inactivated by mercuric chloride and by 15 μM cobaltous sulfate, the latter having been demonstrated to inhibit CT transformation in cultures (Criddle, 1990b; Lewis and Crawford, 1993). These would not be expected to interfere with reduction by Ti^{III}; however, they appear to cause inactivation as rapidly as hydrogen peroxide, which should rapidly oxidize this reductant (not shown).

5. DISCUSSION

Our investigations of the reactive intermediates involved in CT transformation by strain KC clearly show that a substitutive pathway involving sulfur and oxygen is the major route of transformation. These results suggest that the toxicity hazard of CT could be destroyed if such a transformation was successfully implemented, since phosgene and thiophosgene are labile in aqueous environments. How this substitution pattern comes about has not been directly demonstrated; however, a number of indirect observations are consistent with the involvement of trichloromethyl radical. First, the inclusion of molecular oxygen profoundly influences the substitution pattern consistent with its combination with trichloromethyl radical to form phosgene. It is interesting to note that even molecular oxygen with its extremely rapid rate of reaction with trichloromethyl radical did not completely eliminate sulfur substitution. Second, the amount of chloroform produced is related to the amount of particular organics present, consistent with its formation by hydrogen abstraction. The known radical scavenger hydroxyurea gave a substantial increase in the amount of chloroform produced. Molecular oxygen has been seen to reduce the amount of chloroform produced, also consistent with this route of chloroform production and its ability to divert CT carbon atoms into oxygen-substituted products. Chloroform accumulations in all experiments with cultures of strain KC have been less than 4% of the total CT transformed, indicating that other reactions effectively outcompete hydrogen abstraction.

The production of formate from CT is evidence that a net two-electron reduction occurs to some extent. Our data indicate that this is a relatively small fraction of the overall transformation, as this product accounts for less than 5%, and that it is not oxidized to CO_2 to any significant extent. It was also not possible to influence the product spectrum by using the

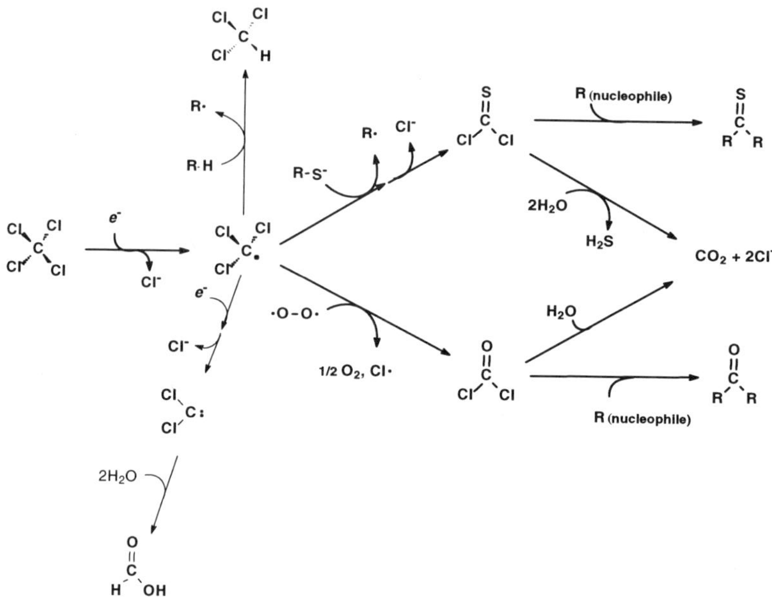

Figure 7. Integrated scheme for CT transformation by strain KC. Bold arrows represent the major routes of transformation shown to account for most of the CT transformation in trapping experiments.

dichlorocarbene-trapping agent dimethylbutene, which has been shown to be effective in studies with rat liver microsomes (Pohl and George, 1983). It is also difficult to envision a pathway by which dichlorocarbene production results in phosgene or thiophosgene formation. The known reactivity of this type of molecule would not lead to this type of product in combination with the reactive sulfur species predominating in biological systems (i.e., thiols, thioloate ions; Kirmse, 1971). A scheme integrating the above data is shown in Figure 7.

Studies with cell-free supernatant have shown a clear dependence on reductant, which is also consistent with a radical substitution pathway. The fact that certain metal ions rapidly inactivate CT transformation points to the involvement of a metal center, which is subject to displacement by inactive species. The sulfur species, which provides the sulfur atom seen in some of the identified products, is unknown and may be directly involved in the transformation or may react fortuitously after initial reduction of CT by the active factor. We hope to use the assay system to purify the active agents from culture supernatants and identify the molecules directly involved in the CT transformation activity and thus further explain some of these phenomena.

ACKNOWLEDGMENTS

This work was supported by grant DE-FG03-96-ER-62273 from the U.S. Department of Energy NABIR (Natural and Attenuated Bioremediation Research) Program.

REFERENCES

Asmus, K-D., Bahnemann, K. D., Krischer, K., Lal, M., Mönig, J. (1985). One-electron induced degradation of halogenated methanes and ethanes in oxygenated and anoxic aqueous solutions. *Life. Chem. Reports* **3**:1–15.

Bouwer, E. J., McCarty, P. L. (1983). Transformations of halogenated organic compounds under denitrification conditions. *Appl. Environ. Microbiol.* **45(4):**1295–1299.

Criddle, C. S., DeWitt, J. T., McCarty, P. L. (1990a). Reductive dehalogenation of carbon tetrachloride by *Escherichia coli* K-12. *Appl. Environ. Microbiol.* **56(11):**3247–3254.

Criddle, C. S., DeWitt, J. T., Grbic-Galic, D., and McCarty, P. L. (1990b). Transformation of carbon tetrachloride by *Pseudomonas* sp. strain KC under denitrification conditions. *Appl. Environ. Microbiol.* **56(11):**3240–3246.

Dybas, M. J., Tatara, G. M., and Criddle, C. S. (1995). Localization and characterization of the carbon tetrachloride transformation activity of *Pseudomonas* sp. strain KC. *Appl. Environ. Microbiol.* **61(2):**758–762.

Egli, C., Tschan, T., Scholtz, R., Cook, A. M., and Leisinger, T. (1988). Transformation of tetrachloromethane to dichloromethane and carbon dioxide by *Acetobacterium woodii*. *Appl. Environ. Microbiol.* **54(11):**2819–2824.

Gantzer, C. J., and Wackett, L. P. (1991). Reductive dechlorination catalyzed by bacterial transition-metal coenzymes. *Environ. Sci. Technol.* **25(4):**715–722.

Gleason, F. K., and Hogenkamp, H. P. (1972). 5'-Deoxyadenosylcobalamin-dependent ribonucleotide reductase: A survey of its distribution. *Biochim. Biophys. Acta* **277:**466–470.

Illman, D. L. (1993). Researchers take up environmental challenge at Hanford. *Chem. Eng. News* June, 9–21.

Jeffers, P. M., Brenner, C., and Wolfe, N. L. (1996). Hydrolysis of carbon tetrachloride. *Environ. Toxicol. Chem.* **15(7):**1064–1065.

Kirmse, W. (1971). Carbene chemistry. 2nd ed. Academic Press, New York.

Kriegman-King, M. R., and Reinhard, M. (1992). Transformation of carbon tetrachloride in the presence of sulfide, biotite, and vermiculite. *Environ. Sci. Technol.* **26(11):**2198–2206.

Krone, U.E., Laufer, K., Thauer, R. K., and Hogenkamp, H. P. C. (1989a). Coenzyme F_{430} as a possible catalyst for the reductive dehalogenation of chlorinated C_1 hydrocarbons in methanogenic bacteria. *Biochemistry* **28**:10061–10065.

Krone, U. E., Thauer, R. K., and Hogenkamp, H. P. C. (1989b). Reductive dehalogenation of chlorinated C_1-hydrocarbons mediated by corrinoids. *Biochemistry* **28**:4908–4914.

Lammers, M., and Follmann, H. (1983). The ribonucleotide reductases—A unique group of metalloenzymes essential for cell proliferation, p. 29–91. In P. H. Connett (ed.), Inorganic elements in biochemistry (Structure and Bonding, 54.) Springer-Verlag, Berlin.

Lewis, T. A., and Crawford, R. L. (1993). Physiological factors affecting carbon tetrachloride dehalogenation by the denitrifying bacterium *Pseudomonas* strain KC. *Appl. Environ. Microbiol.* **59(5):**1635–1641.

Lewis, T. A., and Crawford, R. L. (1995). Transformation of carbon tetrachloride via sulfur and oxygen substitution by *Pseudomonas* sp. strain KC. *J. Bact.* **177(8):**2204–2208.

Mikesell, M. D., and Boyd, S. A. (1990). Dechlorination of chloroform by *Methanosarcina* strains. *Appl. Environ. Microbiol.* **56(4):**1198–1201.

Pohl, L. R., and George, J. W. (1983). Identification of dichloromethyl carbene as a metabolite of carbon tetrachloride. *Biochem. Biophys. Res. Commun.* **117(2):**367–371.

Slater, T. F. (1980). Free radicals as reactive intermediates in tissue injury, p. 575–589. In R. Snyder (ed.), Biological reactive intermediates, chemical mechanisms and biological effects. Plenum Press, New York.

Tatara, G. M., Dybas, M. J., and Criddle, C. S. (1993). Effects of medium and trace metals on kinetics of carbon tetrachloride transformation by *Pseudomonas* sp. strain KC. *Appl. Environ. Microbiol.* **59(7):**2126–2131.

Tatara, G. M., Dybas, M. J., and Criddle, C. S. (1995). Biofactor-mediated transformation of carbon tetrachloride by diverse cell types, p. 69–76. In R. E. Hinchee, A. Leeson, and L. Semprini (eds.), Bioremediation of chlorinated solvents. vol. 3(4), Battelle Press, Columbus, Ohio.

Wackett, L. P., Logan, M. S. P., Blocki, F. A., and Bao-li, C. (1992). A mechanistic perspective on bacterial metabolism of chlorinated methanes. *Biodegradation* **3**:19–36.

2

MICROBIAL TRANSFORMATION OF THE CHIRAL POLLUTANTS MECOPROP AND DICHLORPROP

The Necessity of Considering Stereochemistry

Hans-Peter E. Kohler,* Kathrin Nickel, Monika Bunk,† and Christian Zipper‡

Swiss Federal Institute for Environmental Science and Technology (EAWAG)
Department of Microbiology
Ueberlandstrasse 133, CH-8600 Dübendorf, Switzerland

1. SUMMARY

Mecoprop and dichlorprop, two chiral herbicides, are enantioselectively degraded by *Sphingomonas herbicidovorans* MH. Experiments on the biochemistry of mecoprop and dichlorprop metabolism in this bacterium show that one reason for enantioselectivity is the existence of two distinct α-ketoglutarate-dependent dioxygenases—one that is specific for the *R* enantiomers and one that is specific for the *S* enantiomers of mecoprop and dichlorprop. Uptake experiments with radiolabeled dichlorprop provide evidence that the first step in the degradation of dichlorprop and mecoprop by strain MH is active transport and that two inducible, proton gradient-driven uptake systems exist—one uptake system for (*R*)-dichlorprop and (*R*)-mecoprop and another one for (*S*)-dichlorprop and (*S*)-mecoprop. This indicates that the enantioselective degradation of mecoprop and dichlorprop by the soil isolate *S. herbicidovorans* MH is not only due to enantioselective metabolism but also to enantioselective uptake. Because selective enrichment of one of the enantiomers may occur in the environment it is important to assess the environmental impact of stereoisomers separately. A field study of chiral mecoprop in a contaminated aquifer downstream of a former waste disposal site in Switzerland supports this notion as the

* Corresponding author

† Present address: Oekophil AG, Blegistrasse 23, CH-6340 Baar, Switzerland.

‡ Present address: BMG Engineering, Ifangstrasse 11, CH-8952 Schlieren, Switzerland.

Novel Approaches for Bioremediation of Organic Pollution, edited by
Fass *et al.* Kluwer Academic / Plenum Publishers, New York, 1999.

study clearly shows that the *R* enantiomer enriches during groundwater passage of the landfill leachate that was polluted with racemic mecoprop. We conclude that enantioselective microbial degradation increases the enantiomeric ratio of (*R*)- to (*S*)-mecoprop and suggest that analysis of enantiomeric ratios of chiral contaminants such as mecoprop can provide information about in situ biodegradation processes.

2. INTRODUCTION

Many organic compounds regulated by environmental laws are chiral and are released into the environment as racemates (1, 2). Chemists know well that physico-chemical properties of enantiomers, e.g. boiling points, vapor pressures, and water solubilities are identical. Enantiomers generally behave in an identical way in physicochemical processes and, therefore, have the same environmental behavior with respect to transport and mixing phenomena within a given environmental compartment as well as with respect to transfer processes between different phases or compartments. However, when enantiomers start to interact with other chiral compounds very selective interactions may occur and it is widely recognized that the enantiomers of a chiral drug or pesticide may exert different effects on the biological targets (6, 11, 12).

3. ENANTIOSELECTIVE METABOLISM OF MECOPROP AND DICHLORPROP

Chiral (*RS*)-2-(4-chloro-2-methylphenoxy)propanoic acid (mecoprop) and chiral (*RS*)-2-(2,4-dichlorophenoxy)propanoic acid (dichlorprop) are systemic, post emergence herbicides developed in the fifties for use in the control of broad-leaved weeds in cereal crops and lawns, and commercially are often used in formulations together with other herbicides. Already in 1953, studies about the stereospecificity of plant growth regulators revealed that the herbicidal activity was associated only with the (*R*) enantiomer of mecoprop (7). Our investigations of the degradation of mecoprop and dichlorprop by *S.herbicidovorans* MH demonstrate that both enantiomers are completely degraded by this pure culture, however in an enantioselective manner (14, 15). The (*S*) enantiomers are preferentially degraded and the degradation of the (*R*) enantiomers only starts after a considerable lag period.

4. TWO SEPARATE ENZYME ACTIVITIES FOR THE CONVERSION OF (*R*)- AND (*S*)-MECOPROP

Biochemical investigations focused on the enzyme activities that turned over the two mecoprop isomers. Cells were grown on the pure enantiomers and disrupted. In this way, transport effects at the membrane were excluded from the study, and only the enzymes present in the cell extract were important. The crude cell extracts were tested for the ability to convert either (*R*)- or (*S*)-mecoprop. By enantioselective HPLC we could follow the disappearance of the mecoprop enantiomers as well as the concomitant formation of the achiral metabolite 4-chloro-2-methylphenol. Cell extracts of (*S*)-mecoprop-grown cells effect the selective transformation of (*S*)-mecoprop upon incubation with the racemic mixture, whereas cell extracts of (*R*)-mecoprop-grown cells contain an enzyme activity that specifically transforms the *R* enantiomer (Fig. 1). Both reactions strictly require α-ketoglutarate as the cosubstrate, and the rates of conversion are dependent on the addition of ferrous ions and L-ascorbic acid. From these experiments with crude cell extracts, it is obvious that an *S* enan-

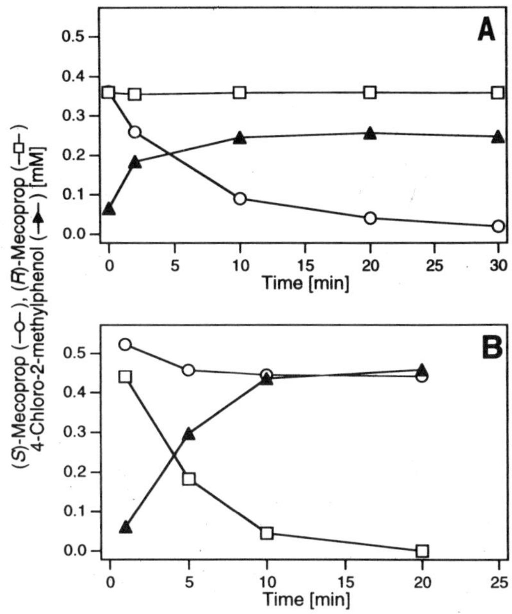

Figure 1. Consumption of (*R*)- or (*S*)-mecoprop and formation of 4-chloro-2-methylphenol in assays with crude cell extracts obtained (A) from (*S*)-mecoprop-grown cells and (B) from (*R*)-mecoprop-grown cells of *S. herbicidovorans* MH.

tiomer-specific enzyme activity is induced upon growth on the *S* enantiomer and an *R* enantiomer-specific enzyme activity when cells grow with the *R* enantiomer (8). Furthermore, we separated the proteins of the crude cell extracts by means of sodium dodecyl sulfate-polyacrylamide gel electrophoresis. In the protein patterns we can easily distinguish a band corresponding to a protein subunit with a molecular mass of 32 kDa in extracts from (*S*)-mecoprop-grown cells, whereas a protein band of 34 kDa is dominant in extracts from (*R*)-mecoprop-grown cells. Both proteins are present in extracts from cells grown on the racemic mixture (8). Nevertheless, each enzyme activity converts its substrate by cleavage of the aliphatic part of the molecule to the common products 4-chloro-2-methylphenol and pyruvate. When (*R*)- and (*S*)-dichlorprop are the substrates of the (*R*)- and (*S*)-enantiomer-specific enzyme, respectively, the common products are 2,4-dichlorophenol and pyruvate (see also chapter 6 and Fig. 3). The cosubstrate α-ketoglutarate is known to be characteristic for a group of dioxygenases that catalyze a rather broad range of reactions. A general reaction mechanism for these enzymes is thought to start with the formation of a ferryl oxidant (Fe^{IV}=O) by the binding of one oxygen molecule to the iron cofactor, a process that is linked to the oxidative decarboxylation of α-ketoglutarate to succinate (3, 4, 9, 10).

5. IDENTIFICATION OF THE REACTION PRODUCTS AND INCORPORATION OF ^{18}O INTO THE PRODUCTS

We identified 4-chloro-2-methylphenol, pyruvate, and succinate as products of both enzyme activities with the respective mecoprop enantiomer as the substrate. In order to confirm the dioxygenation reactions, we followed the incorporation of ^{18}O into the products. The incubations with crude cell extracts were carried out in the presence of a one to one mixture of $^{18}O_2$ and $^{16}O_2$. As the oxygen atom of the keto group of pyruvate readily exchanges with water by the reversible formation of pyruvate hydrate, we fixed the incorporated ^{18}O label by the lactate dehydrogenase-catalyzed reduction of pyruvate to lactate. 4-chloro-2-methylphenol was isolated and its TMS derivative was analyzed by GC-MS.

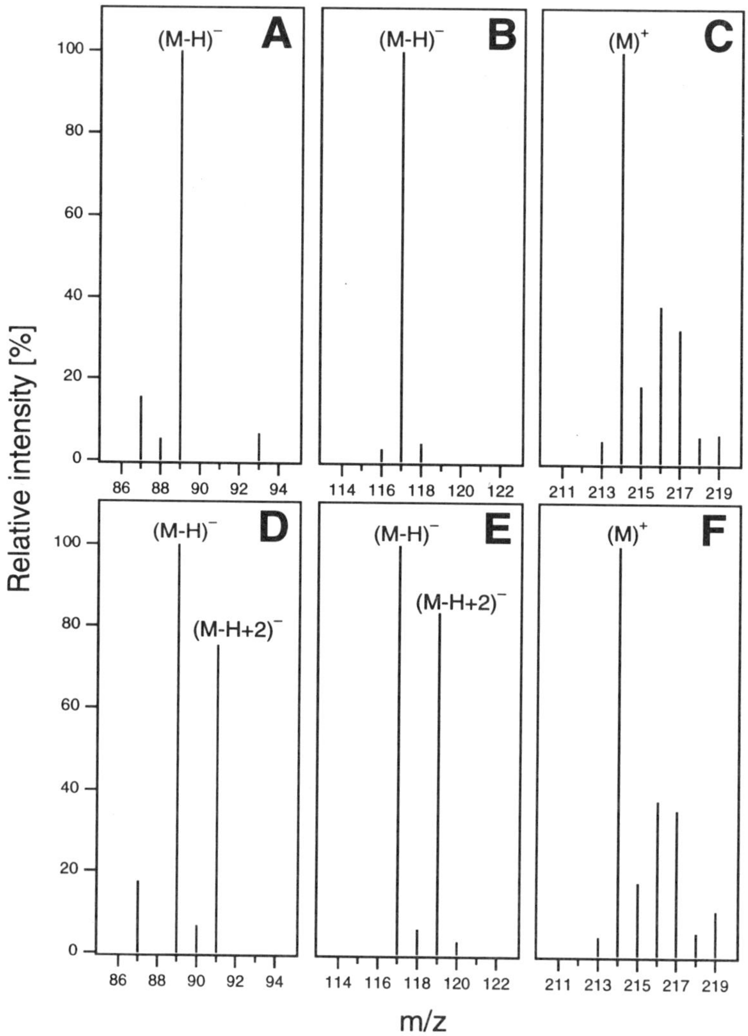

Figure 2. Mass spectral analysis of the products from (*R*)-mecoprop and α-ketoglutarate incubated with cell extracts of (*R*)-mecoprop-grown cells of *S. herbicidovorans* MH in the presence of $^{18}O_2$. According to (8).

Lactate and succinate were directly analyzed by HPLC-MS. The *R* enantiomer-specific and the *S* enantiomer-specific activities produced identical products with identical isotope patterns. Results are shown only for the transformation of (*R*)-mecoprop by the *R* enantiomer-specific enzyme activity. Fig. 2 shows the mass spectra (segment around the signal of the molecular ion) of lactic acid (Fig. 2 A and D), succinic acid (Fig. 2 B and E), and the trimethylsilyl derivative of 4-chloro-2-methylphenol (Fig. 2 C and F) from incubations with $^{16}O_2$ from air (Fig. 2 A, B, C) and with a one to one mixture of $^{16}O_2$ and $^{18}O_2$ (Fig. 2 D, E, F), respectively. The isotope patterns unequivocally show that one atom of molecular oxygen is incorporated into lactate and succinate but none into 4-chloro-2-methylphenol. These results indicate that the two enzyme activities attack the mecoprop enantiomers at the C2-atom of the aliphatic side chain, i.e. at the asymmetric C-atom. Thus, there are two different enzymes catalyzing the initial steps of mecoprop degradation in *S. herbicidovorans* MH—each specific for one enantiomer of mecoprop—but sharing the overall

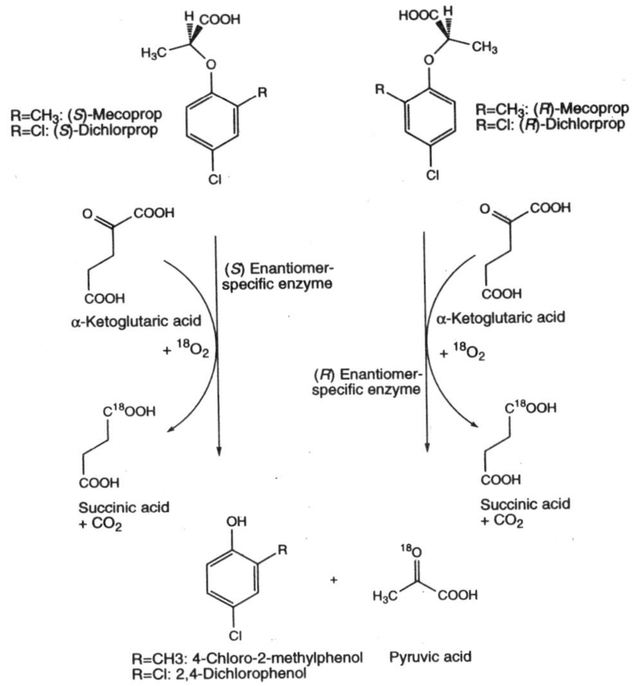

Figure 3. Proposed scheme for the initial steps in the degradation of the enantiomers of mecoprop in *S. herbicidovorans* MH. Modified after (5).

chemistry of the reaction (Fig. 3). Future research projects will address the question about what kind of structural elements determine the stereospecificity of the two enzymes.

6. SUBSTRATE SPECIFICITIES OF THE TWO MECOPROP-DEGRADING ENZYMES AND OF TFDA

Cell extracts from *S. herbicidovorans* MH grown on different substrates were examined for their ability to transform mecoprop and other phenoxyalkanoic acid herbicides to the corresponding phenols. For this comparison, we exploited a colorimetric method to measure the amounts of the phenolic compounds formed during the assay. The activities of the cell extracts are highly dependent on the growth substrate. Growth on the *R* enantiomer of mecoprop effects a high activity in the extract for the transformation of the *R* enantiomers of mecoprop and dichlorprop. When the cells grow with (*S*)-mecoprop as the carbon and energy source, the prevalent enzyme activity in the extract is specific for the *S* enantiomers of those two herbicides. Cell extract of cells grown on complex medium contain the enzyme activity that preferentially turns over the *S* enantiomers of mecoprop and dichlorprop (8). Cell extract of cells grown on 2,4-D also have the activity for the *S* enantiomers and do not appear to contain an activity specific for 2,4-D. This is in contrast to cell extract of *R. eutropha* JMP 134 grown on 2,4-D. The extract we prepared from such cells act on 2,4-D, 4-chloro-2-methylphenoxyacetic acid (MCPA), and (*S*)-dichlorprop, but not on (*S*)-mecoprop. These data indicate that the two α-ketoglutarate-dependent dioxygenase activities in *S. herbicidovorans* MH are distinct from TfdA, the α-ketoglutarate-dependent dioxygenase that initiates the degradation of 2,4-D in *R. eutropha* JMP 134. All cell extracts of *S. herbicidovorans* MH convert 4-chloro-2-methylphenoxyacetic acid. The activity was highest in those cell extracts that were active with the *S* enantiomer of mecoprop. Both, the *R* and the *S* enantiomer-specific enzyme activity, seem to share the overall chemistry with TfdA, but not the substrate specificity. Whereas TfdA preferentially transforms 2,4-D, the enzyme activities present in cell extracts of *S. herbicidovorans* MH specifically turn over either the *R* or the *S* enantiomer of mecoprop and of dichlorprop. They only have small activities with 2,4-D as the substrate (8). Again, the configuration and substitution at the C2-atom of the alkanoic acid side chain determines the ability of the two enzymes to attack the substrates. Differences in substitutions on the aromatic part of the molecule did not seem to play such an important role in substrate specificity.

7. ENANTIOSELECTIVE UPTAKE OF MECOPROP AND DICHLORPROP

Comparing the behavior of whole cells of *S. herbicidovorans* MH with the enzyme activities actually present in cell extracts, we found another interesting feature of the degradation pattern. Cells grown on complex medium without mecoprop do not take up oxygen when exposed to (*R*)- or (*S*)-mecoprop, but the enzyme activity for the transformation of (*S*)-mecoprop is rather high in cell extracts prepared from such cells. Oxygen uptake of whole cells and enzyme activities measured in cell extracts are consistent only when the cells grow on the respective mecoprop enantiomer. We hypothesized that an uptake system that catalyzes the entry of the compounds through the cell membrane must exist. In order to test this hypothesis, transport studies were carried out with pure ^{14}C-labeled dichlorprop enantiomers. Uptake of (*R*)-dichlorprop, (*S*)-dichlorprop, and 2,4-D (2,4-dichlorophenoxyacetic acid) was inducible. Initial uptake rates of cells grown on the respective substrate show substrate saturation kinetics with apparent affinity constants (K_t) of 108, 93, and 117 μM and maximal velocities (V_{max}) of 19, 10, and 21 nmol min^{-1} mg of protein^{-1} for (*R*)-dichlorprop, (*S*)-dichlorprop, and 2,4-D, respectively. Transport of (*R*)-dichlorprop, (*S*)-dichlorprop, and 2,4-D is completely inhibited by various uncouplers and by nigericin but is only marginally inhibited by valinomycin and by the ATPase inhibitor *N,N'*-dicyclohexylcarbodiimine. Experiments on the substrate specificity of the putative transport systems revealed that (*R*)-dichlorprop uptake is inhibited by (*R*)-mecoprop but not by (*S*)-mecoprop, (*S*)-dichlorprop, or 2,4-D. On the other hand, the (*S*)-dichlorprop transport is inhibited by (*S*)-mecoprop but not by (*R*)-mecoprop, (*R*)-dichlorprop, or 2,4-D. These results provide evidence that the first step in the degradation of dichlorprop, mecoprop, and 2,4-D by *S. herbicidovorans* MH is active transport and that three inducible, proton gradient-driven uptake systems exist; one uptake system for (*R*)-dichlorprop and (*R*)-mecoprop, another one for (*S*)-dichlorprop and (*S*)-mecoprop, and a third one for 2,4-D (14).

8. ENANTIOSELECTIVE DEGRADATION OF MECOPROP DURING GROUNDWATER PASSAGE

In a recent field study, we determined enantiomeric ratios of (*R*)-mecoprop to (*S*)-mecoprop in leachates from the waste disposal site in Kölliken (Switzerland) as well as in ground water from locations 20 to 50 meters downstream of this landfill. We successfully applied a sensitive enantiomer-specific analytical method (based on gas chromatography-mass spectrometry (13)) to the leachate as well as to the ground water samples in order to

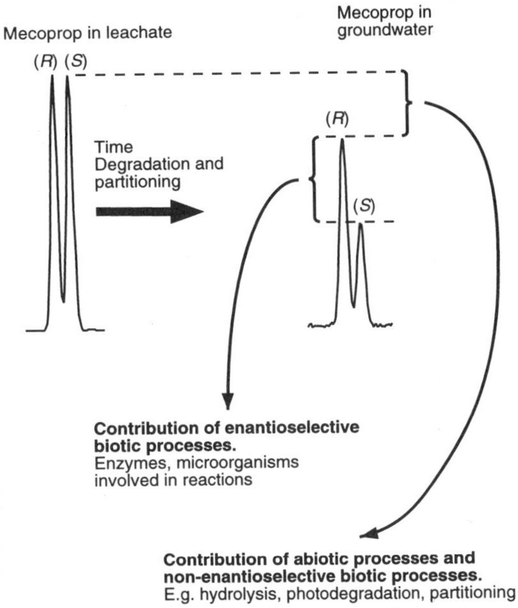

Figure 4. Illustration of the change in the enantiomeric ratio of mecoprop during groundwater passage of leachate that is polluted with a racemic mixture. The contribution of biotic and abiotic processes to such changes is shown.

determine the enantiomeric ratios of (*R*)- to (*S*)-mecoprop. The enantiomeric ratios in the leachate samples from within the landfill turned out to be one. In contrast, the concentrations of (*R*)-mecoprop were elevated up to sevenfold as compared to those of (*S*)-mecoprop in some of the ground water samples taken from locations downstream. Sorption of (*R*)- and (*S*)-mecoprop to aquifer matrix and to various reference minerals is generally low and does not discriminate between the mecoprop enantiomers. We conclude that enantioselective microbial degradation increases the enantiomeric ratio of (*R*)- to (*S*)-mecoprop during groundwater passage of the landfill leachate and that alterations in the ratio unequivocally indicate the occurrence of biological degradation processes, as physicochemical processes do not lead to discrimination of enantiomers (Fig. 4). Most leachate affected groundwaters are subject to a very complex input pattern of contaminants in time and space. Thus, observations of concentration changes of contaminants with time or space are rarely indicative of specific transformation processes in such aquifers. Under such conditions, analysis of the enantiomeric ratio of chiral contaminants like mecoprop can provide information about in situ biodegradation processes. Changes in the enantiomeric ratios do not give any information regarding absolute depletion of a chiral compound, but they are indicators for in situ biological transformations (Fig. 4). Especially in complex groundwater environments – e.g. at landfills, such as the Kölliken site – data with regard to in situ biotransformation of organic pollutants is very valuable for assessing their long term fate in the aquifer and can not be easily obtained otherwise (16).

9. CONCLUSIONS

- Not only may the enantiomers of a chiral drug or pesticide exert different effects on their biological targets, but also their biodegradation and environmental fate may differ.
- Chiral compounds can be degraded enantioselectively. Therefore, selective enrichment of a single enantiomer is possible in the environment.

- The studies on mecoprop and dichlorprop show that even though both enantiomers were degraded by strain MH – on the biochemical level – enzymes and carrier systems were specific for only one enantiomer.
- Changes in the values of enantiomeric ratios are indicators for in situ biological transformations. Especially in complex groundwater environments such data is valuable for assessing long term behavior of the pollutants and can not be easily obtained otherwise.
- The degradation and fate of stereoisomers has to be studied separately with enantiomer-specific tools.

ACKNOWLEDGMENTS

K. Nickel was supported by the Swiss National Sciences Foundation grant 21-44303.95

REFERENCES

1. Ariëns, E. J. 1989. Racemates—an impediment in the use of drugs and agrochemicals, p. 31–80. *In* A. M. Krstulovic (ed.), Chiral separations by HPLC. Ellis Horwood Limited, Chichester.
2. Armstrong, D. W., G. L. Reid, M. L. Hilton, and C.-D. Chang. 1993. Relevance of enantiomeric separations in environmental science. Environ. Pollut. **79:**51–58.
3. Feig, A. L., and S. J. Lippard. 1994. Reactions of non-heme iron(II) centers with dioxygen in biology and chemistry. Chem. Rev. **94:**759–805.
4. Hanauske-Abel, H. M., and V. Günzler. 1982. A stereochemical concept for the catalytic mechanism of prolylhydroxylase. Applicability to classification and design of inhibitors. J. Theor. Biol. **94:**421–455.
5. Kohler, H.-P. E., W. Angst, W. Giger, C. Kanz, S. Müller, and M. J.-F. Suter. 1997. Environmental fate of chiral pollutants-the necessity of considering stereochemistry. Chimia **51:**947–951.
6. March, J. 1985. Advanced Organic Chemistry: Reactions, Mechanisms, and Structure, 3rd ed. John Wiley & Sons, New York.
7. Matell, M. 1953. Stereochemical studies on plant growth regulators. VII. Optically active α-(2-methyl-4-chlorophenoxy)propionic acid and α-(2,4-dichlorophenoxy)-n-butyric acid and their steric relations. Arkiv foer Kemi **6:**365–373.
8. Nickel, K., M. J.-F. Suter, and H.-P. E. Kohler. 1997. Involvement of two α-ketoglutarate-dependent dioxygenases in the enantioselective degradation of (*R*)- and (*S*)-mecoprop by *Sphingomonas herbicidovorans* MH. J. Bacteriol. **179:**6674–6679.
9. Prescott, A. G. 1993. A dilemma of dioxygenases (or where biochemistry and molecular biology fail to meet. J. Exp. Bot. **44:**849–861.
10. Roach, P. L., I. J. Clifton, V. Fülöp, K. Harlos, G. Barton, J. Hajdu, I. Andersson, C. J. Schofield, and J. E. Baldwin. 1995. Crystal structure of isopenicillin *N* synthase is the first from a new structural family of enzymes. Nature **375:**700–704.
11. Sheldon, R. A. 1993. Chirotechnology: industrial synthesis of optically active compounds. Marcel Dekker, Inc., New York.
12. Simoni, S., S. Klinke, C. Zipper, W. Angst, and H.-P. E. Kohler. 1996. Enantioselective metabolism of chiral 3-phenylbutyric acid, an intermediate of linear alkylbenzene degradation, by *Rhodococcus rhodochrous* PB1. Appl. Environ. Microbiol. **62:**749–755.
13. Zipper, C. 1998. Ph.D. thesis, ETHZ No. 12543, Zürich, Switzerland.
14. Zipper, C., M. Bunk, A. J. B. Zehnder, and H.-P. E. Kohler. 1998. Enantioselective uptake and degradation of the chiral herbicide dichlorprop [(*RS*)-2-(2,4-dichlorophenoxy)propanoic acid] by *Sphingomonas herbicidovorans* MH. J. Bacteriol. **180:**3368–3374.
15. Zipper, C., K. Nickel, W. Angst, and H.-P. E. Kohler. 1996. Complete microbial degradation of both enantiomers of the chiral herbicide mecoprop [(*RS*)-2-(4-chloro-2-methylphenoxy)propionic acid] in an enantioselective manner by *Sphingomonas herbicidovorans* sp. nov. Appl. Environ. Microbiol. **62:**4318–4322.
16. Zipper, C., M. J.-F. Suter, S. B. Haderlein, M. Gruhl, and H.-P. E. Kohler. 1998. Changes in the enantiomeric ratio of (*R*)- to (*S*)-mecoprop indicate in situ biodegradation of this chiral herbicide in a polluted aquifer. Environ. Sci. Technol. **32:**2070–2076.

3

THE USE OF FUNGAL LACCASE FOR OXIDATION OF PHOSPHOROTHIOLATES

G. Amitai,[1,*] R. Adani,[1] G. Sod-Moriah,[1] I. Rabinovitz,[1] Y. Chai,[1] A. Vincze,[1] H. Leader,[1] B. Chefetz,[2] L. Leibovitz-Persky,[2] D. Friesem,[2] and Y. Hadar[2]

[1]Israel Institute for Biological Research
Ness Ziona 74100, Israel
[2]Faculty of Agriculture, Hebrew University
Rehovot 76100, Israel

SUMMARY

Enzymatic degradation of organophosphorus (OP) insecticides and nerve agents involves the hydrolytic breakdown of the bond between the phosphorus atom and either O-alkyl, O-aryl (P-OR) or halogen (P-X) moieties. These moieties also serve as leaving groups during irreversible inhibition of acetylcholinesterase (AChE) by these OP's. However, the P-S bond in phosphorothiolates is more resistant to enzymatic hydrolysis. So far, phosphorothiolates were decomposed rapidly only by chemical oxidation. In the current work we have successfully employed enzymatic oxidation for the rapid degradation of P-S containing OP's. Degradation kinetics was followed by measuring the residual inhibitory activiy of AChE by the OP compound at specified time intervals. Purified laccase isolated from the white rot fungus *Pleurotus ostreatus (Po)* together with as 2,2' azinobis (3-ethylbenzthiazoline - 6-sulfonate) (ABTS) as a mediator caused rapid degradation of the persistent nerve agent O-ethyl S-[N,N-diisopropylaminoethyl] methylphosphonothiolate (VX) ($t_{1/2}$=10 min with 6μg/ml enzyme, k_{sp} = 2200 nmole min^{-1} mg^{-1}) and its structural analog O,O-diethyl S-[N,N-diisopropylaminoethyl] phosphorothiolate (DiPr-Amiton) (k_{sp} =1833 nmole $min^{-1}mg^{-1}$). The optimal pH for VX degradation by *Pleurotus* laccase and ABTS was 7.4 whereas DiPr-Amiton was decomposed at a higher rate at pH 8. The maximal rate of VX and DiPr-Amiton degradation by *Po* laccase and ABTS was obtained with ABTS:OP molar ratio of 20:1 and 10:1, respectively. The complete degradation of both optical isomers of VX and identification of oxidative biodegradation products were dis-

[*] Corresponding author: Department of Pharmacology, Fax: 972-8-9381559, Tel: 972-8-9381565, e-mail: amitai@iibr.gov.il

Novel Approaches for Bioremediation of Organic Pollution, edited by Fass *et al.* Kluwer Academic / Plenum Publishers, New York, 1999.

played by ^{31}P NMR and GC/MS analysis. A thermostable laccase purified from the fungus *Chaetomium thermophylium (Ct)* in the presence of the mediator ABTS (ABTS:VX, 20:1) caused a 52 fold slower degradation of VX than by *Po* laccase ($t_{1/2}$ =16 min, with 1 mg/ml enzyme in phosphate buffer 0.05M, pH 7.4, 37°C). The specific activity (k_{sp}) of laccase from *Ct* for the degradation of VX was 42 nmole min^{-1} mg^{-1}. Thus, fungal laccases combined with the appropriate mediator can be used for the enhanced oxidative degradation of P-S-containing toxic OP compounds.

INTRODUCTION

Enzymatic degradation of organophosphorus (OP) insecticides and nerve agents is carried out either by OP acid hydrolases (OPH) (EC 3.1.8.1) or OP acid anhydrolase (OPAA, EC 3.1.8.2) (LaDu, 1992). Certain OPH's show selectivity toward the insecticide parathion and particularly toward its oxo metabolite O,O-diethyl O-*p*-nitrophenol phosphate (paraoxon). Various OPH's were isolated either from bacterial or mammalian origins and catalyze the breakdown of the P-O bond of paraoxon with a bimolecular rate constant of $5x10^7$ $M^{-1}sec^{-1}$ (LaDu, 1992). OPH enzymes with high paraoxonase activity were purified from *Pseudomonas diminuta* MG, *pseudomonas sp.* and *Flavobacterium* bacterial species ATCC 27551 (Dumas et al. 1990; Cheng et al., 1996a) and also from human plasma (Furlong et al; 1988). OPAA isolated from the thermophilic bacteria *Altermonas sp.* JD6.5 displayed high hydrolytic activity toward the nerve agent soman that contain a P-F bond in its chemical structure (Cheng et al., 1996b and 1996c).

However, it was observed that the P-S bond in certain phosphorothiolates (e.g. Amiton and VX) is usually more resistant to OPH enzymatic hydrolysis. Since VX is a persistent nerve agent and its decontamination is critical for survival and remediation it became a challenge to find an enzymatic system for its environmentally-safe degradation. It was previously noted that certain phosphorothiolate nerve agents such as VX could rapidly be degraded by chemical oxidation of the P-S bond using various peroxides such as hydrogen peroxide (Yang et al., 1993) and monomagnesium perphthalate (MPP) (Magnaud et al., 1998). As shown in scheme 1, the oxidative hydrolyis of phosphonothiolates such as VX leads to the formation of O-ethyl methylphosphonic acid and N,N-diisoprpylaminoethanesulfonate as compared to the corresponding alkylthiol formed via the hydrolysis pathway. The alkylthiol hydrolytic product is notorious for its extremely unpleasant smell and mod-

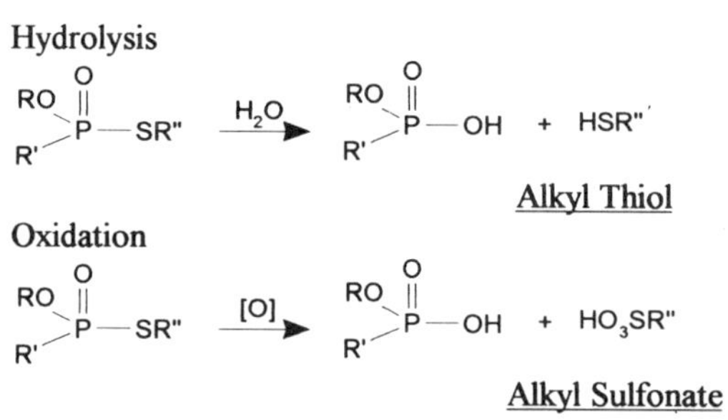

VX. R= C_2H_5, R'= CH_3, R''= $CH_2CH_2N(iPr)_2$
DiPr-Amiton: R= C_2H_5, R'= OC_2H_5, R''= $CH_2CH_2N(iPr)_2$

Scheme 1. Degradation pathways of VX and DiPr-Amiton by hydrolysis and oxidation.

erate toxicity. Thus, the oxidative pathway forms less toxic and more environmentally benign degradation products.

Therefore, it is conceivable that enzymatic oxidative breakdown of the P-S bond in certain phosphorothiolates (e.g. VX), rather than hydrolysis, will eventually lead to more efficient detoxification. The degradative properties of phenol oxidase (laccase), lignin peroxidase and Mn peroxidases isolated from various white rot fungi have been extensively studied (Tien 1987; Kerem et al., 1998; Kirk and Farrell 1987). These enzymes are applicable for biodegradation of lignin as a more environmentally safe degradation method than the chlorine-based bleaching process. Laccase isolated from various fungi such as *Trametes versicolor*, and *Pleurotus ostreatus* degrade efficiently lignin using various mediators such as 2,2' azinobis (3-ethylbenzthiazoline-6-sulfonate) (ABTS), 1-hydroxy benztriazole (HOBT) and several phenothiazines such as promazine and chlorpromazine (Call and Mucke, 1997). The combination of fungal laccase with certain mediators was proved to be active in deligninification of kraft pulp and created considerable interest in the paper industry (Bourbonnais et al., 1997). Since oxidative biodegradation of lignin proceeds without the formation of a direct Michaelis complex between laccase and the substrate but rather by using a diffusible oxidizing mediator, the scope of laccase substrates may be expanded. Indeed, we have successfully employed fungal laccase together with several mediators for degradation of toxic phosphorothiolates. We report here the rapid oxidative degradation O,O diethyl S-[N,N-diisopropylaminoethyl] phosphorothiolate (DiPr-Amiton), and the nerve agent O-ethyl S-[N,N-diisopropylaminothyl] methylphosphonothiolate (VX) by fungal laccase using ABTS as a mediator.

MATERIALS AND METHODS

OP Substrates

The Organophosphorus Compounds O-Ethyl S-[N,N'-Diisoprpyl Aminoethyl] Methylphosphonothiolate (VX) and O,O-Diethyl S-[N,N'-Diisoprpylaminoethyl] Phosphorothiolate (DiPr-Amiton) were synthesized according to previously described procedures (Tammelin, 1957; Amitai et al., 1976). the structure and purity of the OP compounds were elucidated by ^{1}H and ^{31}P NMR and GC/MS analysis

Growth of Fungus and Purification of *Pleurotus ostreatus* Laccase

Culture of Organisms Pleurotus spp. "Florida" F6. The fungus was grown in a basidomycete synthetic medium (BSM) that contained the following per liter of culture: glucose (5 gr), K_2HPO_4 (1 gr), L-aspargine (0.6 gr), KCl (0.5 gr), $MgSO_4.7H_2O$ (0.5 gr), yeast extract (Difco, 0.2 gr), Zn NO_3 (0.003 gr), $CuSO_4.5H_2O$ (0.003 gr), $Ca(NO_3)_2.4H_2O$ (0.006 gr) at pH 5.5. Stock cultures were maintained on 2% agar BSM. A typical preparation consisted of 20 discs (5mm diameter) in 250 Erlenmeyer flasks containig 60ml of liquid BSM. After 5 days of growth, the mycelium was homogenized by Ultra-Turrax (TP18/10, IKI-WERK, Staufen, Germany) for 15 seconds. Cultures were cultivated (40 ml) by shaking in 500 ml Erlenmeyer flasks containing 200 ml liquid BSM. All of the flasks were incubated for 2 days at 28°C in the dark in a rotary shaker at 120 rpm and then were used (12ml) for inoculation in 250 ml Erlenmeyer flasks containing 60 ml of liquid BSM.

Protein Purification. After 8 days of growth the extracellular medium was collected by filtration through glass wool followed by 0.45 m filters. The fluids were concentrated 50 fold relative to the original concentration, using an Amicon 10K membrane filter and dialysed against 10mM sodium acetate buffer , pH 6. The proteins were separated on FPLC AKTA explorer (Pharmacia Biotech, Sweden). The concentrated extracellular medium was loaded onto a HiTrap Q anion exchange column (HiTrap Q 5ml column, Pharmacia Biotech) previuosly equilibrated with 10mM sodium acetate, pH 6. After washing with 6 column volumes (cv) of 10mM sodium acetate buffer, the enzyme was eluted by two linear NaCl gradients, 0-0.3M in 30cv 0.3-1.0M in 10cv. Enzyme activity was monitored using ABTS as a substrate and the active fractions were collected, concentrated and loaded onto a gel filtration column (Superdex 200HR 10/30. Pharmacia Biotech). Prior to elution, the column was equilibrated with 10mM sodium acetate buffer pH 6. After elution, the active fractions were collected. These fractions showed a single band on PAGE (10% acrylamide) as visualized by silver stain. Laccase from *Chaetomium thermophilium* was purified according to Chefetz et al. (1998).

Enzyme Activity. Laccase activity was monitored using ABTS as a substrate. The reaction mixture contained 0.05ml of enzyme, 0.1ml of ABTS (5mM), 0.1M phosphate buffer (pH 6) and incubated for 3 minutes at 30°C. Absorbance at 436nm was recorded.

Enzymatic Assay for Degradation Kinetics of OP Substrates. The enzymatic degradation mixtrure contained 50 μl of OP substrate at a final concentration of 10μM (stock solutions of VX or DiPr-Amiton are at a concentration of 0.2 mM in citrate buffer pH 5.6), 20 μl of ABTS (10mM) and 20 μl of purified *Pleurotus ostreatus* laccase (6ug/ml) in 1ml of phosphate buffer (50mM pH 7.4) incubated at 37°C. Three control solutions that were monitored parallel to the degradation solution contained the OP substrate in the presence of the following reagents: control 1: phosphate buffer, control 2: phosphate buffer with ABTS and control 3: phosphate buffer with laccase. Degradation of OP compounds in all control solutions was less than 10% at the time enzymatic degradation was complete. The residual OP level was monitored by 5×10^3 fold dilution of the degradation solution for measuring the initial rate of inhibition of fetal bovine serum acetylcholinesterase (FBS-AChE) for 5 minutes at 25° C. FBS-AChE activity was determined using the Ellman method (Ellman et al., 1961).

GC/MS Analysis. Samples of solutions after enzymatic oxidative degradation were evaporated to dryness using Speed Vac centrifuge under reduced pressure. The solid residue was acidified with 100 μl 5mM HCl and 0.25ml of chloroform was added. Diazomethane in ether was added to the separated organic phase until the yellow color was sustained in solution. Aliquots of 1 μl were injected into the GC injector. GC/MS analysis was performed using VG model 70VSEQ mass spectrometer directly interfaced to a GC model HP 5890. Chemical ionization (CI) gas was isobutane. GC conditions: HP5 column 30m length, 0.22 mm i.d., 1μm film, head pressure: 55kPa. and temperature programming: 40°C for 1 minute, heating rate: 10°C/min up to 280 °C and remains at 280°C for 10 minutes.

NMR Spectra. NMR spectra were recorded with a GN 300WB NMR instrument (General Electric, USA) at 121.65 MHz for ^{31}P and 300.0 MHz for ^{1}H. The NMR spectra were determined for purity and structure elucidation of the synthesized OP compounds and analysis of their enzymatic degradation products.

RESULTS AND DISCUSSION

The time-course of enzymatic degradation of VX and DiPr-Amiton by purified *Pleurotus ostreatus* laccase in the presence of ABTS as a mediator is described in Figure 1A. Figure 1B describe the linear plots of Ln(%OP) as a function of initial degradation time. The semilogarithmic plots in figure 1B were used for the calculation of first order rate constants k_{obs} (min^{-1}) of OP degradation. The k_{obs} values for VX and DiPr-Amiton are: 0.066 and 0.055 min^{-1}, respectively (for 6 μg/ml enzyme protein), and the corresponding specific activity values (k_{sp}) are: 2200 and 1833 nmole min^{-1} mg^{-1}, respectively (Table 1). As shown in figure 1A, the oxidative enzymatic degradation of these phosphorothiolates proceeds to completion. Complete degradation of both optical isomers of VX was shown by the enzymatic AChE assay (figure 1) and confirmed by ^{31}P-NMR spectrum following 6 hours incubation of 0.5mM VX with laccase/ABTS in ammonium bicarbonate buffer (0.05M, pH 7.5) (not shown). The mass chromatograms and mass spectra obtained from the GC/MS analysis of VX is presented in Figure 2. The formation of O-ethyl methyphosphonic acid is displayed in the mass chromatogram and mass spectra of the methyl esters of the degradation products of VX following methylation by diazomethane. The mass spectrum of the compound eluting with a retention time of 17:07 min in the gas chromatogram of degraded VX (see total ion current in lower part of figure 2A) has an $M+1]^+$ of m/z =139 calculated to be the quasimolecular ion of O-ethyl O-methyl methylphosphonate (figure 2B, where M is the molecular ion). The appearance of the molecular ion peak of methyl ester of the corresponding O-ethyl methylphosphonic acid (figure 2) together with

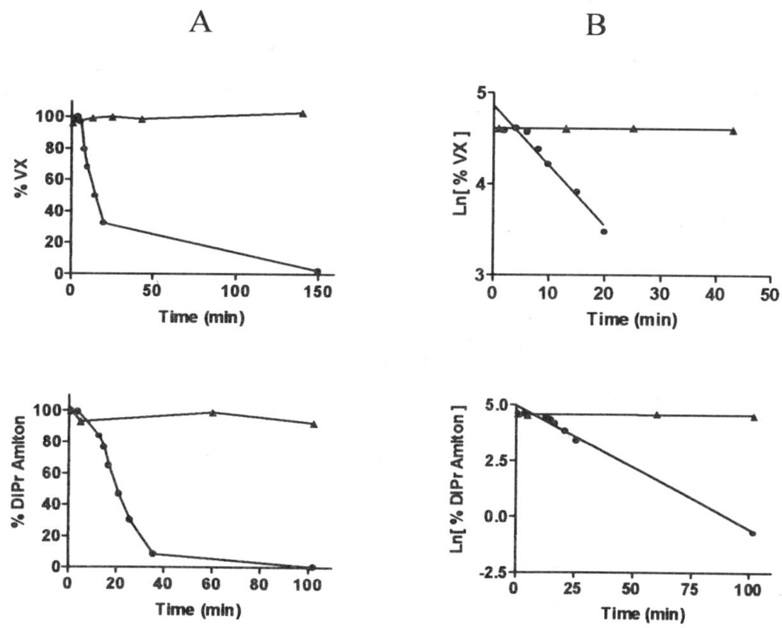

Figure 1. Degradation kinetics of VX and DiPr Amiton by *Pleurotus ostreatus* laccase in the presence of ABTS ([OP]=10μM, [ABTS]=200μM, pH=7.4, 50mM phosphate, 37°C). A. Time-course of OP enzymatic degradation B. Ln[%OP] as a function of initial degradation time (the slopes of these lines are k_{obs}, min^{-1}) Enzymatic degradation (●) Buffer control (▲).

Table 1. First order rate constants (k_{obs}, min^{-1}) and specific activity values (k_{sp}, $nmole/min^{-1}mg^{-1}$) of VX and DiPr-Amiton degradation by fungal laccase

Enzyme (substrate)	k_{obs} (min^{-1})	k_{sp} ($nmole\ min^{-1}\ mg^{-1}$)
Pleurotus ost. Laccase (VX)	0.066	2200
Pleurotus ost. Laccase (Amiton)	0.055	1833
Chaetomium th. Laccase (VX)	0.0013	42

k_{obs} values were calculated for 6μg/ml purified laccase, [OP]=10μM, [ABTS]=200μM, pH=7.4, 50mM phosphate, 37°C.

the absence of the molecular ions of VX and its hydrolytic products N,N'-diisopropylaminoethanethiol, its corresponding sulfide or disulfide in the GC/MS chromatogram indicate that both optical isomers of VX were degraded via the oxidative hydrolysis pathway.

The dependence of degradation rate of VX and DiPr Amiton on pH, buffer concentration and ABTS:OP ratio are described in Figure 3A and 3B, respectively. The optimal pH for degradation of VX is 7.4 as compared to DiPr Amiton that is degraded at a higher rate at pH 8. The most rapid degradation rate of VX is obtained at a concentration of 0.05 M phosphate whereas the rate of DiPr Amiton degrdation is higher at 0.2M phosphate. The optimal stoichiometric ratio of ABTS:OP for VX degradation is 20:1 whereas the rate of DiPr-Amiton degradation reaches it maximum at a molar ratio of 10:1.

The oxidative degradation of VX and DiPr by *Po* laccase together with ABTS as mediator (table 1) displays relatively rapid reaction rates that are unprecedented so far for enzymatic degradadtion of phosphorothiolates. It is pertinent to note that the cobalt-metalo enzyme form of recombinant *Pseudomonas diminuta* OPH displayed hydrolytic activity toward VX and the Russian VX (RVX) (Rastogi et al., 1997). Furthermore, a very slow (14-16 days) and only partial degradation of the P-S containing insecticides Fonofos and Terbufos was caused by nitrogen-limited cultures of the fungus *Phanerochaete chrysosporium* (Bampus et al., 1993). This white rot fungus caused slow mineralization of Fonofos and Terbufos only to a level of 12.2 and 26.6%, respectively (Bampus et al., 1993). The exact mechanism of oxidative degradation of phosphorothiolate by laccase/mediator is as yet unknown and it is assumed that the sulfur atom is oxidized and then the P-S bond is cleaved (Yang et al., 1990).

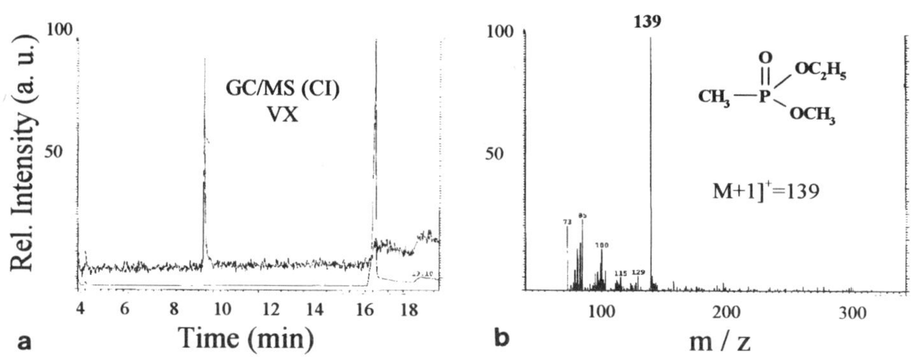

Figure 2. GC/MS of degraded VX (A) Total ion current mass chomatograms (B) mass spectrum showing $M+1]^+$ of O-methyl O-ethyl methylphosphonate in extract of degraded VX obtained after 1h incubation with *Pleurotos ostreatus* laccase and ABTS at 37°C.

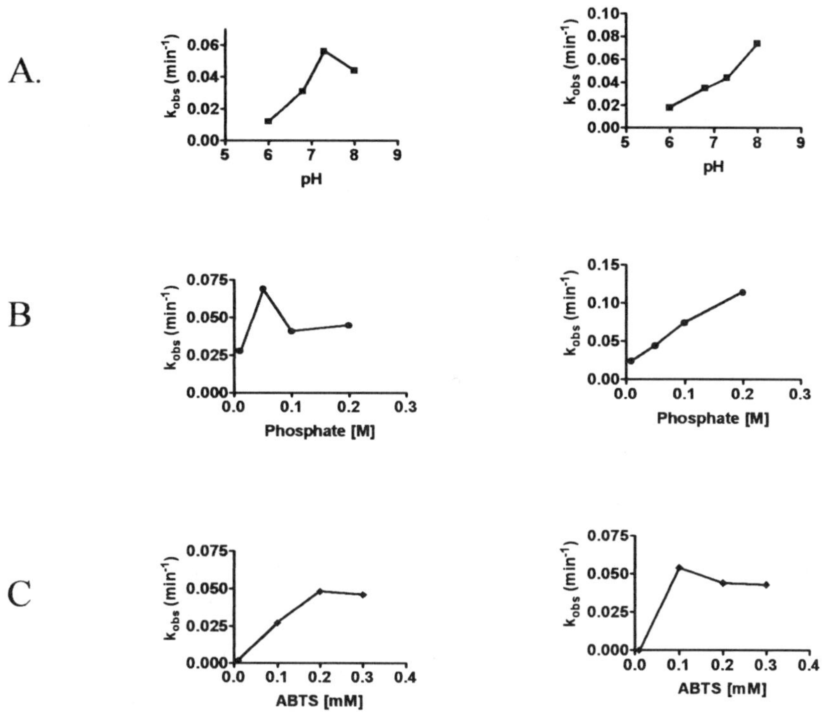

Figure 3. Dependence of VX (A) and DiPr-Amiton (B) degradation rate by *Pleurotus ostreatus* laccase on pH, phosphate buffer concentration and ABTS concentration ([OP]=10μM, 37°C) A. k_{obs} (min^{-1}) as a function of pH (ABTS 0.2mM, phosphate 50mM) B. k_{obs} (min^{-1}) as a function of phosphate concentration (ABTS 0.2mM, pH=7.4) C. k_{obs} (min^{-1}) as a function of ABTS concentration (pH=7.4, phosphate 50mM).

The nitrogen atom at β position to the carbon bound to the sulfur atom also plays an important role in the enzymatic oxidation reaction. One possible pathway is the formation of an N-oxide intermediate in the N,N- dialkylaminoethyl moiety at alkaline pH that may affect the cleavage of the P-S bond (Yang et al., 1990). We have shown, that oxidative biodegradation of both optical enantiomers of VX by laccase/ABTS was verified by the enzymatic AChE assay, ^{31}P NMR and GC/MS analysis. It was previously noted by Hall et al. (1977) that the P(-) enantiomer of VX is 6.4 fold more toxic than its P(+) antipode. However, in contrast to other chiral OP compounds where the P(+) isomer is far less toxic than its P(-) counterpart, the P(+) enantiomer of VX is highly toxic (LD_{50} = 0.21 μmole/kg, subcutaneous in rats) and the apparent bimolecular rate constants of AChE inhibition in vitro by these enantiomers are similar ($4.5x10^7$ and $2.8x10^6$ M^{-1} min^{-1}, respectively) (Hall et al., 1977). Therefore, oxidative biodegradation of both optical isomers of VX by laccase/ABTS is important since it leads to complete detoxification of this toxic nerve agent.

REFERENCES

Amitai, G., Ashani, Y., Grunfeld, Y., Kalir A. and Cohen, S., (1976), The Synthesis and Properties of 2-oxo-2-S[2'(N,N dialkylaminoethyl)] -thio 1,3,2-dioxaphosphorinane and the corresponding quaternary derivatives as potential antiglaucoma agents. J. Med. Chem. **19:** 810.

Bampus, J.A., Kakar, S.N. and Coleman, R.D., (1993), Fungal degradation of organophosphorus insecticides, Appl. Biochem. Biotech. **39/40:** 715-726.

Bourbonnais, R. M., Paice, G., Freiermuth, B., Bodie, E. and Borneman, S., (1997) Reactivites of various mediators and laccase with kraft pulp and lignin model compounds, Applied and Environ. Microbiol. **63:** 4627-4632.

Call, H.P. and Mucke, L., (1997), History, overview and applications of mediated lignolytic systems, especially laccase-mediator-systems (Lygnozym-process), J. Biotechnol. **53:** 163-202.

Chefetz, B., Chen Y. and Hadar. Y., (1998), Laccase from *Chaetomium thermophilium*: purification characterization and role in humification, Appl. and Environ. Microbiol.. (in press)

Cheng, T-c. and J.J. Calomiris, (1996 a), A cloned bacterial enzyme for nerve agent decontamination. Enzyme and Micro Technol **18:** 597-601.

Cheng, T-c., Harvey, SP. and Chen, GL., (1996 b), Cloning and expression of a gene encoding a bacterial enzyme for decontamination of organophosphorus nerve agents and nucleotide sequence of the enzyme. Appl. Environ. Microbiol. **62:** 1636-1641.

Cheng, T-c., Liu, L., Wang, B., Wu, J., DeFrank, JJ., Anderson, DM., Rastogi, VK. and Hamilton., AB., .(1996 c) Nucleotide sequence of a gene encoding an organophosphorus nerve agent degrading enzyme from Alteromonas haloplanktis. J. Indus Microbiol. **17:** 554-563.

Dumas, D. P., Durst, H.D., Landis, W.G., Raushel, F.M. and Wild, J.R. (1990), Inactivation of Organophosphorus nerve agents by the phosphotriesterase from *Pseudomonas diminuta,* Arch. Biochem. Biophys. **277:** 155-159.

Ellman, GC, Courtney, KD, Anders, V and Featherstone, RM, (1961), A new and rapid colorimetric determination of acetylcholinesterase activity. *Biochem. Pharmacol.* **7:** 88-95.

Furlong, C.E., Richter, R.J., Seidel S. and Motulsky, A.G., (1988), Role of genetic polymorphism of human plasma paraoxonase/arylesterase in hydrolysis of the insecticide chlorpyrifos oxon and paraoxon. Am. J. Hum. Genet. **43:** 230-238.

Hall, C.R., Inch, T.D., Inns, R.H., Muir, A.W., Sellers, D.J. and Smith, A.P., (1977), Differences between some biological properties of enantioms of alkyl S-alkyl methylphosphonothioates, J. Pharm. Pharmac., **29:** 574-576.

Kerem Z. and Hadar, Y., (1998), Lignin-degrading fungi: Mechanisms and utilization, in: Agricultural Biotechnology, pp.351-365 Ed. A. Altman Marcel Dekker, Inc.

Kirk, T.K. and Farrell, R.L., (1987) Enzymatic "combustion": the microbial degradation of lignin, Annu. Rev. Microbiol. **41:** 465.

LaDu, B.N., (1992), in: Pharmacogenetics of drug metabolism, (Kalow, W. Ed.) pp 51-91 Pergamon Press, New York

Magnaud, G., Lion, C., Delmas G. and Reynaud, A., (1998), Thorough decontamination system, Proceedings of the 6th CBW protection symposium, Stockholm, Sweden, p. 309-312, May 10-15.

Rastogi, V.K., DeFrank, J.J., Cheng, T,-c. and Wild, J.R., (1997), Enzymatic hydrolysis of Russian VX by organophophsorus hydrolase, Biochem. Biophys. Res. Comm. **241:** 294-296.

Tammelin, T.E., (1957), Dialkoxy-phosphorothiocholines, alkoxy methyl phosphorylthiocholine esters, Syntheses, pKa, of tertiary homologues and cholinesterase inhibition, Acta Che. Scand. **11:** 1340-1349.

Tien, M., (1987), Properties of ligninase from Phanerochaete chrysosporium and their possible applications, CRC Crit. Rev. Microbiol. **15:** 141-168.

Yang, Y.c., Szafraniec, L.L and Beaudry, W. T., (1993), Perhydrolysis of Nerve Agent VX, J. Org. Chem. **58:**6964-6965.

Yang, Y.c., Szafraniac, L., Beaudry, W.T. and Bohrbaugh, D.K., (1990), Oxidative detoxification of VX, J. Am. Chem. Soc. **112:** 6621-6624.

4

MSA MONOOXYGENASE

An Enzyme from Terrestrial and Marine Bacteria which Degrades the Natural Sulfonate Methanesulfonate

Wolfram Reichenbecher,[1,*] Paolo De Marco,[2] Julie Scanlan,[1] Nardia Baxter,[1] and J. Colin Murrell[1]

[1]Department of Biological Sciences
University of Warwick
Coventry CV4 7AL, UK
[2]Department of Microbiology
University of Porto, Portugal

1. OCCURRENCE AND DEGRADATION OF SULFONATE COMPOUNDS

1.1. Sulfonates

Sulfonates are organosulfur compounds with sulfur in the oxidation state +5 generally linked to a terminal carbon atom ($R-CH_2-SO_3H$). Naturally occuring sulfonates are usually non-aromatic and include taurine (2-aminoethanesulfonate), which is found abundantly in mammals (Huxtable, 1992), isethionate (2-hydroxyethanesulfonate), which occurs in red algae (Holst, 1994), the squid axon (Koechlin, 1954) and in mammals as a result of taurine conversion, cysteate (DL-2-amino-3-sulfopropionate), which is derived from cysteine, the archaebacterial coenzyme M (2-mercaptoethanesulfonate), the membrane constituents sulfonolipids, and methanesulfonate, which is discussed below. Synthetic sulfonates commonly have an aromatic nucleus, for example in the linear alkylbenzene sulfonate surfactants (LAS) or the sulfonated dyestuffs (Kertesz *et al.* 1994). Some of the artificial buffers used in the laboratory are sulfonates like 3-(cyclohexylamino)-1-propanesulfonic acid (CAPS), 4-(2-hydroxyethyl)piperazine-1-ethanesulfonic acid (HEPES) and 3-morpholinopropanesulfonic acid (MOPS).

* Correspondence should be send to J. C. Murrell. Phone: +44 1203 523553, fax: 523568, e-mail: cm@dna.bio.warwick.ac.uk.

Novel Approaches for Bioremediation of Organic Pollution, edited by Fass *et al.* Kluwer Academic / Plenum Publishers, New York, 1999.

Sulfonate compounds can be used as a source of sulfur, or carbon and energy, by aerobic bacteria (Locher *et al.* 1989; Seitz *et al.* 1993; Kertesz *et al.* 1994) and, as was found more recently, also by anaerobic bacteria (e.g. Chien *et al.* 1995; Denger *et al.* 1997). Utilization of the simplest sulfonate, methanesulfonate, so far is known to occur only in aerobic bacteria, where it is used as a source of carbon and energy (Baker *et al.* 1991; Thompson *et al.* 1995).

1.2. Methanesulfonate

Methanesulfonate is derived from atmospheric dimethyl sulfide (DMS) which itself is produced by enzymatic hydrolysis of dimethylsulfoniopropionate (DMSP). DMSP is used as osmolyte or cryoprotectant by algae and marsh grass. A large proportion of the volatile compound DMS enters the atmosphere, making it the principal organic sulfur compound derived from the biosphere. In the atmosphere, 25–70 % of DMS is oxidized photochemically to methanesulfonic acid (MSA) which is equivalent to an annual rate of 20–50 Tg MSA formed (see Kelly *et al.* 1993). In the atmosphere, the stable compound MSA is important in cloud condensation nuclei (Charlson *et al.* 1987), but does not undergo further degradative transformation. Instead it falls on the Earth's surface in rain and snow, thus re-entering the terrestrial and marine biosphere. Studies of Antarctic ice cores (Saigne and Legrand 1987) suggest that MSA has been continuously deposited on the Earth for thousands of years without apparent accumulation, indicating its mineralization in the biosphere thereby, recycling sulfur and carbon from marine-derived DMS. From both the terrestrial and the marine environment, MSA-degrading-bacteria have been isolated.

2. ISOLATION AND CHARACTERIZATION OF MSA-UTILIZING BACTERIA

2.1. Isolation Procedures

Aerobic chemostat and batch culture enrichments with MSA as sole carbon and energy source, together with a methylotrophic co-substrate such as methanol or methylamine, led to growth of various bacteria from several inocula including soil, freshwater, rainwater, and marine samples (Baker *et al.* 1991; Thompson *et al.* 1995). Without the methylotrophic co-substrates, most of the bacteria enriched from non-marine inocula were small rods, resembling the well described strain M2 (Holmes *et al.* 1997). Anaerobic enrichments with MSA as sole carbon and energy source did not lead to bacterial growth (Reichenbecher and Murrell, unpublished), and similar observations were made with MSA or higher non-substituted alkanesulfonates as sulfur or energy source (Chien *et al.* 1995; Denger *et al.* 1997). The failure so far to enrich anaerobic MSA-utilizing bacteria is probably due to the enormous stability of the C-S bond in the unsubstituted alkanesulfonate. Since MSA is probably derived from the atmosphere and enters the biosphere via rain and snow, aerobic habitats are always reached first, giving anaerobic bacteria less chance to develop the ability to utilize it.

2.2. The MSA-Utilizing Strains M2 and PSCH4

The rod shaped *Methylosulfonomonas methylovora* strain M2 isolated from garden soil and the rod shaped, often rosette-forming marine *Marinosulfonomonas methylotropha* strain PSCH4 isolated from coastal water are both Gram negative, restricted facultative methylotrophic, non-autotrophic bacteria with the unique capability to grow on MSA. Furthermore both

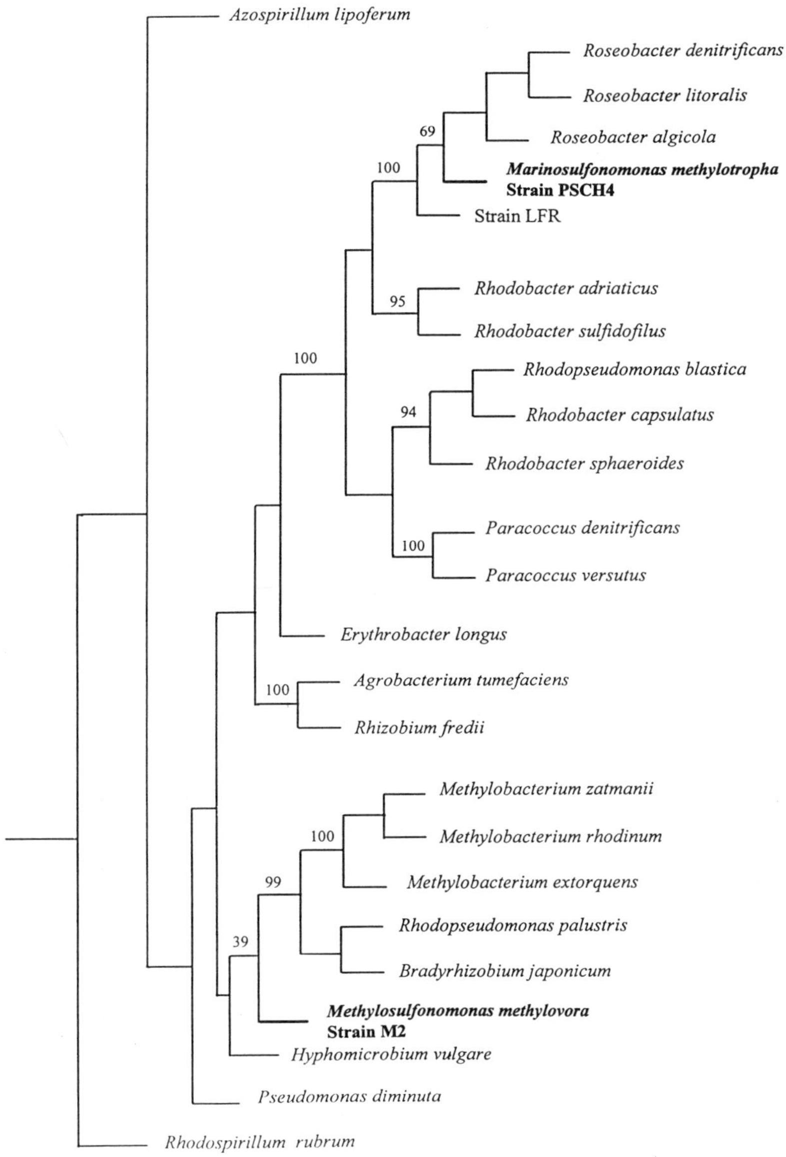

Figure 1. Phylogenetic tree showing the position of the terrestrial *Methylosulfonomonas methylovora* strain M2 and the marine *Marinosulfonomonas methylotropha* strain PSCH4 within the α subgroup of the Proteobacteria (taken from Holmes *et al.* 1997). Numbers at the nodes indicate the number of times the species to the right occurred in 100 bootstrap replicates.

grow on methanol, methylammonium chloride, formaldehyde and formate, but not on methane or dimethyl sulfide (Holmes *et al.* 1997; Thompson *et al.* 1995). With MSA, the doubling time is 7–8 hours for strain M2 and 10–15 hours for strain PSCH4. The marine strain has an obligate requirement for NaCl (0.5–3.5 % w/v), and it grows poorly on solid media.

Analysis of the 16S rDNA sequences revealed that the terrestrial and the marine strain belong to different subgroups of the α-Proteobacteria and that each of them forms a new, separate genus (Figure 1).

Table 1. Substrate specificity for the oxidation of different sulfonates by whole cells (*in vivo*) or cell-free extract (*in vitro*) of strain M2

Substrate	Relative *in vivo* oxidation rate [%]*	Relative *in vitro* oxidation rate [%]**
MSA	100	100
Amino-MSA	86	36
Ethanesulfonate	36	79
Propanesulfonate	8	29
Butanesulfonate	0	0
Pentanesulfonate	0	0
Taurine	0	34
Isethionate	11	46
Cysteate	0	23

*100 % was 28.4 nmol oxygen per min per mg dry weight cells.
** 100 % was 58.9 nmol NADH per min per mg protein.
Modified after Kelly *et al.* 1994 and Higgins *et al.* 1996.

3. DEGRADATION OF MSA

3.1. Oxidation of MSA by a Monooxygenase

Whole cells of strain M2 oxidized MSA with oxygen, indicating the involvement of an oxygenase in the breakdown of MSA. This was confirmed by measuring an MSA-stimulated oxidation of NADH in cell-free extracts, and the enzyme responsible was described as methanesulfonic acid monooxygenase (MSAMO). It became clear from studies with the methanol dehydrogenase inhibitor cyclopropanol and with growth-deficient mutants that the enzyme converts the carbon moiety of MSA into formaldehyde rather than methanol suggesting equation (1) for the reaction of the MSAMO (Higgins *et al.* 1996).

$$CH_3SO_3H + O_2 + NADH + H^+ \rightarrow HCHO + H_2SO_3 + H_2O + NAD^+ \quad (1)$$

The enzyme is located in the cytoplasm, induced only during growth on MSA, and has a very narrow substrate spectrum. Experiments with whole cells and with cell-free extracts showed that the enzyme oxidized MSA, amino-MSA, ethanesulfonate, propanesulfonate, and isethionate, but not butanesulfonate or pentanesulfonate. Cell-free extract also oxidized taurine and cysteate (Table 1). When the partially purified, reconstituted enzyme (see section 4.) was tested for substrate specificity, it only oxidized MSA and ethanesulfonate, but not the other substrates mentioned above. The reconstituted enzyme was also unable to oxidize methane, dimethyl sulfide, methylsulfoxide, methanethiol, methylamine, methanol, formaldehyde or formate, indicating a pronounced restriction towards short-chain alkylsulfonates (Kelly *et al.* 1994; Higgins *et al.* 1996). Oxidation of isethionate, taurine and cysteate by cell-free extract but not by the reconstituted MSAMO suggests that there might be other sulfonate-oxidizing oxygenases present in M2.

3.2. Pathway for Degradation of MSA

Formaldehyde, which is produced in the MSAMO reaction, is further oxidized to formate and carbon dioxide thereby generating reducing equivalents for the reduction of oxygen or regeneration of NADH (Figure 2). Alternatively, formaldehyde can be assimilated via the Serine pathway involving hydroxypyruvate reductase. The absence of any ribulose bisphosphate carboxylase was demonstrated by several different means (Kelly *et*

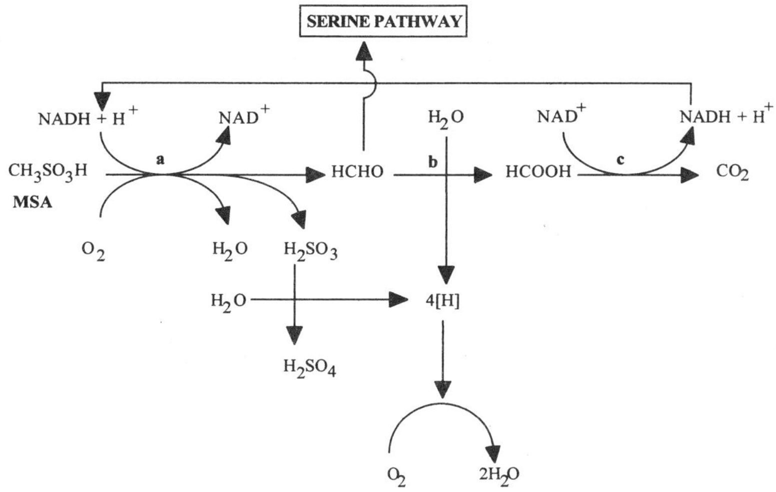

Figure 2. Proposed pathway for the degradation and assimilation of methanesulfonic acid (MSA) by the methylotrophic bacteria *Methylosulfonomonas methylovora* strain M2 and *Marinosulfonomonas methylotropha* strain PSCH4 (based on Kelly *et al.* 1994, Thompson *et al.* 1995, and Higgins *et al.* 1996). (a) methanesulfonic acid monooxygenase, (b) formaldehyde dehydrogenase, (c) formate dehydrogenase. The monooxygenase reaction breaks MSA into sulfite, which is subsequently oxidized to sulfate, and formaldehyde. The latter is either assimilated via the Serine pathway or further oxidized to formate and carbon dioxide. The regeneration of NAD/NADH and the generation of reducing power for oxygen respiration is indicated.

al. 1994) showing that M2 is a strict heterotrophic bacterium. The second product of the MSAMO reaction, sulfite, is oxidized to sulfate probably in a purely chemical manner The whole pathway proposed for the oxidation and degradation of MSA is outlined in Figure 2. The mineralization of MSA to carbon dioxide proceeds via equation (2),

$$CH_3SO_3H + 2O_2 \rightarrow CO_2 + H_2O + H_2SO_4 \qquad (2)$$

which is consistent with the stoichiometry of 1.96 oxygen:MSA found for the utilization of MSA (Kelly *et al.* 1994). Other C_1 substrates like methylamine, methanol, formaldehyde and formate were also completely oxidized to carbon dioxide. In addition to the pathway in figure 2, oxidation of methanol involves a constitutively expressed pyrroquinoline quinone-linked methanol dehydrogenase. Methylamine was only oxidized by cells grown on methylamine indicating that the pathway for the degradation of this C_1 compound differs at least in the initial step from the one for MSA.

The enzymes methanesulfonic acid monooxygenase, formate dehydrogenase, methanol dehydrogenase, but not ribulose bisphosphate carboxylase were also found in strain PSCH4 (Holmes *et al.* 1997) suggesting that the terrestrial and the marine strain degrade C_1 compounds via similar pathways.

4. PURIFICATION AND CHARACTERIZATION OF THE METHANESULFONIC ACID MONOOXYGENASE (MSAMO)

When crude extract of strain M2 was separated on a Q-Sepharose column, MSA-dependent NADH-oxidation could only be restored by the addition of three different fractions designed A, B, and C, indicating that the MSAMO is a three component monooxygenase. Fraction A (reductase) eluted at 200 mM NaCl, fraction B (hydroxylase) at 300 mM NaCl, and fraction C (ferredoxin) at 440 mM NaCl (Higgins *et al.* 1996). Addition of FAD and ferrous ions to the reconstituted enzyme complex resulted in a five-fold increase in enzyme activity, suggesting the loss of these factors from at least one of the components. When chromatographic fractions were tested for activity, a rapid colorimetric assay was applied, based on the reaction of sulfite derived from MSA with Ellman's reagent to give a yellow dye (Johnston *et al.* 1975). Components A, B or C were further purified individually and activity was checked by adding fractions back to the two other components obtained from the Q-Sepharose column.

4.1. The Reductase Component

Further purification of the reductase-containing fraction A was attempted by a variety of chromatographic techniques including anion-exchange (Mono Q, Pharmacia), gel filtration (Sephadex 75 and Sephadex 200, Pharmacia), hydrophobic interaction (Toyo pearl starter kit, ToshoHaas), hydroxyapatite (Macro-Prep Ceramic Type I, BioRad), and affinity (HiTrap Blue, Pharmacia, and NAD-Agarose, Sigma). In all cases, this was accompanied by loss of activity. Chromatograms and SDS-gels indicated that the best separation was obtained with hydroxyapatite and Blue Sepharose. Activity was also lost after a Blue Sepharose run under conditions which helped maintain activity of reductases from other oxygenases (Batie *et al.* 1987; Haigler and Gibson 1990) : a high sample concentration (20 mg/ml), a chilled and light-protected column, a short process time (30 min.), addition of 1 mM dithiothreitol. No activity was recovered when the total fractions from a Blue Sepharose run were pooled and added to the other components of the MSAMO. The reductase fraction from the Q-Sepharose column was active for at least two hours when kept on ice, but lost activity immediately during subsequent purification steps. Similar difficulties have been reported for reductases from other oxygenases (A. M. Cook, personal communication).

4.2. The Hydroxylase Component

Further purification of the hydroxylase-containing fraction B from the Q-Sepharose column was attempted with anion-exchange (Mono Q), hydrophobic interaction (Phenyl-Superose, Pharmacia), affinity (HiTrap Blue), gel filtration (Superdex 200), hydroxyapatite (Macro-Prep Ceramic, Typ I) and cation-exchange (HiTrap SP, Pharmacia) chromatographic methods. HiTrap Blue and HiTrap SP both have a sulfonate group as part of their protein-binding ligand. When hydroxylase-containing extract was eluted from the HiTrap Blue column, a polypeptide was retained which later turned out to be the large subunit of the hydroxylase, indicating that this subunit binds the sulfonate substrate MSA. When loaded onto HiTrap SP neither of the subunits bound to the column.

Purification of the hydroxylase component was achieved with a three step chromatographic procedure consisting of Q-Sepharose, hydroxyapatite and Mono Q. According to SDS-PAGE the hydroxylase is an $\alpha\beta$-protein with a molecular mass of 43.7 kDa for the α-subunit, and 23.4 kDa for the β-subunit. Active fractions from the hydroxyapatite column and the Mono Q column contained an additional band of 21 kDa, which was not present in fractions from the first column. This band is likely to be a degradation product of the small subunit. When run on a Superdex 200 column, the hydroxylase eluted with a mass of around 200 kDa, indicating an $\alpha_3\beta_3$ structure for the native protein. The oxidized spectrum of the hydroxylase exhibited maxima at 330 nm, 465 nm and 570 nm, which is typical for ironsulfur centres. According to these preliminary results, the hydroxylase component of the MSAMO appears to be similar to hydroxylases from several hydroxylating dioxygenases which have ironsulfur centres as cofactors and an $\alpha_2\beta_2$ or an $\alpha_3\beta_3$ structure.

4.3. The Ferredoxin Component

The component C of the MSAMO was purified to homogeneity by a three step procedure including Q-Sepharose, Superdex 75 and Mono Q chromatography (Higgins *et al.* 1997). Evidence for a Rieske [2Fe-2S] center was found by absorbance spectrometry and element analysis. The protein has a mass of 16.6 kDa (SDS-PAGE) or 13,752 Da (mass spectrometry) respectively, and formed homodimers. The gene encoding protein C was cloned and identified by N-terminal sequencing. Predictions made from the gene sequence matched well with the biochemical results. A BLAST research showed high similarity of protein C to ferredoxin components of several oxygenases, identifying protein C as the electron transfer component of the MSAMO. In particular protein C was closely related to ferredoxin components of two toluene monooxygenases (Higgins *et al.* 1997). The hydroxylase component of some toluene-monooxygenases has been shown to contain a μ-oxo bridged binuclear iron center (Byrne *et al.* 1995; Yen *et al.* 1991) like in the soluble methane monooxygenase (Nordlund *et al.* 1992), something which has yet to be demonstrated for the hydroxylase component of the MSAMO.

5. MOLECULAR BIOLOGY OF THE MSAMO

5.1. The MSAMO Gene Cluster

Two MSA-induced polypeptides, designated C and D, were purified and sequenced at their N-termini. This amino acid sequence information was then used to design and synthesize oligonucleotide probes for the identification of genes on the chromosome of M2 which were involved in the degradation of MSA. A clone with a 6.5 kbp *Sph*I insert was chosen from a gene library and sequenced. It contained the genes for all three components of the MSAMO enzyme, organized in a cluster. The first two genes (B1 and B2) encode for the large and the small subunit of the hydroxylase, the second gene (C) encodes for the ferredoxin, and the third (gene A) for the reductase component. Gene C (Higgins *et. al.* 1997) and the genes B1 and B2 were confirmed by comparison with amino acid sequences at the N-termini of the ferredoxin and the hydroxylase polypeptides. Gene A has yet to be confirmed as the reductase component of the MSAMO. Sequence analysis reveals a high degree of similarity between this gene and the genes encoding reductases from other oxygenases. The upstream region of gene B1 has open reading frames for two or possibly three more proteins which, on the basis of preliminary sequence analysis, are believed to encode for an ABC-type transport system that might transport MSA into the cell. A portion of the corresponding gene cluster has also been cloned from the marine methanesulfonate-utilizer *Marinosulfonomonas methylotropha* strain PSCH4. Preliminary DNA sequence analysis suggests that the methanesulfonate monooxygenase genes of this organism are also linked on the chromosome and also, unlike in strain M2, there appear to be two copies of this gene cluster.

5.2. Ecology of Methanesulfonate-Utilizing Bacteria

A number of methanesulfonate-utilizing enrichment cultures have been established using a variety of different environmental samples. Preliminary analysis indicates a diversity of bacteria in these enrichments. Using DNA sequence information from the MSAMO gene clusters of strain M2 and strain PSCH4, PCR primers have been designed which are being used to amplify specific DNA fragments encoding the larger subunit of the hydroxylase component of MSAMO. In this way the genetic diversity in methanesulfonate-degraders in the environment can be assessed.

6. HIGHER ALKANE SULFONATE-UTILIZING BACTERIA

From different soils, rainwater and compost aerobic, rod-shaped bacteria were enriched and isolated with ethane-, propane-, and butanesulfonate as sole carbon and energy source. Apart from one isolate, the bacteria stained Gram positive and accumulated sulfite in the medium during growth on sulfonates. Growth was good with propane-, butane-, and pentanesulfonate, but poor with ethanesulfonate. Other sulfonates like taurine, isethionate and cysteate also supported growth, but benzenesulfonate did not. The different isolates are very similar with respect to their substrate spectrum, pH optimum, temperature optimum and salt requirements. According to an RFLP analysis, they fall into three different groups. A more detailed analysis will reveal the phylogenetic relationship of the new isolated bacteria. Physiological studies will concentrate on the question, of whether breakdown of higher alkanesulfonates involves a monooxygenase enzyme like that responsible for the breakdown of MSA.

REFERENCES

Baker, S.C., Kelly, D.P., and J.C. Murrell (1991) Microbial degradation of methanesulfonic acid: a missing link in the biogeochemical sulfur cycle. Nature 350:627–628.

Batie, C.J., LaHaie, E., and D.P. Ballou (1987) Purification and characterization of phthalate oxygenase and phthalate oxygenase reductase from *Pseudomonas cepacia*. J. Biol. Chem. 262:1510–1518.

Byrne, A.M., Kukor, J.J., and R.H. Olsen (1995) Sequence-analysis of the gene cluster encoding toluene-3-monooxygenase from *Pseudomonas picketti* PK01. Gene 154:65–70

Charlson, R.J., Lovelock, J.E., Andrea, M.O., and S.G. Warren (1987) Oceanic phytoplankton, atmospheric sulphur cloud albedo and climate. Nature 326:655–661.

Chien, C.-C., Leadbetter, E.R., and W. Godchaux III (1995) Sulfonate-sulfur can be assimilated for fermentative growth. FEMS Microbiol. Letters 129:189–194.

De Marco, P. (1996) Molecular biology and genetics of methanesulfonic acid-utilising bacteria. PhD Thesis, University of Warwick.

Denger, K., Laue, H., and A.M Cook (1997) Anaerobic taurine oxidation: A novel reaction by a nitrate-reducing Alcaligenes sp. Microbiology 143:1919–1924.

Haigler, B.E. and D.T. Gibson (1990) Purification and properties of NADH-Ferredoxin$_{NAP}$Reductase, a component of naphthalene dioxygenase from *Pseudomonas* sp. strain NCIB 9816. J. Bacteriol. 172:457–464.

Higgins, T.P., Davey, M., Trickett, J., Kelly, D.P., and J.C. Murrell (1996) Metabolism of methanesulfonic acid involves a multicomponent monooxygenase enzyme. Microbiology 142:251–260.

Higgins, T.P., De Marco, P., and J.C. Murrell (1997) Purification and molecular characterization of the electron transfer protein of methanesulfonic acid monooxygenase. J. Bacteriol. 179:1974–1979.

Holmes, A.J., Kelly, D.P., Baker, S.C., Thompson, A.S., De Marco, P., Kenna, E.M., and J.C. Murrell (1997) *Methylosulfonomonas methylovora* gen. nov., sp. nov., and *Marinosulfonomonas methylotropha* gen. nov., sp. nov.: novel methylotrophs able to grow on methanesulfonic acid. Arch. Microbiol. 167:46–53.

Holst, P.B., Nielsen, S.E., Anthoni, U., Bisht, K.S., Christophersen, C., Gupta, S., Parmar, V.S., Nielsen, P.H., Sahoo, D.B., and A. Singh. (1994) Isethionate in certain red algae. J. Appl. Phycol. 6:443–446.

Huxtable, R.J. (1992) Physiological actions of taurine. Physiol. Rev. 72:101–163.

Johnston, J.B., Murray, K., and R.B. Chain (1975) Microbial metabolism of aryl sulphonates. A reassessment of colorimetric methods for the determination of sulphite and their use in measuring desulphonation of aryl and alkylbenzene sulphonates. Antonie Leuweenhoek. 41:493–511.

Kelly, D.P., Malin, G., and A.P. Wood (1993) Microbial transformations and biogeochemical cycling of one-carbon substrates containing sulphur, nitrogen or halogens. In: Microbial growth on C1 compounds (Murrell, J.C. and D.P. Kelly eds.) pp 47–64. Intercept, Andover, UK.

Kelly, D.P., Baker, S.C., Trickett, J., Davey, M., and J.C. Murrell (1994) Methanesulphonate utilization by a novel methylotrophic bacterium involves an unusual monooxygenase. Microbiology 140:1419–1426.

Kertesz, M.A., Cook, A.M., and T. Leisinger (1994) Microbial metabolism of sulfur-and phosphorus-containing xenobiotics. FEMS Microbiol. Reviews 15:195–215.

Koechlin, B.A. (1954) The isolation and identification of the major anion fraction of the axoplasm of squid giant nerve fibers. Proc. Natl. Acal. Scie. USA 40:60–62.

Locher, H.H., Thurnheer, T., Leisinger, T., and A.M. Cook. (1989) 3-nitrobenzenesulfonate, 3-aminobenzenesulfonate, and 4-aminobenzenesulfonate as sole carbon-sources for bacteria. Appl. Environ. Microbiol. 55:492–494.

Nordlund, P.H., Dalton, H., and E. Eklund (1992) The active site structure of methane monooxygenase is closely related to the binuclear iron center of ribinucleotide reductase. FEBS Lett. 307:257–262.

Saigne, C. and Legrand (1987) Measurements of methanesulphonic acid in Antarctic ice. Nature 330:240–242.

Seitz, A.P., Leadbetter, E.R., and W. Godchaux III (1993). Utilization of sulfonates as sole sulfur source by soil bacteria including *Comamonas acidovorans* Arch. Microbiol. 159:440–444.

Thompson A.S., Owens, N.J.P., and J.C. Murrell (1995) Isolation and characterization of methanesulfonic acid-degrading bacteria from the marine environment. Appl. Environ. Microbiol. 61:2388–2393.

Yen, K.-M., Karl, M.R., Blatt, L.M., Simon, M.J., Winter, R.B., Fausset, P.R., Lu, H.S., Harcourt, A.A., and K.K. Chen (1991) Cloning and characterization of a *Pseudomonas mendocina* KR1 gene cluster encoding toluene-4-monooxygenase. J. Bacteriol. 173:5315–5327.

5

BACTERIAL FLUORANTHENE DEGRADATION

Indication for a Novel Degradation Pathway

Klaus Rehmann,[1,2] Norbert Hertkorn,[1] and Antonius A. Kettrup[1,2]

[1]GSF-National Research Center for Environment and Health
Institute of Ecological Chemistry
Ingolstädter Landstraße 1, D-85764 Neuherberg, Germany
[2]Technical University Munich
Chair of Ecological Chemistry and Environmental Analytics
D-85350 Freising-Weihenstephan, Germany

SUMMARY

During the last fifteen years it became evident that Polycyclic Aromatic Hydrocarbons (PAHs) encompassing more than three rings despite their low water solubility may serve as growth substrates for several kinds of soil bacteria.

Mycobacterium sp. strain KR20 was isolated from a PAH contaminated soil of a former gasworks by its ability to utilise fluoranthene as sole source of carbon and energy. Five novel fluoranthene metabolites were isolated and identified from fluoranthene grown cultures of this isolate, namely *cis*-2,3-fluoranthene dihydrodiol, z-9-carboxymethylene-fluorene-2-carboxylic acid, *cis*-1,9a-dihydroxy-1-hydrofluoren-9-one-8-carboxylic acid, 4-hydroxy-benzochromene-6-one-7-carboxylic acid, and benzene-1,2,3-tricarboxylic acid. They possibly indicate a novel metabolic pathway for the bacterial metabolisation of fluoranthene.

INTRODUCTION

Among other, artificial wetland wastewater treatment plants are applied within pilot scale bioremediation projects in Germany to clean ground water at former carbonising plant or gasworks sites contaminated by Polycyclic Aromatic Hydrocarbons (PAHs). Due to the mutagenic and carcinogenic properties of this class of compounds and their respective degradation products (Harvey, 1991) there is a reasonable interest in the behaviour and fate of this substances during and after the cleanup process. Thus the behaviour and

Novel Approaches for Bioremediation of Organic Pollution, edited by
Fass *et al.* Kluwer Academic / Plenum Publishers, New York, 1999.

fate of selected PAHs (phenanthrene and fluoranthene) and a PAH contaminated groundwater in a model artificial wetland plant type Aquaplant® was investigated at the Institute of Ecological Chemistry in co-operation with the Umweltschutz Nord company over a three year period (Machate *et al.*, 1997). As the microbial flora contributes to a large extent to the degradation processes in such systems, detailed accompanying studies were performed to gain information on microbial PAH-metabolism and possible breakdown products to be expected during the cleaning process.

Here we describe the identification of several novel bacterial metabolites of fluoranthene produced by *Mycobacterium* sp. strain KR20 and discuss the implications of our results for a novel bacterial degradation pathway for fluoranthene. Detailed spectroscopic data of all metabolites will be given elsewhere (Rehmann *et al.*, in preparation).

MATERIAL AND METHODS

Chemicals and microbiological cultivation techniques as well as analytical procedures were essentially the same as recently described (Rehmann *et al.*, 1998). Further information concerning the structure elucidation of the fluoranthene metabolites, e. g. special NMR methods applied, will be published soon (Rehmann *et al.*, in preparation). Briefly proceeding was as follows:

Using samples of PAH contaminated soil from a former gasworks site a mixed culture was set up to enrich for microorganisms capable of utilising fluoranthene as their sole source of carbon and energy. Fluoranthene degrading bacteria from this enrichment cultures were visualised by the PAH-spray-agar plate method of Kiyohara *et al.*, 1982, isolated by conventional microbiological techniques and preliminary characterised by selected physiological tests. The fluoranthene degrading isolate KR20 was grown at 20°C, in a mineral salt medium (Weißenfels *et al.*, 1991) without organic additives, supplying 0.5 mg/mL of fluoranthene as the sole source of carbon and energy. Cultures were transferred by inoculating 10 % (v/v) every 14 d.

Determination of the fluoranthene and fluoranthene metabolite content of the cultures was performed by RP-HPLC analysis of ethyl acetate extracts from acidified cultures. Starting from this extracts preparative isolation of metabolites was carried out by RP-LC separation and, if necessary, a second purification step based on Silica gel LC or DC. Purified metabolites were characterised by UV-Vis spectroscopy (spectra recorded by a diodearray detector during HPLC analysis), NMR-spectroscopy (^{1}H-NMR spectra, ^{13}C-NMR spectra, and combined two-dimensional spectra), mass spectroscopy (GC-MS analysis of trimethyl-silylated derivatives and direct infusion LC-MS analysis) and by HPLC co-chromatographic comparison to reference substances if they were commercially available.

RESULTS

A bacterium capable of utilising fluoranthene as growth substrate was isolated from a bacterial consortium enriched from a PAH contaminated soil of a former gasworks area by supplying fluoranthene as the sole source of carbon and energy. The strain, isolate KR20, did not require any organic supplements besides fluoranthene and multiplied at least 30-fold (CFU) under the growth conditions selected within 7 d when inoculated at 10% (v/v) from a 14 d old culture pregrown on fluoranthene.

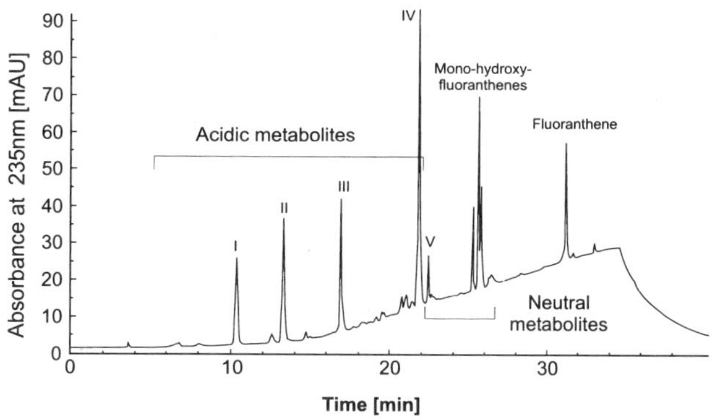

Figure 1. Combined RP-HPLC elution profiles of the neutral and acidic ethyl acetate extract of a 14 d old fluoranthene grown culture of *Mycobacterium* sp. strain KR20. For experimental details see Rehmann *et al.*, 1998.

The rod shaped (diameter 0.5 µm - 0.8 µm, length 1.5 µm - 2 µm), non motile bacterium was scotochromogenic and Gram negative. Due to its strong acid fastness, its mycolic acid pattern, which was similar to that of *Mycobacterium phleii* DSM 43239 (Minnikin *et al.*, 1975) and its ability to utilise Tween 80® (Wayne *et al.*, 1974) strain KR20 was assigned to the genus *Mycobacterium*.

RP-HPLC analysis of ethyl acetate extracts of fluoranthene grown cultures revealed the presence of at least five major metabolites (I - V, Figure 1) four of which (I - IV) could only be extracted from acidified cultures.

Comparison of their UV-Vis spectra (Figure 2) to UV-Vis spectra of PAH metabolites already published (LaVoie *et al.*, 1982; Grifoll *et al.*, 1994) gave first hints on the chemical structures of metabolites V, II, and III.

By its UV-Vis spectroscopic properties metabolite V was assigned the structure of a 2,3-fluoranthene dihydrodiol (LaVoie *et al.*, 1982). The structural assignment was confirmed by GC-MS analysis of the trimethyl-silylated derivative which gave the predicted M^+ peak at 380 m/z. A decision on the *cis*- or *trans*-configuration of the hydroxy groups was made based on the ^{1}H-NMR data of the compound: As the 2H/3H ^{3}J coupling constant was 5.4 Hz the hydroxyl groups occupied the *cis*-position since the corresponding ^{3}J of the *trans*-isomer was reported to be 8 Hz (Rice *et al.*, 1983).

Metabolite II had a UV-Vis spectrum almost identical to the UV-Vis spectrum of *cis*-1,9a-dihydroxy-1-hydrofluorene-9-one which was demonstrated to be a bacterial fluorene metabolite (Grifoll *et al.*, 1994). However the mass spectrum of the trimethyl-silylated derivative of metabolite II did not show a M^+ signal at 358 m/z as expected for trimethyl-silylated *cis*-1,9a-dihydroxy-1-hydrofluorene-9-one but at 416 m/z. The occurrence of a carboxy-signal in the ^{13}C-NMR spectrum related to ring position N° 8 of the latter compound as well as the presence of a methyl-proton singulett in the ^{1}H-NMR spectrum suggested that the observed mass difference might be attributable to a carboxy methyl ester group. Hence metabolite II was assigned the structure of *cis*-1,9a-dihydroxy-1-hydrofluorene-9-one-8-carboxylic acid methyl ester. The methoxy group probably represented an artefact of the metabolite cleanup procedure which encompassed a concentration of an acidic methanolic solution of the raw metabolite.

The UV-Vis and NMR spectroscopic properties of metabolite III also strongly reminded of a fluorene metabolite described by Grifoll *et al.*, 1994: 1-hydroxy-3,4-benzocoumarin. But again the ^{13}C-NMR spectrum revealed the presence of an additional carboxy group, which could be located at ring position N° 8 of the molecule mentioned. The resulting structure of 4-hydroxy-benzochromene-6-one-7-carboxylic acid was con-

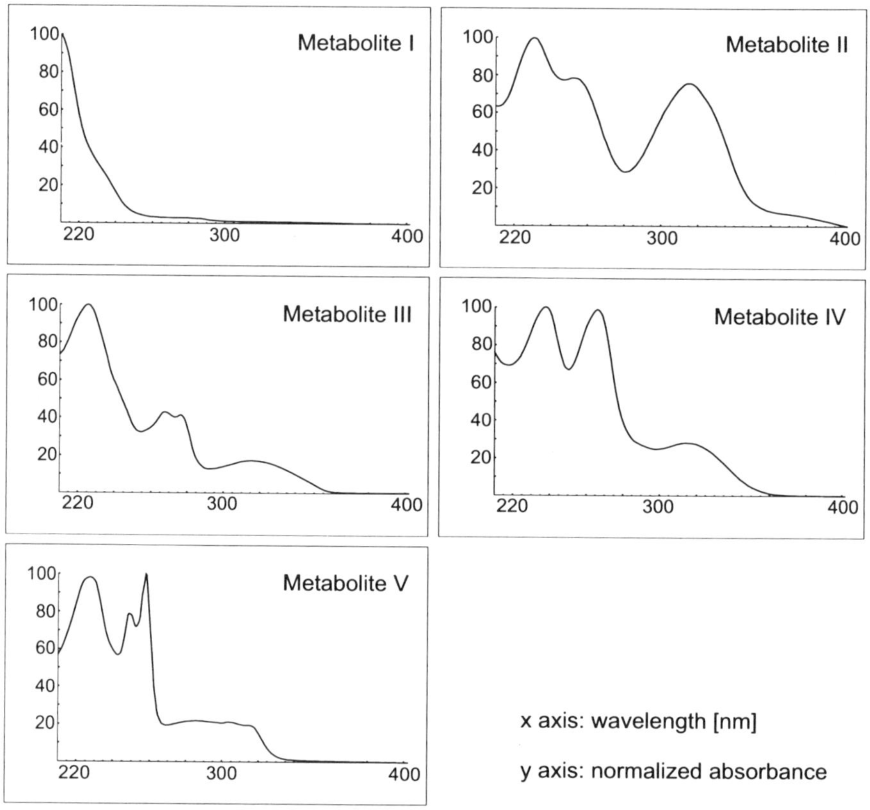

Figure 2. UV-Vis spectra of the fluoranthene metabolites produced by *Mycobacterium* sp. strain KR20 as recorded during RP-HPLC analysis of ethyl acetate culture extracts by the diodearray detector. For experimental details see Rehmann *et al.*, 1998.

firmed by GC-MS of the trimethyl-silylated derivative which showed a M^+ signal at 400 m/z.

The ^{1}H-NMR spectrum of metabolite I showed a signal pattern typical for 1,2,3-symmetrically substituted benzene derivatives. The corresponding ^{13}C-NMR spectrum encompassed seven signals, two of which showed a chemical shift characteristic for carboxylic C-atoms. As the intensities of this two signals were very different we assumed that the stronger of both resulted from two carboxy-C-atoms. Thus metabolite I was assigned the structure of benzene-1,2,3-tricarboxylic acid (hemimellitic acid). A HPLC-MS analysis confirmed the structural assignment as metabolite I produced the predicted M-H molecule ion peak at 209 m/z. Finally the identity of metabolite I and hemimellitic acid was verified by HPLC-co-chromatography of commercially available hemimellitic acid and metabolite I.

The structure of metabolite IV, *Z*-9-carboxymethylene-1-fluorene carboxylic acid, was completely deduced from its ^{1}H- and ^{13}C-NMR data. The M^+ signal of the trimethyl-silylated derivative at 410 m/z and its fragmentation pattern (Zink and Lorber, 1995) was completely in accordance with this structure. However, attempts to further prove the identity of the compound by synthesising it by oxidative ring cleavage of 2,3-fluoranthene quinone, which is easily accessible from fluoranthene, were not successful (Kneifel, H. 1997, personal communication).

DISCUSSION

Currently the database on bacterial fluoranthene metabolites and the corresponding degradation pathway(s) is very small and, as can be seen from the data published by Weißenfels *et al.*, 1991 and Kelley and Cerniglia, 1991, especially suffers from missing proofs for early metabolic intermediates. Whereas Weissenfels *et al.* proposed a initial attack on the benzylic moiety of the fluoranthene skeleton, Kelley and Cerniglia, proposed two different primary dioxygenation positions: the 7,8-region according to the former authors but additionally in the naphthoic substructure of the fluoranthene molecule in 1,2-position. The structures of the five fluoranthene metabolites identified from cultures of *Mycobacterium* sp. strain KR20 however suggest a third possible degradation route (Figure 3, A).

Fluoranthene metabolism by this hypothetical pathway starts with a dioxygenation of the fluoranthene molecule in 2,3-position, yielding *cis*-2,3-fluoranthene dihydrodiol (metabo-

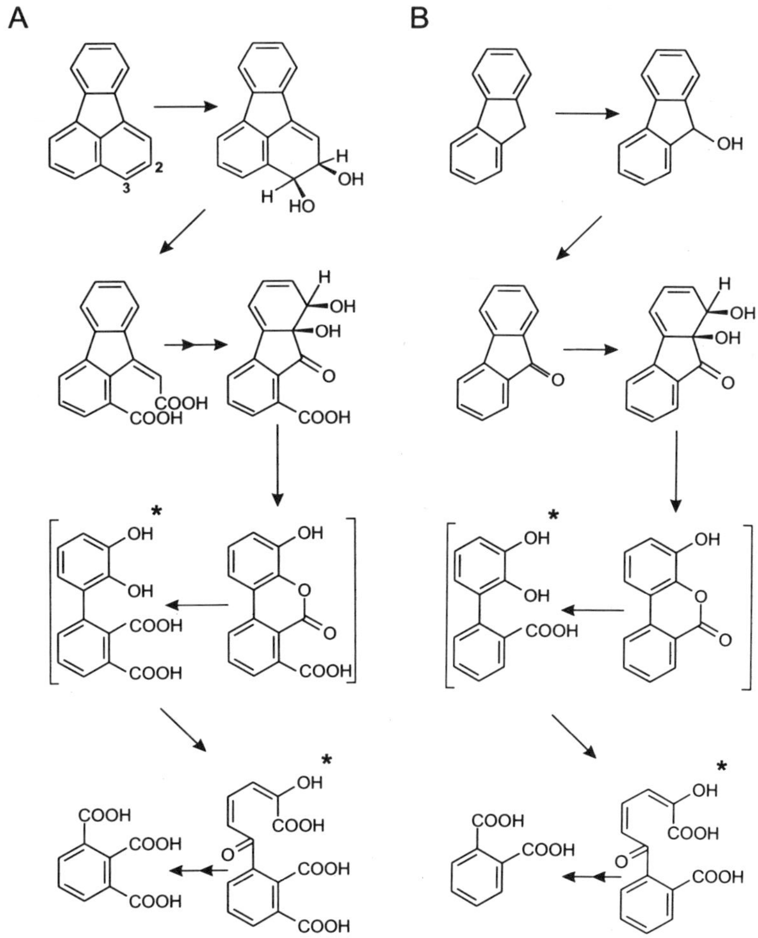

Figure 3. Degradation pathways for fluoranthene in *Mycobacterium* sp. strain KR20 (A), and fluorene in *Pseudomonas* sp. strain F274 (B) (as reported by Grifoll *et al.*, 1994.) Compounds not detected in the corresponding cultures are marked with an asterisks.

lite V). The expected follow up product 2,3-dihydroxyfluoranthene could not be detected yet. However the same holds for the corresponding *ortho*-dihydroxy derivatives in the degradation pathways of anthracene (Fernley *et al.*, 1964) and pyrene respectively (Heitkamp *et al.*, 1988; Walter *et al.*, 1991; Dean-Ross and Cerniglia, 1996; Rehmann *et al.*, 1998) probably because of a strong association of the dihydroxy metabolites with the enzymes involved in the formation and / or cleavage of this intermediates. Ortho ring cleavage of 2,3-dihydroxyfluoranthene leads to the formation of *Z*-9-carboxymethylene-fluorene-1-carboxylic acid (metabolite IV) which subsequently looses a C_2-unit producing 1-carboxyfluoren-9-one. However this compound in contrast to its successor *cis*-1,9a-dihydroxy-2-hydrofluorene-9-one-8-carboxylic acid (metabolite II) could also not be detected in cultures of isolate KR20 but was found to be a prominent metabolic intermediate of fluoranthene degradation in cultures of *Mycobacterium* sp. strain PYR1 (Kelley *et al.*, 1993). *Cis*-1,9a-dihydroxy-2-hydrofluorene-9-one-8-carboxylic acid forms the precursor of the second ring opening reaction which might proceed in one of two alternative ways. Either by a biological Baeyer-Villiger reaction as suggested in a similar form by Weißenfels *et al.*, 1991 or by dehydrogenation of the secondary hydroxy group at C2 resulting in the formation of a spontaneously hydrolysing 1,3-diketone whose hydrolysis product under acidic conditions (as given during metabolite cleanup) would easily dehydrate to yield the corresponding δ-lactone (Grifoll *et al.*, 1994). Anyhow, the resulting metabolic product 4-hydroxy-benzochromene-6-one-7-carboxylic acid (metabolite III) after hydrolysis of the lactone is most likely metabolised analogous to biphenyl (Higson, 1992) resulting in the formation of hemimellitic acid (metabolite I) as the final aromatic intermediate. The fate of this latter compound is unclear at the moment.

Two facts support the reality of the pathway proposed: 1. None of the metabolites identified accumulated in cultures of strain KR20 after 480 h of incubation. Only intermediary concentration maxima could be observed (data not shown). 2. The metabolic route suggested bears a striking similarity to a fluorene degradation pathway found in a *Pseudomonas* sp. (Grifoll *et al.*, 1994, Figure 3, B). Exactly spoken, both pathways are completely analogous beginning from the second dihydrodiol with only one difference: The fluoranthene metabolites carry an additional carboxy group in one of the aromatic rings.

As the results described were completely obtained with laboratory cultures the question remained whether this metabolic route is of any significance under *in situ* conditions. Analysis of water samples taken from the artificial wetland plant (see introduction) during a fluoranthene degradation experiment revealed the presence of two of the fluoranthene metabolites described. Z-9-carboxymethylene-fluorene-1-carboxylic acid and 1-hydroxy-3,4-benzocoumarin-8-carboxylic acid could unambiguously be identified by comparison of HPLC retention times and UV-Vis spectra (data not shown).

ACKNOWLEDGMENTS

We thank Mr. A. Papaderos for the preparative isolation of some fluoranthene metabolites, Dr. T. Machate (Umweltschutz Nord GmbH) and Dr. H. Noll for supplying data from the artificial wetland wastewater treatment plant, Mrs. L. Lattmann and Mr. S. Achatz for performing the GC-MS and HPLC-MS-measurements respectively, and Prof. Dr. H. Kneifel (ICG 6, FZ Jülich) for helpful discussion.

REFERENCES

Dean-Ross, D., and Cerniglia, C. E. 1996. Degradation of Pyrene by *Mycobacterium flavescens*. *Appl. Microbiol. Biotechnol.* **46**: 307–312.

Fernley, H. N., Griffiths, E., and Evans, W. C. 1964. Oxidative Metabolism of Phenanthrene and Anthracene by Soil Bacteria: The Initial Ring-Fission Step. *Biochem. J.* ***91***: 15p-16p.

Grifoll, M., Selifonov, S. A., and Chapman, P. J. 1994. Evidence for a Novel Pathway in the Degradation of Fluorene by *Pseudomonas* sp. Strain F274. *Appl. Environ. Microbiol.* **60**: 2438–2449.

Harvey, R. G. 1991. Polycyclic Aromatic Hydrocarbons: Chemistry and Carcinogenicity. New York, Port Chester, Melbourne, Sydney, Cambridge University Press.

Heitkamp, M. A., J. P. Freeman, J. P., Miller, D. W., and Cerniglia, C. E. 1988. Pyrene Degradation by a *Mycobacterium* sp. : Identification of Ring Oxidation and Ring Fission Products. *Appl. Environ. Microbiol.* **54**: 2556–2565.

Higson, F. 1992. Microbial Degradation of Biphenyl and Its Derivatives. *Adv. Appl. Microbiol.* **37**: 135–164.

Kelley, I., Freeman, J., Evans, F. E., and Cerniglia, C. E. 1993. Identification of Metabolites from the Degradation of Fluoranthene by *Mycobacterium* sp. Strain PYR-1. *Appl. Environ. Microbiol.* **59**: 800–806.

Kiyohara, H., Nagao, K. and Yana, K. 1982. Rapid Screen for Bacteria Degrading Water-Insoluble, Solid Hydrocarbons on Agar Plates. *Appl. Environ. Microbiol.* **43**: 454–457.

LaVoie, E. J., Hecht, S. S., Bedenko, V., and Hoffmann, D. 1982. Identification of the Mutagenic Metabolites of Fluoranthene, 2-Methylfluoranthene, and 3-Methylfluoranthene. *Carcinogenesis* **3**: 841–846.

Machate, T., Noll, H., Behrens, H. and Kettrup, A. 1997. Degradation of Phenanthrene and Hydraulic Characteristics in a Constructed Wetland. *Wat. Res.* **31**: 554–560.

Minnikin, D. E., Alshamaony, L., and Goodfellow, M. 1975. Differentiation of *Mycobacterium*, *Nocardia*, and Related Taxa by Thin-layer Chromatographic Analysis of Whole-Organism Methanolysates. *J. Gen. Microbiol.* **88**: 200–204.

Rehmann, K., Noll, H. P., Steinberg, C. E. W., and Kettrup, A. A. 1998. Pyrene Metabolism by *Mycobacterium* sp. Strain KR2. *Chemosphere* **36**: 2977–2992.

Rice, J. E., LaVoie, E. J., and Hofmann, D.1983. Synthesis of the Isomeric Phenols and the *trans*-2,3-Dihydrodiol of Fluoranthene. *J. Org. Chem.* **48**: 2360–2363.

Walter, U., Beyer, M., Klein, J., and Rehm, H.-J. 1991. Degradation of Pyrene by *Rhodococcus* sp. UW1. *Appl. Microbiol. Biotechnol.* **34**: 671–676.

Wayne, L. G., Engbœk, C., Engel, H. B. W., Froman, S., Gross, W., Hawkins, J., Käppler, W., Karlson, A. G., Kleeberg, H. H., Krasnow, I., Kubica, G. P., McDurmont, C., Nel, E. E., Pattyn, S. R., Schröder, K. H., Showalter, S., Tarnok, I., Tsukamura, M., Vergmann, B., and Wolinsky, E. 1974. Highly Reproducible Techniques for Use in Systematic Bacteriology in the Genus *Mycobacterium*: Tests for Pigment, Urease, Resistance to Sodium Chloride, Hydrolysis of Tween80, and β-Galactosidase. *Int. J. Syst. Bacteriol.* **24**: 412–419.

Weißenfels, W. D., Beyer, M., Klein, J., and Rehm, H.-J. 1991. Microbial Metabolism of Fluoranthene: Isolation and Identification of Ring Fission Products. *Appl. Microbiol. Biotechnol.* **34**: 528–535.

Zink, G., and Lorber, K. E. 1995. Mass Spectral Identification of Metabolites Formed By Microbial Degradation of Polycyclic Aromatic Hydrocarbons (PAH). *Chemosphere* **31**: 4077–4084.

6

STRATEGIES FOR THE AEROBIC BIODEGRADATION OF DINITROTOLUENES

Shirley F. Nishino and Jim C. Spain*

Air Force Research Laboratory/MLQR
139 Barnes Dr., Ste. 2, Tyndall AFB, Florida 32403-5323

1. INTRODUCTION

Dinitrotoluenes (DNT) are intermediates in the production of the explosive 2,4,6-trinitrotoluene (TNT) and precursors of toluenediisocyanate used in the manufacture of polyurethane foams (Hartter, 1985) Disposal practices associated with TNT manufacturing dating back to World Wars I and II, have resulted in an enormous contamination problem at ammunition production and handling facilities in the US and worldwide. Although TNT is no longer produced in the US, the current annual worldwide production of DNT is 2.7 billion pounds. Both 2,4- and 2,6-DNT are listed as US EPA priority pollutants in the Clean Water Act, the Safe Drinking Water Act, the RCRA, and the CERCLA.

1.1. Rationale for Bioremediation of DNT

To date there are no biological methods to effectively mineralize TNT under aerobic conditions without the production of unwanted byproducts. As a result very costly incineration (\$200–300 per ton soil, (Bartell, 1989)) and more recently, composting have been the favored methods for treatment of soils contaminated with TNT, its precursors, and by-products. Sorption to activated carbon has been the standard for treatment of contaminated water.

At the turn of the century, German scientists developed the sequential batch method for production of TNT. Under acidic conditions, toluene undergoes mono-, di- and trinitration (Figure 1). Each nitration step in the sequence was physically isolated to reduce the danger of explosion, and as a consequence mononitrotoluene (MNT) and DNT contamination of soil and groundwater are often not congruent with TNT contamination. Because MNT and DNT are more soluble than TNT, groundwater also frequently contains much

* Corresponding Author: Phone (850) 283-6058, FAX (850) 283-6090, JSpain@mlg.afrl.af.mil

Novel Approaches for Bioremediation of Organic Pollution, edited by Fass *et al.* Kluwer Academic / Plenum Publishers, New York, 1999.

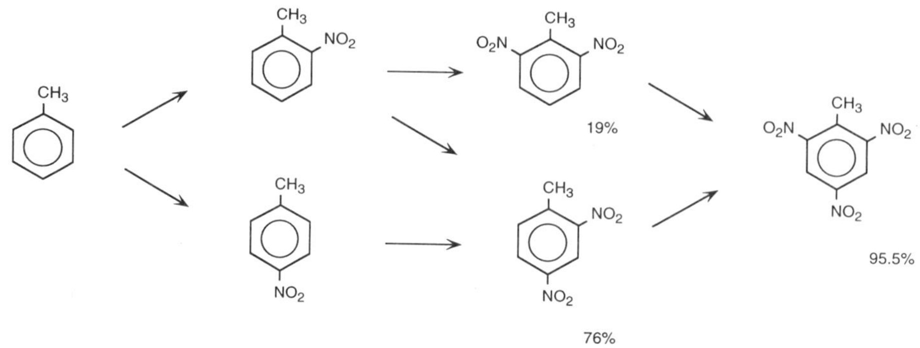

Figure 1. Production of TNT. Percentages indicate yield of intermediates and products at each step.

higher levels of the lesser nitrated compounds. Specific bacteria capable of using mono- and dinitroaromatic compounds as growth substrates appear to be widespread at explosives contaminated sites (Nishino and Spain, 1996). The physical separation of nitrotoluene contamination coupled with our current understanding of MNT and DNT mineralization, presents the opportunity to reduce the overall expense of explosives remediation by segregation of the contaminated materials according to the nitroaromatic compounds involved, and treatment with the appropriate biotechnology.

1.2. Current Knowledge about DNT Degradation

1.2.1. Anaerobic Studies. DNT can undergo a variety of transformations under environmental conditions (Figure 2). One of the most commonly reported transformations has been nonspecific reduction of one or more of the nitro groups to form aminonitro- and diaminotoluenes via hydroxylamino- and nitroso- intermediates (Liu *et al.*, 1984; McCormick *et al.*, 1978). More recently, acetylation has been reported as a significant reaction of arylamines produced by reduction of DNT (Noguera and Freedman, 1996). The reduced intermediates are believed to be responsible for much of the toxicity attributed to DNT.

The nonspecific reductions are generally the result of a process in which the transformation of the chemical is incidental to the degradation of another chemical. Bacteria do not seem to derive any metabolic benefit from such transformations of DNT and the products are not further degraded. Cometabolic systems are difficult to control and do not completely destroy the contaminants. Thus, continued transformation of DNT requires input of another carbon source to provide energy to drive the reaction.

Two studies have shown that mixed cultures can reduce 2,4-DNT when ethanol is provided under anaerobic conditions as the primary carbon and energy source (Berchtold *et al.*, 1995; Cheng *et al.*, 1996). In both instances aromatic amines accumulated. In one of the systems an aerobic reactor in series removed the aminotoluenes (Berchtold *et al.*, 1995), but the process by which the aminotoluenes were degraded remains unknown. To date, efficient destruction of diaminotoluenes has not been demonstrated. The major strength of the anaerobic system is the ability to transform 2,4-DNT in the presence of other carbon sources.

1.2.2. Composting. Composting systems for explosives-contaminated soils have received much attention in recent years (Breitung *et al.*, 1996; Griest *et al.*, 1993; Griest *et al.*,

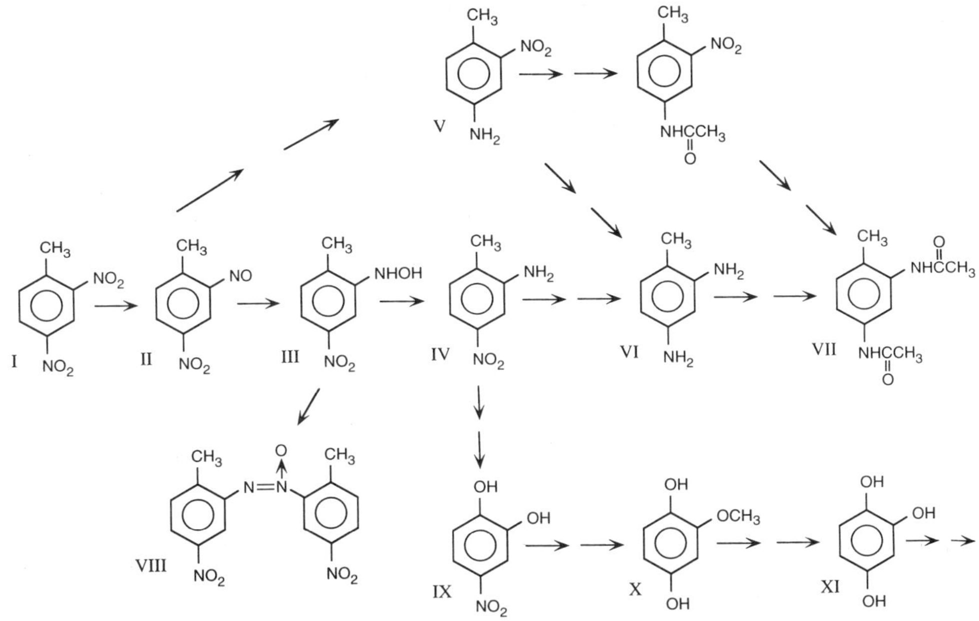

Figure 2. Cometabolic transformations of 2,4-dinitrotoluene. 2,4-DNT (I) can be reduced at either nitro group to nitroso- (II) and hydroxylamino- (III) nitrotoluenes (only one pathway is shown) to form aminonitrotoluene (IV, V). Aminonitrotoluene may be further reduced to diaminotoluene (VI). Any amino group may be acetylated (VII) to dead end products. Hydroxylaminonitrotoluenes can dimerize in any combination to form azoxynitrotoluene (VIII, only one shown), also dead end products. Fungal exoenzymes can produce the partially reduced intermediates (II - IV), then further attack ring substituents to form benzenetriol (XI) which is productively metabolized. Modified from (McCormick *et al.*, 1978; Valli *et al.*, 1992).

1994; Gunderson *et al.*, 1997; Thorne and Leggett, 1997), however the mechanisms of DNT degradation during composting have not been established. At present composting presents two major drawbacks. First, no more than 30% of the compost pile can be made up of the contaminated soil which results in an increase in the amount of product that must eventually be landfilled. Second, despite concerted efforts, residual toxicity of finished compost is still a possibility.

1.2.3. Mineralization. Under aerobic conditions some strains of bacteria can use DNT as a source of carbon and energy for growth. There is little or no toxicity associated with the endproducts of mineralization. Because the bacteria derive a benefit, mineralization of DNT results in a controllable self sustaining process.

Bausum et al. (Bausum *et al.*, 1992) reported mineralization of 2,4- and 2,6-DNT at concentrations up to 10 mg L^{-1} by natural river water populations collected downstream of a TNT-manufacturing plant following lag periods of up to 3 weeks. Neither the pathways involved nor the organisms responsible for the mineralization were characterized. Based upon rate studies they concluded that at low DNT concentrations, microbial degradation played only a minor role in the removal of DNT from surface waters. A series of recent papers (Bradley *et al.*, 1995; Bradley *et al.*, 1997; Bradley *et al.*, 1994) reports mineralization of 2,4- and 2,6-DNT by mixtures of microorganisms indigenous to a shallow aquifer at an explosives-contaminated site. In all cases DNT concentrations were less than 20 mg

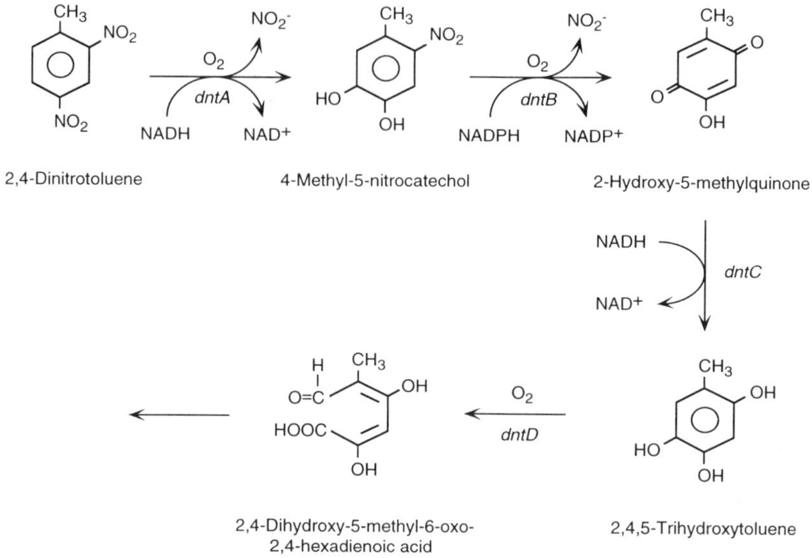

Figure 3. Pathway and genes for degradation of 2,4-DNT by *Burkholderia* sp. DNT. Modified from (Haigler *et al.*, 1994a).

L^{-1}. After 28 days 28% of initial 2,4-DNT was mineralized, 20% was undegraded, 28% was recovered as aminonitrotoluenes and 24% was in unidentified metabolites. For 2,6-DNT the percentages were 8, 67, 14, and 11%. The picture that emerges from the investigations of DNT degradation by natural assemblages is one of slow mineralization in competition with nonspecific transformation reactions.

Fungi mineralize DNT under ligninolytic conditions (Valli *et al.*, 1992). However, the ligninolytic systems do not appear to be involved in the initial reduction of DNT which requires live mycelia. The ligninolytic system is necessary for the mineralization of the aminoaromatic products of DNT reduction. The limiting factor in the fungal systems has been the toxicity of DNT at concentrations found in contaminated soils (Barr and Aust, 1994).

1.2.4. DNT-Degradation Pathways. Bacterial strains capable of mineralizing 2,4-DNT have been isolated and characterized and the degradation pathway determined (Haigler *et al.*, 1994a; Spain, 1995; Spanggord *et al.*, 1991). All 2,4-DNT-degrading isolates studied to date utilize the same pathway for mineralization of 2,4-DNT (Figure 3). The pathway proceeds through dioxygenation of 2,4-DNT to 4-methyl-5-nitrocatechol; monooxygenation of 4-methyl-5-nitrocatechol then yields 2-hydroxy-5-methylquinone which is subsequently reduced to 2,4,5-trihydroxytoluene prior to meta ring cleavage.

Strains that mineralize 2,4-DNT can degrade DNT in continuous, controlled systems. *Burkholderia* sp. DNT, the earliest reported DNT-degrading strain (Spanggord *et al.*, 1991), grew on 2,4-DNT in a fixed bed bioreactor at maximal 2,4-DNT load of 95 mg L^{-1} h^{-1} (Heinze *et al.*, 1995). When *Pseudomonas* sp. PR7 was immobilized on diatomaceous earth, the strain removed 2,4-DNT at a maximum rate of 59 mg L^{-1} h^{-1} (Reardon and Spain, 1993). These early isolates of 2,4-DNT-degrading bacteria were not, however, able to degrade the mixtures of DNT isomers (80:20 2,4-DNT:2,6-DNT) found at contaminated sites, and the presence of high concentrations of 2,6-DNT inhibited 2,4-DNT degradation (Nishino and Spain, 1995).

Recently discovered strains are able to use 2,6-DNT as the sole growth substrate (Nishino and Spain, 1996). In all strains examined to date, the initial transformation of 2,6-DNT produces 3-methyl-4-nitrocatechol via a dioxygenase attack similar to that used by 2,4-DNT degrading strains. During the initial dioxygenation, one of the nitro groups is released (Nishino and Spain, 1996). The aromatic ring of 3-methyl-4-nitrocatechol is opened by a catechol-2,3-dioxygenase. The remaining nitro group is released in as yet undescribed reactions subsequent to ring cleavage (Nishino and Spain, 1997).

Short-term (1–2 day) studies with radiolabeled DNT isomers and pure cultures of specific DNT-degrading bacteria demonstrated the release of 39 to 55% of the initial radiolabel as $^{14}CO_2$, as well as release of both nitro groups and 2 to 4 doublings in protein which indicates growth of the bacteria. DNT-degrading strains removed DNT from aqueous solutions containing as much as 1 mM DNT, the approximate limit of solubility of DNT in water at room temperature.

2. TREATMENT OF DNT CONTAMINATED WATER

The discovery of strains able to mineralize 2,6-DNT and more effective strains for the degradation of 2,4-DNT, has enabled the combined use of 2,4- and 2,6-DNT degrading bacteria for the simultaneous degradation of mixed DNT isomers. The use of mixtures of DNT-degrading strains to remediate groundwater contaminated with MNT and DNT was demonstrated in a fluidized bed reactor (Lendenmann *et al.*, 1998) in aqueous solutions containing 50 mg L^{-1} total DNT.

Both isomers of DNT were degraded effectively at high rates (Figure 4). Nitrate as well as nitrite was recovered because of the action of nitrite oxidizing bacteria present in the bioreactor. The sum of the nitrogen recovered as nitrite and nitrate represented greater than 90% of theoretical nitrogen released from DNT. Allowing for nitrogen used for growth, the total nitrogen released was stoichiometric with DNT degradation. Transient breakthrough of DNT occurred on days 57 and again on day 78 due to mechanical malfunctions. Upon resumption of aeration and circulation a significant amount of biomass was washed out, however performance was recovered within 4 days. Figure 5 shows the removal efficiencies in percentage of inflowing medium and the resulting effluent DNT concentrations at various hydraulic residence times. The FBR met the U.S. EPA treatment standards for DNT (40CFR 268.48) for all but the shortest hydraulic residence time.

Subsequent to the operation with DNT in tap water, contaminated ground water from Volunteer Army Ammunition Plant (VAAP) was supplied to the bioreactor as the sole influent. The groundwater contained 2,4-DNT (2.1 mg L^{-1}), 2,6-DNT (1.6 mg L^{-1}), 2-nitrotoluene (2-NT, 2.5 mg L^{-1}), and 4-nitrotoluene (4-NT, 1.7 mg L^{-1}). After 10 days operation at a hydraulic residence time of 3 h, nitrotoluene removal was 99.6, 62, 98 and 99.5% (2,4-DNT, 2,6-DNT, 2-NT, and 4-NT, respectively). The results indicated that bacteria able to degrade 2-NT and 4-NT could coexist with the DNT degrading biomass and extended the approach to include mixtures of isomers of both MNT with DNT. The mechanisms for mineralization of 2-NT (Haigler *et al.*, 1994b) and 4-NT (Haigler and Spain, 1993; Rhys-Williams *et al.*, 1993; Spiess *et al.*, 1998) have been worked out in detail. Discussion of the pathways is beyond the scope of this paper.

The success with the laboratory scale studies has led to a field study which is currently underway at VAAP near Chattanooga, TN. VAAP is a U.S. government owned 6700 acre TNT manufacturing facility where 2.9 billion pounds of TNT were produced while it was in active status. No TNT has been produced there since 1977, yet groundwater moni-

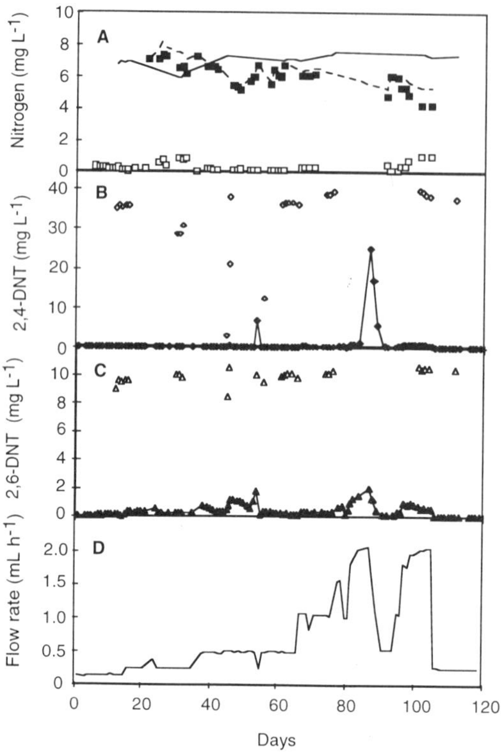

Figure 4. Degradation of mixed DNT isomers in a fluidized bed bioreactor. Starting with the bottom panel are flow rates, influent and effluent concentrations of 2,4-DNT and 2,6-DNT, and the released nitrogen species (nitrite and nitrate) (Lendenmann *et al.*, 1998).

toring wells at the site currently contain up to 76 mg L^{-1} of DNT and up to 57 mg L^{-1} MNT ((USAEC), 1995). Contaminated groundwater from the site is being supplied to a pilot scale FBR. The biomass is supported on a granular activated carbon bed. Preliminary results indicate that the pilot scale system functions as efficiently as the laboratory scale unit. Typical removal efficiencies are: 2,4-DNT, > 99%; 2,6-DNT, 70%; 2-NT, > 99%; 4-NT, > 99%; and TNT, 70%.

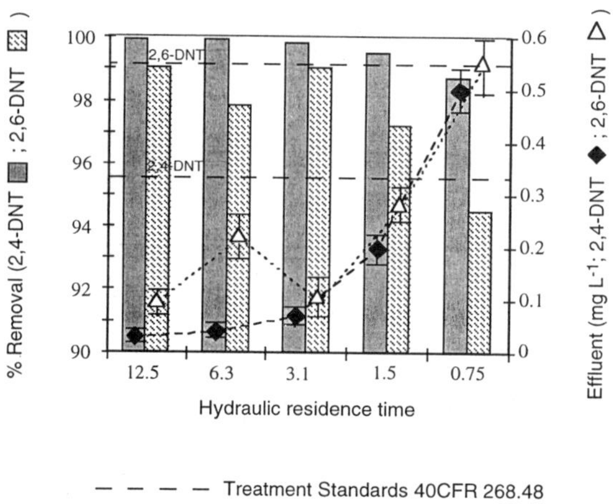

Figure 5. Efficiency of DNT removal as a function of hydraulic residence time in a fluidized bed reactor. Modified from (Lendenmann *et al.*, 1998).

3. BIODEGRADATION OF DNT IN CONTAMINATED SOIL

3.1. Radioisotope Studies in Recently Contaminated Soil

The majority of contamination at former explosives manufacturing sites is in soil where sorption can reduce the bioavailability of the contaminant. We therefore investigated whether bacteria could mineralize DNT associated with soil. *Burkholderia cepacia* JS872, a 2,4-DNT degrading strain, and *Hydrogenophaga palleronii* JS863, a 2,6-DNT degrading strain, were combined in non-sterile 10% (w/v) soil slurries containing 2,4-DNT (800 μM) and 2,6-DNT (200 μM). The nonsterile soils were a composite from a variety of sources with no history of DNT contamination. Both 2,4- and 2,6-DNT became nondetectable in the culture fluid and in organic extracts of the soil within 48 h. Two mol of nitrite were released per mol of DNT degraded indicating complete degradation of DNT. Over 40% of the initial radiolabel was trapped as $^{14}CO_2$, indicating levels of mineralization well within the norm for growth substrates (Figure 6). Mercury killed control cultures released no NO_2^- or $^{14}CO_2$, and DNT concentrations in the culture fluids remained unchanged. The preliminary results indicated clearly that bacteria can rapidly mineralize high concentrations of DNT in soil.

3.2. Remediation of Historically Contaminated Soil in a Draw and Fill Reactor

Sequestration in historically contaminated soils can dramatically reduce the availabilty of nitroaromatic compounds in the aqueous phase (Grant *et al.*, 1995). The extraction kinetics of explosives varies considerably between field-contaminated soils and spiked soils as well as among various types of soils (Jenkins *et al.*, 1989). Experiments with aged field-contaminated soil were conducted in a bench scale slurry reactor (Figure 7). The soil was nonsterile, dried and sieved clay and sand fractions from the former TNT-manufacturing plant at Hessisch Lichtenau, Germany which was operated during WWII. Despite exposure to environmental conditions for 50 years, the soil contained 3.6 g of 2,4-DNT and 2.5 g of 2,6-DNT per kg. The reactor was inoculated with a mixture of specific 2,4- and 2,6-DNT-degrading strains to initiate the trial. When DNT was biodegraded, 90%

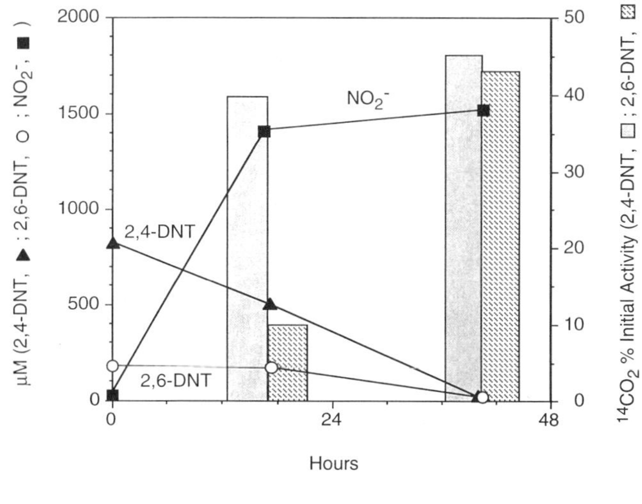

Figure 6. Mineralization of DNT in a spiked soil slurry. The inoculum included *Burkholderia cepacia* JS782, a 2,4-DNT-degrading bacterium, and *Hydrogenophaga palleronii* JS863, a 2,6-DNT-degrading strain (Nishino and Spain, submitted to ES&T).

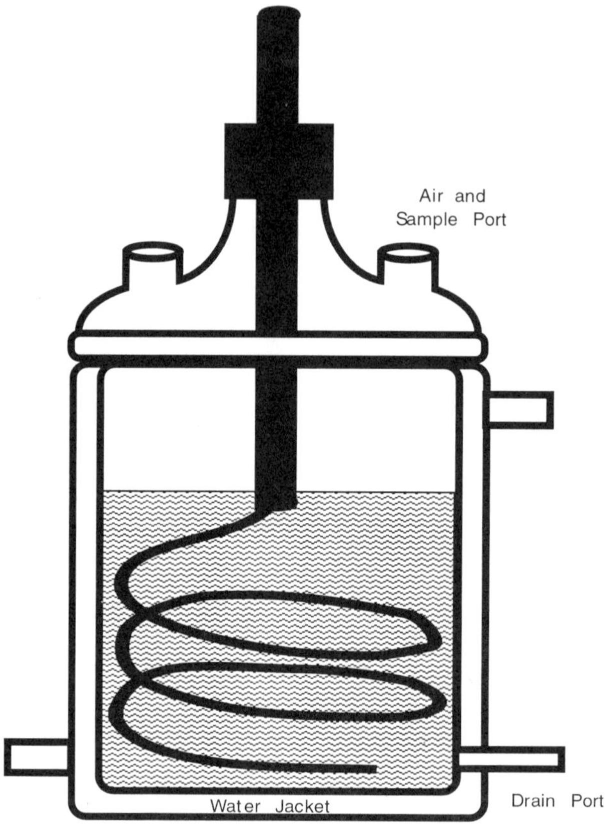

Figure 7. Soil slurry reactor with a helical ribbon impeller.

of the slurry was drawn from the reactor and replaced with fresh soil and phosphate buffer, no further additions of bacteria were made. During 600 hours of operation, 9 draw and fill cycles were completed (Figure 8). In each cycle, both DNT isomers disappeared with the 2,4-DNT disappearing in half the time of the 2,6-DNT. Two moles of nitrite were released per mole of DNT degraded, which clearly indicated mineralization of DNT.

Early in each cycle DNT was found in both the aqueous phase and in the organic extract of the soil. By midway through the cycle, DNT disappeared from the aqueous phase but continued to be extractable from the soil by acetonitrile. The concentration of DNT in the organic solvent decreased throughout the cycle from initial values of 350 and 150 μM (for 2,4- and 2,6-DNT, respectively) to a residual level of 15 - 20 μM (for each DNT, equivalent to 30 - 35 μg kg^{-1} soil). The residual contamination remained bound to the soil and was unavailable to the degradative bacteria even after extended incubation. Tests have been initiated to determine the toxicity of the remaining DNT.

4. CONCLUSION

Environmental contamination by synthetic chemicals is an enormous problem and costs for cleanup are staggering. Where bioremediation can provide an effective treatment

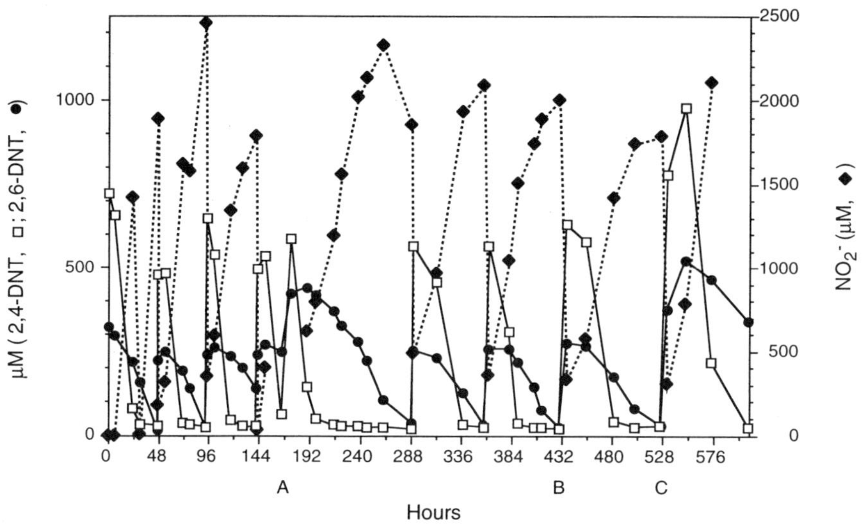

Figure 8. Field contaminated soil slurry inoculated with DNT-degrading bacteria. At (A) additional 10% soil (by weight) added when 2,4-DNT was gone but 2,6-DNT remained; (b) temperature of reactor reduced to 23 - 25 °C; and (C) temperature returned to 30 °C, soil concentration doubled to 20% (Nishino and Spain, submitted to ES&T).

strategy it is generally much less costly than more conventional strategies such as carbon sorption or incineration. Most organic chemicals are susceptible to attack by microorganisms. Some chemicals such as TNT can only be transformed by microorganisms in processes that require energy input and yield no benefit to the organisms. Such compounds can be treated by bioremediation, but the processes are not self sustaining and are therefore inherently expensive and difficult to control.

In contrast, some compounds can be completely degraded and serve to support the growth of specific bacteria. In such instances the degradation process will be self sustaining and will have a high potential for successful bioremediation or natural attenuation. It is clear from the work described here that the isomeric MNT and DNT can serve as growth substrates for bacteria and are, thus, excellent candidates for bioremediation. In situations where MNT and DNT are the predominant contaminants of soil or groundwater the remediation strategies described here should provide substantially more cost effective cleanup than traditional strategies. The bacteria that degrade DNT can tolerate moderate amounts of TNT and seem robust in both FBR and slurry phase reactors.

Because bacteria able to degrade the nitrotoluenes are widely distributed at contaminated sites it is not clear why the contaminants have persisted for so long. The influence of bioavailability, toxicity, and nutrient limitation must be evaluated to allow an understanding of the factors that limit natural attenuation of nitrotoluenes in soil and groundwater.

ACKNOWLEDGMENTS

This work was supported in part by the Air Force Office of Scientific Research, the Strategic Environmental Research and Development Program and the Swiss National Science Foundation.

REFERENCES

U.S.Army Environmental Center. (1995). Draft final site investigation and remedial investigation/feasibility study, Volunteer Army Ammunition Plant, Prepared by IT Corporation.

Barr, D.P., and Aust, S.D. (1994). Mechanisms white rot fungi use to degrade pollutants. *Environ. Sci. Technol.* **28**:78–87.

Bartell, R. (1989). Introduction/overview. Proceedings, Workshop on composting of explosives contaminated soils: U.S. Army Toxic and Hazardous Materials Agency, Edgewood, MD.

Bausum, H.T., Mitchell, W.R., and Major, M.A. (1992). Biodegradation of 2,4- and 2,6-dinitrotoluene by freshwater microorganisms. *J. Environ. Sci. Health A* **27**:663–695.

Berchtold, S.R., VanderLoop, S.L., Suidan, M., and Maloney, S.W. (1995). Treatment of 2,4-dinitrotoluene using a two-stage system: Fluidized-bed anaerobic granular activated carbon reactors and aerobic activated sludge reactors. *Water Environ. Res.* **67**:1081–1091.

Bradley, P.M., Chapelle, F.H., and Landmeyer, J.E. (1995). Degradation of 2,4-DNT, 2,6-DNT and 2,4,6-TNT by indigenous aquifer microorganisms, p. 267–271. In Hinchee, R.E., Anderson, D.B., and Hoeppel, R.E. (eds.), Bioremediation of Recalcitrant Organics: International In Situ On Site Bioreclamation Symposium, Battelle Press, Columbus.

Bradley, P.M., Chapelle, F.H., Landmeyer, J.E., and Schumacher, J.G. (1997). Potential for intrinsic bioremediation of a DNT-contaminated aquifer. *Ground Water* **35**:12–17.

Bradley, P.M., Chapelle, R.H., Landmeyer, J.E., and Schumacher, J.G. (1994). Microbial transformation of nitroaromatics in surface soils and aquifer materials. *Appl. Environ. Microbiol.* **60**:2170–2175.

Breitung, J., Bruns-Nagel, D., Steinbach, K., Kaminski, L., Gemsa, D., and von Löw, E. (1996). Bioremediation of 2,4,6-trinitrotoluene-contaminated soils by two different aerated compost systems. *Appl. Microbiol. Biotechnol.* **44**:795–800.

Cheng, J., Kanjo, Y., Suidan, M.T., and Venosa, A.D. (1996). Anaerobic biotransformation of 2,4-dinitrotoluene with ethanol as primary substrate: mutual effect of the substrates on their biotransformation. *Water Res.* **30**:307–314.

Grant, C.L., Jenkins, T.F., Myers, K.f., and McCormick, E.F. (1995). Holding-time estimates for soils containing explosives residues: comparison of fortification vs. field contamination. *Environ. Toxicol. Chem.* **14**:1865–1874.

Griest, W.H., Stewart, A.J., Tyndall, R.L., Caton, J.E., Ho, C.-H., Ironside, K.S., Caldwell, W.M., and Tan, E. (1993). Chemical and toxicological testing of composted explosives-contaminated soil. *Environ. Toxicol. Chem.* **12**:1105–1116.

Griest, W.H., Tyndall, R.L., Stewart, A.J., Caton, J.E., Vass, A.A., Ho, C.-H., and Caldwell, W.M. (1994). Chemical characterization and toxicological testing of windrow composts from explosives-contaminated sediments. *Environ. Toxicol. Chem.* **14**:51–59.

Gunderson, C.A., Kostuk, J.M., Gibbs, M.H., Napolitano, G.E., Wicker, L.F., Richmond, J.E., and Stewart, A.J. (1997). Multispecies toxicity assessment of compost produced in bioremediation of an explosives-contaminated sediment. *Environ. Toxicol. Chem.* **16**:2529–2537.

Haigler, B.E., Nishino, S.F., and Spain, J.C. (1994a). Biodegradation of 4-methyl-5-nitrocatechol by *Pseudomonas* sp. strain DNT. *J. Bacteriol.* **176**:3433–3437.

Haigler, B.E., and Spain, J.C. (1993). Biodegradation of 4-nitrotoluene by *Pseudomonas* sp. strain 4NT. *Appl. Environ. Microbiol.* **59**:2239–2243.

Haigler, B.E., Wallace, W.H., and Spain, J.C. (1994b). Biodegradation of 2-nitrotoluene by *Pseudomonas sp.* Strain JS42. *Appl. Environ. Microbiol.* **60**:3466–3469.

Hartter, D.R. (1985). The use and importance of nitroaromatic chemicals in the chemical industry, p. 1–13. In Rickert, D.E. (ed.)^(eds.), Toxicity of nitroaromatic compounds, Hemisphere Publishing Corp., Washington, D. C.

Heinze, L., Brosius, M., and Wiesmann, U. (1995). Biologischer Abbau von 2,4-Dinitrotoluol in einer kontinuierlich betriebenen Versuchsanlage und kinetische Untersuchungen. *Acta Hydrochim. Hydrobiol.* **23**:254–263.

Jenkins, T.F., Walsh, M.E., Schumacher, P.W., Miyares, P.H., Bauer, C.F., and Grant, C.L. (1989). Liquid chromatographic method for determination of extractable nitroaromatic and nitramine residues in soil. *J. Assoc. Off. Anal. Chem.* **72**:890–899.

Lendenmann, U., Spain, J.C., and Smets, B.F. (1998). Simultaneous biodegradation of 2,4-dinitrotoluene and 2,6-dinitrotoluene in an aerobic fluidized-bed biofilm reactor. *Environ. Sci. Technol.* **32**:82–87.

Liu, D., Thomson, K., and Anderson, A.C. (1984). Identification of nitroso compounds from biotransformation of 2,4-dinitrotoluene. *Appl. Environ. Microbiol.* **47**:1295–1298.

McCormick, N.G., Cornell, J.H., and Kaplan, A.M. (1978). Identification of biotransformation products from 2,4-dinitrotoluene. *Appl. Environ. Microbiol.* **35**:945–948.

Nishino, S.F., and Spain, J.C. Mineralization of 2,4- and 2,6-dinitrotoluene in soil slurries. *Submitted to Environmental Science and Chemistry.*

Nishino, S.F., and Spain, J.C. (1995). The effect of 2,6-dinitrotoluene on 2,4-dinitrotoluene degradation, abstr. PW253, p. 277. In Abstracts, Second SETAC World Congress, Society of Environmental Toxicology and Chemistry, Pensacola.

Nishino, S.F., and Spain, J.C. (1996). Degradation of 2,6-dinitrotoluene by bacteria, abstr. Q-380, p. 452. In Abstracts, 96th General Meeting of the American Society for Microbiology, American Society for Microbiology, Washington D.C.

Nishino, S.F., and Spain, J.C. (1997). Ring-cleavage of 3-methyl-4-nitrocatechol by 2,6-dinitrotoluene-degrading bacteria, abstr. Q-348, p. 513. In Abstracts, 97th Annual Meeting American Society for Microbiology., American Society for Microbiology, Washington, DC.

Noguera, D.R., and Freedman, D.L. (1996). Reduction and acetylation of 2,4-dinitrotoluene by a *Pseudomonas aeruginosa* strain. *Appl. Environ. Microbiol.* **62**:2257–2263.

Reardon, K.F., and Spain, J.C. (1993). Immobilized cell bioreactor for 2,4-dinitrotoluene degradation, abstr. BIOT 86. In Abstr. 205th ACS National Meeting, American Chemical Society, Washington, D.C.

Rhys-Williams, W., Taylor, S.C., and Williams, P.A. (1993). A novel pathway for the catabolism of 4-nitrotoluene by *Pseudomonas. J. Gen. Microbiol.* **139**:1967–1972.

Spain, J.C. (1995). Bacterial degradation of nitroaromatic compounds under aerobic conditions, p. 19–35. In Spain, J.C. (ed.), Biodegradation of nitroaromatic compounds, Plenum Publishing Corp., New York.

Spanggord, R.J., Spain, J.C., Nishino, S.F., and Mortelmans, K.E. (1991). Biodegradation of 2,4-dinitrotoluene by a *Pseudomonas* sp. *Appl. Environ. Microbiol.* **57**:3200–3205.

Spiess, T., Desiere, F., Fischer, P., Spain, J.C., Knackmuss, H.-J., and Lenke , H. (1998). A new 4-nitrotoluene degradation pathway in a *Mycobacterium* strain. *Appl. Environ. Microbiol.* **64**:446–452.

Thorne, P.G., and Leggett, D.C. (1997). Hydrolytic release of bound residues from composted soil contaminated with 2,4,6-trinitrotoluene. *Environ. Toxicol. Chem.* **16**:1132–1134.

Valli, K., Brock, B.J., Joshi, D.K., and Gold, M.H. (1992). Degradation of 2,4-dinitrotoluene by the lignin-degrading fungus *Phanerochaete chrysosporium*. *Appl. Environ. Microbiol.* **58**:221–228.

7

THE DEGRADATION OF NITRATE ESTER EXPLOSIVES AND TNT BY *Enterobacter cloacae* PB2

Christopher E. French[1,*], Peter R. Binks, Deborah A. Rathbone, Susan J. Rosser, Richard E. Williams, Stephen Nicklin[2] and Neil C. Bruce[1,†]

[1.]Institute of Biotechnology, University of Cambridge
Tennis Court Road, Cambridge CB2 1QT, United Kingdom
[2.]Defence Evaluation and Research Agency, Fort Halstead
Sevenoaks, Kent TN14 7BP, United Kingdom

1. INTRODUCTION

One of the major environmental problems facing the military establishment is the considerable amount of land and water that is contaminated with explosives, particularly TNT (2,4,6-trinitrotoluene). Explosives are found in the environment as a result of waste water produced by manufacturing plants and from the disposal of off-specification material or demilitarisation of out-of-date munitions. Concerns have arisen regarding the toxicity and environmental fate of TNT and nitrate ester explosives such as PETN (pentaerythritol tetranitrate) and GTN (glycerol trinitrate) due to their relative toxicity. PETN and GTN are produced in large amounts for use as explosives in blasting caps and detonators and as vasodilators for the control of angina. Nitrate esters are extremely rare in nature and multiply-substituted nitrate esters are not known to occur naturally. The environmental fate of such compounds which are produced in large quantities by the chemical industries, is therefore of considerable interest. In this paper we describe the isolation and characterisation of a strain of *Enterobacter cloacae* from explosives contaminated soil

* Present address: Institute of Cell, and Molecular Biology, University of Edinburgh, Darwin Building, King's Buildings, Mayfield Road, Edinburgh EH9 3JR, United Kingdom.

† Corresponding author. Mailing address: Institute of Biotechnology, University of Cambridge, Tennis Court Road, Cambridge, CB2 1QT, United Kingdom. Phone: 44 (0) 1223 334168. Fax: 44 (0) 1223 334162. E-mail: n. bruce@biotech. cam. ac. uk

Novel Approaches for Bioremediation of Organic Pollution, edited by
Fass *et al.* Kluwer Academic / Plenum Publishers, New York, 1999.

that is capable of utilising nitrate ester explosives and TNT as sole nitrogen sources for growth.

2. RESULTS AND DISCUSSION

2.1. Isolation of *Enterobacter cloacae* PB2

Enrichment cultures of explosives-contaminated soil with glycerol as the sole carbon source and PETN as sole nitrogen source led to the isolation of a bacterial strain designated PB2 (Binks *et al*, 1996). This organism, a gram negative rod, was identified on the basis of physiological characteristics as a strain of *Enterobacter cloacae* by the National Collection of Industrial and Marine Bacteria (NCIMB, Aberdeen, UK). Subsequently, the 16S rRNA genes of PB2 were cloned by polymerase chain reaction (PCR) and sequenced. Alignments of these sequences with other 16S rRNA sequences in the Gen-EMBL database supported the relationship of PB2 to the family *Enterobacteriaceae*.

Following growth of *E. cloacae* PB2 with PETN as sole nitrogen source, metabolites were extracted from culture supernatants using ethyl acetate. Three metabolites were isolated and separated by thin layer chromatography and found to have lower R_f values than PETN, indicating that the unknown metabolites were more polar than PETN. The identities of the unknown metabolites were investigated using electron impact (EI) mass spectrometry. The EI mass spectrum of metabolite 1 revealed a molecular ion of m/z 227 which is consistent with the empirical formula $C_5H_{10}N_2O_8$. The mass spectrum contained a fragment ion at m/z 209. We propose that this metabolite is pentaerythritol dinitrate (2,2-*bis*[nitrooxy)methyl]-1,3-propanediol), a denitration product of PETN. The EI mass spectrum of metabolite 2 revealed a molecular ion of m/z 222 which corresponds to the empirical formula $C_5H_8N_20_8$. The mass spectrum contained fragment ions at m/z 207, 170, 141, 99, 94, 76, 74, 158, 46 and 44. This is believed to represent 1-hydroxy-2,2-*bis*[(nitroxy)methyl]propanal. The EI mass spectrum of metabolite 3 revealed a molecular ion of m/z 222 which corresponds to the empirical formula $C_5H_6N_2O_8$. The mass spectrum contained fragment ions at m/z 208, 177, 91. This is believed to represent 2,2-*bis*[(nitroxy)methyl]propanedial. This indicates that PB2 removes two nitrate groups from PETN leaving hydroxyl groups, which are subsequently oxidized to aldehydes.

2.2. Biodegradation of PETN by Crude Cell-Free Extracts

Crude cell-free extracts prepared from cultures of *E. cloacae* on PETN as the sole nitrogen source were added to reaction mixtures containing PETN and 50 mM bis-Tris propane buffer (pH 7.0) at 30 °C. In the presence of NADPH the PETN was found to disappear rapidly. No activity was detected in the presence of the cofactors NAD^+, NADH or glutathione. A small amount of activity was detected in the presence of $NADP^+$, presumably due to endogenous reduction of this cofactor. PETN degradation could also be measured by the liberation of nitrite using Griess reagent (Rosenblatt et al,. 1991). A constititutive NADPH-dependent PETN reductase activity was detected in cell extracts from PB2. This activity was present in extracts from cells grown in the presence or absence of PETN. The products of the action of PETN reductase with PETN were extracted into ethyl acetate and analysed by mass spectrometry. It was found that PETN reductase liberates two nitrite ions from PETN to produce pentaerythritol trinitrate and pentaerythritol dinitrate (Figure 1) (Binks *et al*, 1996).

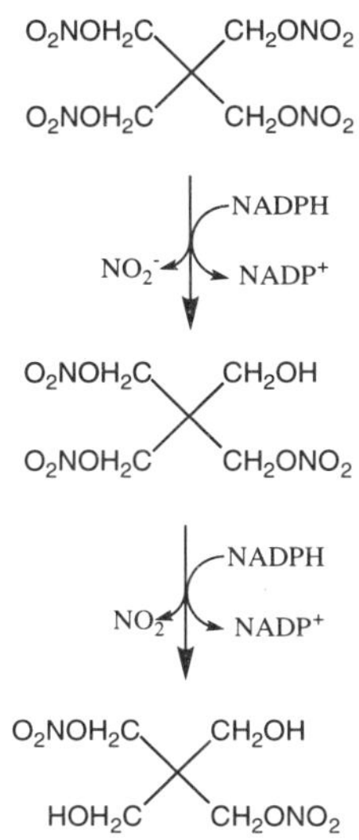

Figure 1. Reactions catalysed by PETN reductase.

2.3. Purification and Characterization of PETN Reductase

PETN reductase was purified from cell extracts by ion exchange and affinity chromatography using Mimetic Orange 2 A6XL (Binks *et al*, 1996). Analysis by sodium dodecylsulphate polyacrylamide gel electrophoresis (SDS-PAGE) showed a single protein band with M_r approximately 40 000. The native M_r of purified PETN reductase was investigated by gel filtration and was estimated as 40 000, suggesting that it is active as a monomer.

The purified PETN reductase was yellow and showed an absorption spectrum typical of an oxidized flavoprotein. The flavin was liberated by boiling and analysed by thin layer chromatography. In two different TLC systems, the liberated flavin comigrated with authentic flavin mononucleotide (FMN) and not with authentic flavin adenine dinucleotide (FAD) or riboflavin. It was concluded that PETN reductase is a monomeric flavoprotein of M_r 40 000, binding FMN non-covalently.

PETN reductase was also active with GTN and with EGDN (ethylene glycol dinitrate). The stoichiometry of nitrite release from GTN and EGDN has not been examined thoroughly; preliminary results suggest that one nitrite is released rapidly from GTN with at least one more released at a much lower rate. The products of PETN reductase activity with GTN and EGDN have not been characterized but are presumed to be the denitrated alcohols. PETN reductase was also found to catalyse NADPH-dependent reduction of 2-cyclohexen-1-one and 2,4,6-trinitrotoluene. As in most flavoproteins, a background oxidation rate was detected due to oxidation of the reduced enzyme by molecular oxygen.

The activity against various substrates was compared in an assay system consisting of 0.2 mM NADPH in 50 mM phosphate buffer (pH 7) at 30°C. GTN, EGDN and 2-cyclohexen-1-one were sufficiently soluble that the substrate concentration could be varied over a sufficiently wide range to estimate kinetic parameters. Kinetic parameters could not be estimated for PETN or TNT due to the poor solubility of these compounds. Results are shown in Table 1. It is clear that activity is highest against PETN and GTN. Much higher concentrations of EGDN are required to achieve comparable rates, as indicated by the high K_m for EGDN.

Table 1. Specificity of PETN reductase

A.Substrates for which kinetic parameters could be estimated		
Compound	V_{max} (U/mg)	K_m (mM)
GTN	12.1 ± 0.5	0.023 ± 0.002
EGDN	9.5 ± 0.5	2.4 ± 0.3
2-cyclohexen-1-one	3.2 ± 0.1	0.74 ± 0.06
B. Substrates for which kinetic parameters could not be estimated		
Compound	Concentration	Activity (U/mg)
PETN	0.05 mM	14
TNT	0.05 mM	0.4
RDX	0.05 mM	No activity
Oxygen	Saturated	0.1

2.4. Cloning and Analysis of the PETN Reductase Gene

The N-terminal amino acid sequence of purified PETN reductase was determined by automated Edman degradation, and was found to be:

S-A-E-K-L-F-T-P-L-K-V-G-A-V-T-A-P-N-R-V-F-M-A-P-L-T-

Two degenerate oligonucleotide probes, matching the predicted nucleic acid sequence encoding residues 2 to 11 and 18 to 26 of this protein, were synthesized:

i. AC-TTT-(G/C)AG-(G/C)GG-(G/C)GT-GAA-(G/C)AG-TTT-TTC-(G/C)GC
ii. GT-(G/C)AG-(G/C)GG-(G/C)GC-CAT-GAA-(G/C)AC-(G/C)CG-GTT

Southern blots using standard procedures (Sambrook *et al*, 1989) showed that both probes bound to the same region of PB2 genomic DNA. A 1525 bp *Nco* I/ *Cla* I genomic DNA fragment was cloned in pBluescript SK+ (Stratagene) to give pONR1. Sequencing indicated the presence of an open reading frame beginning with codons matching the known N-terminal sequence of PETN reductase. The sequence of *onr* was obtained and the deduced amino acid sequence of PETN reductase is shown in Figure 2. The sequence predicts a protein of 364 residues with a M_r of 39 358 excluding the N-terminal methionine, consistent with the M_r of approximately 40 000 estimated for PETN reductase by SDS-PAGE.

The deduced amino acid sequence of PETN reductase was compared to sequences in protein and nucleic acid databases using the BLAST program of the GCG package (Genetics Computer Group,1994). The most similar proteins found were morphinone reductase from *Pseudomonas putida* M10 (French and Bruce, 1994, 1995) and other members of the Old Yellow Enzyme family of α/β-barrel flavoprotein oxidoreductases (French *et al*, 1996). These include Old Yellow Enzyme of *Saccharomyces carlsbergensis* and *Saccharomyces cerevisiae* and homologs from the yeasts *Kluyveromyces lactis* and *Candida albicans*, the protozoan *Trypanosoma cruzii*, and the plants *Arabidopsis thaliana*, *Oryza sativa* and *Brassica campestris*. Other than morphinone reductase, which is believed to be involved in the breakdown of morphine alkaloids, the physiological functions of these enzymes are unknown. PETN reductase and morphinone reductase share 53% sequence identity and 71% sequence similarity as determined by the GAP program of the GCG package. An alignment of the deduced amino acid sequences of these proteins is shown in Figure 2.

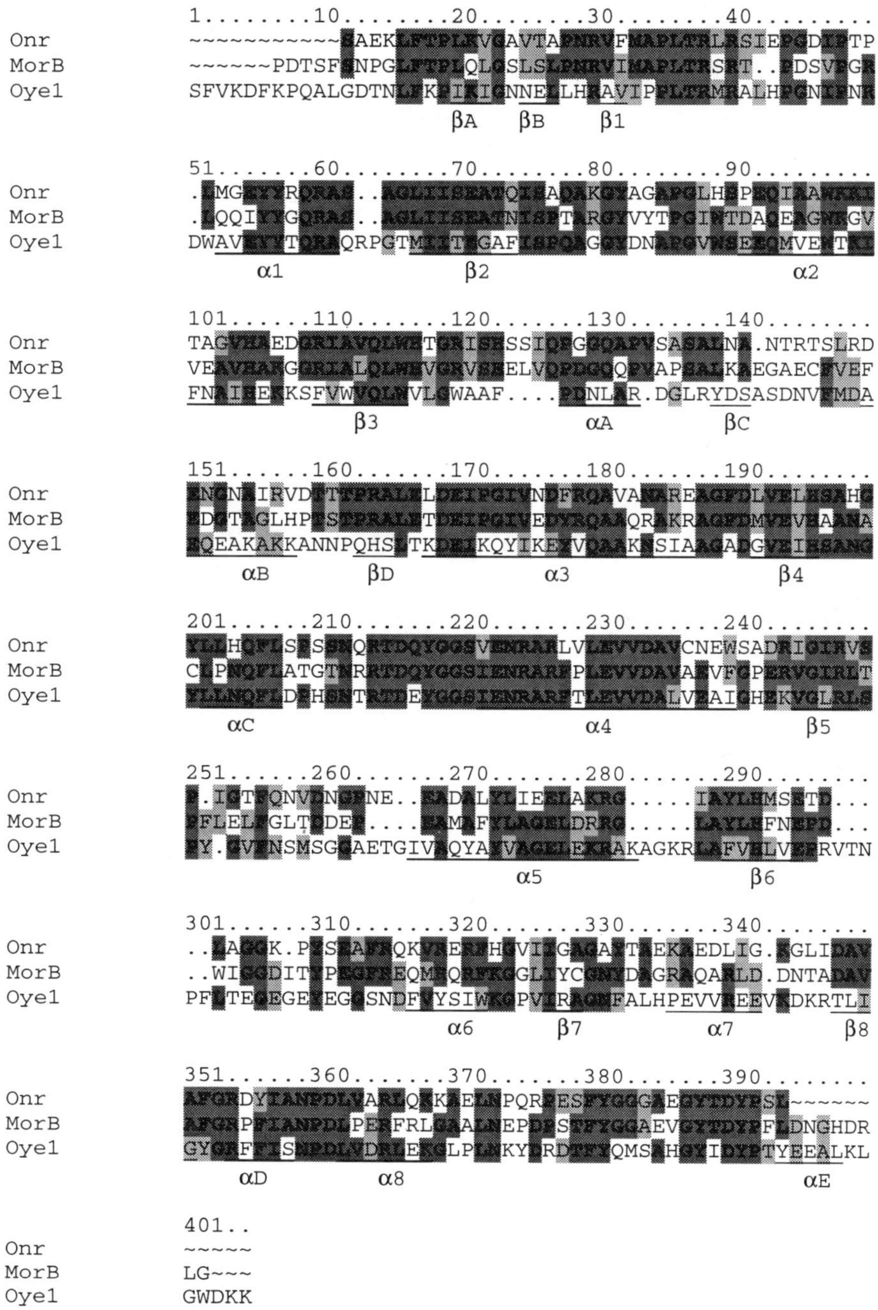

Figure 2. PETN reductase and related enzymes.

The physical properties of PETN reductase are consistent with a close relationship to this family: like all of the known members of this group, it is a simple flavoprotein with FMN as a prosthetic group, its subunit M_r is approximately 40 000, and it is an oxidoreductase using a reduced pyridine nucleotide cofactor as an electron donor.

The structure of Old Yellow Enzyme has been shown to be an eight-stranded α/ß-barrel (Fox and Karplus, 1994). The flavin is hydrogen-bonded by side chains from residues T37, Q114, R243 and R348. All of these residues are conserved in both morphinone reductase and PETN reductase (Figure 2). Of the residues which hydrogen-bond with the flavin through the peptide backbone in Old Yellow Enzyme, G324 and G347 are conserved, whilst several others are replaced by conservative substitutions such as glycine to alanine. Residues F374 and Y375, which provide a hydrophobic pocket for the dimethylbenzene ring of the flavin, are also conserved, although F296 is not. This suggests that all three enzymes have similar structures.

2.5. Overexpression and Purification of Recombinant PETN Reductase

The insert of pONR-1 is immediately downstream of the *lac* promoter of pBluescript SK+. It was found that *E. coli* JM109/pONR-1 grown in a rich medium such as SOB in the presence of 0.4 mM isopropyl-β-D-thiogalactoside (IPTG), an inducer of the *lac* promoter, expressed high levels of PETN reductase activity, suggesting that PETN reductase formed 30% to 50% of soluble cell protein. In the absence of IPTG, expression was approximately half of that observed with IPTG, suggesting that expression is driven both by the *lac* promoter of the vector and by a promoter in the insert. Recombinant PETN reductase was purified by affinity chromatograpy using Mimetic Orange 2 A6XL (French *et al*, 1996). The purified enzyme appeared homogeneous by SDS-PAGE and comigrated with PETN reductase purified from *E. cloacae* PB2. The N-terminal 10 amino acid residues of the recombinant enzyme were determined and were found to be identical to those of the enzyme purified from *E. cloacae* PB2. The purified enzyme showed no detectable loss in activity after several months of storage at -20°C.

2.6. Kinetic Mechanism and Steroid Inhibition of PETN Reductase

The availability of large amounts of purified PETN reductase allowed an investigation of the kinetic mechanism (French et al., 1996). PETN reductase activities were measured at varying concentrations of glycerol trinitrate (GTN) and NADPH. GTN was used as a substrate rather than PETN since its greater solubility allows variation of the substrate concentration over a much wider range. GTN concentration was varied between 10 mM and 100 mM and NADPH concentration between 40 mM and 150 mM. Apparent V_{max} and K_m values were determined at each NADPH concentration by non-linear regression using the GraFit software package (Leatherbarrow, 1990). Figure 3 (A) shows a double-reciprocal plot including the lines fitted by non-linear regression. Convergence occurs to the right of the y-axis. This is characteristic of a ping-pong mechanism with competitive inhibition by the fixed substrate, in this case NADPH (Cleland, 1970). In physical terms this implies that NADPH is capable of binding unproductively to the reduced form of the enzyme. The general equation for such behaviour is:

$$v = \frac{V_{max}[A][B]}{[A][B] + K_B[A](1 + [A]/K_{AR}) + K_A[B]}$$

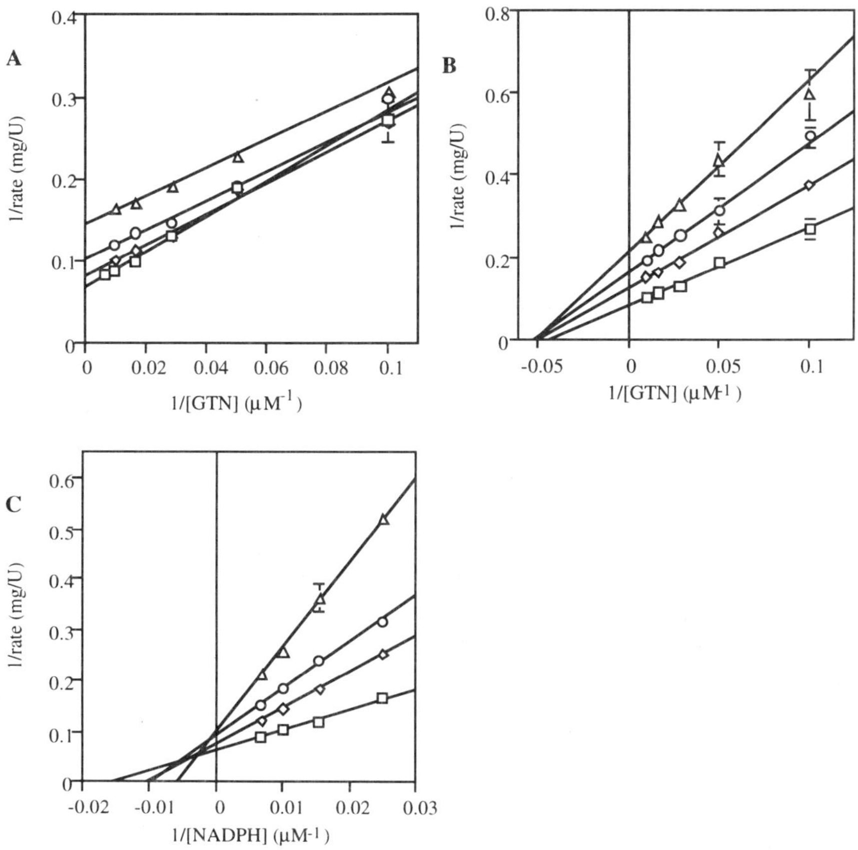

Figure 3.

where [A] is the concentration of NADPH, [B] is the concentration of GTN, v is the observed rate, V_{max} is the theoretical rate at saturating substrate concentrations ignoring substrate inhibition, K_A is the apparent K_m for NADPH at saturating levels of GTN, K_B is the apparent K_m for GTN at saturating levels of NADPH and K_{AR} is the dissociation constant for unproductive binding of NADPH to the reduced form of the enzyme.

Parameters for this equation were estimated by linear regression analysis of the slopes and intercepts of the lines shown in Figure 2.6A. The following values were obtained: $V_{max} = 25.2 \pm 1.5$ U/mg; $k_{cat} = 16.6 \pm 1.0$ s^{-1}; $K_A = 107 \pm 10$ mM; $K_B = 39.3 \pm 3.5$ mM; $K_{AR} = 400 \pm 58$ mM.

Errors indicated are one standard error based on the error of the linear regression.

A partial F-test was performed to determine whether the substrate inhibition term was statistically significant in describing the data. The F statistic was calulated as 30.1 with 1 numerator and 68 denominator degrees of freedom. This value is significant at the 1% level of confidence.

Similar kinetic behaviour has been described for morphinone reductase, in which substrate inhibition was observed at high levels of NADH, and product inhibition patterns also suggested that the reduced product hydrocodone could bind unproductively to the reduced form of the enzyme (French and Bruce, 1994, 1995). Old Yellow Enzyme and morphinone reductase are known to bind steroids and to reduce the double bond of 2-cyclo-hexen-1-one (French and Bruce, 1994; Schopfer and Massey, 1991). Several steroids were tested as inhibitors of PETN reductase and were found to be potent inhibitors of GTN reduction. The concentrations of progesterone, testosterone and cortisone required to halve the observed rate at 100 mM GTN and 100 mM NADPH were approximately 0.022 mM, 0.12 mM and 0.31 mM respectively.

Inhibition by testosterone was investigated in more detail. Assays were performed with testosterone concentrations ranging from 0.05 mM to 0.20 mM, either with NADPH fixed at 100 mM and GTN varied between 10 mM and 100 mM, or with GTN fixed at 100 mM and NADPH varied between 40 mM and 150 mM. Apparent V_{max} and K_m for each testosterone concentration were determined by non-linear regression using the GraFit software package. Double reciprocal plots showing data and fitted lines are shown in Figures 3 (B) and (C). Convergence to the left of the y-axis occurs with both GTN and NADPH as varied substrates, implying that testosterone binds both to the oxidized and reduced forms of the enzyme. The general equation describing this behaviour is:

$$v = \frac{V_{max}[A][B]}{[A][B](1+[A]/K_{AR}+[I]/K_{IR})+K_A[B](1+[I]/K_{IO})}$$

where K_{IO} is the dissociation constant for the binding of the inhibitor to the oxidized form of the enzyme, K_{IR} is the dissociation constant for the binding of the inhibitor to the reduced form of the enzyme and other parameters are as previously described.

Using the values for V_{max}, K_A, K_B and K_{FA} previously determined, K_{IO} and K_{IR} were estimated from linear slope replots of the lines fitted for each inhibitor concentration. K_{IO} was estimated as 70 ± 6 nM and K_{IR} was estimated as 136 ± 7 nM. Errors indicated are one standard error based on the error of the linear regression in each case. By contrast, apparent K_m values for GTN were in the order of 10–30 mM depending on the concentration of NADPH. It therefore appears that PETN reductase, like morphinone reductase, has an extremely high affinity for steroids, suggesting that these enzymes may have evolved from steroid reductases.

2.7. TNT Degradation by *E. Cloacae* PB2

Interestingly, *E. cloacae* PB2 was capable of very slow growth with TNT as sole nitrogen source in a mineral salts medium with glucose as the carbon source (French et al., 1998). The gowth curves obtained are shown in Figure 4. Growth estimated by protein concentration was observed in the presence of TNT or $NaNO_2$ and was proportional to the amount of nitrogen present.

During early characterization of PETN reductase, it was noted that the enzyme showed a low level of activity against TNT, as observed by increased oxidation of NADPH in the presence of TNT. To determine whether PETN reductase might play a role in the dergradation of TNT by *E. cloacae* PB2, PETN reductase was added to reaction mixtures containing TNT and NADPH. The reaction mixtures developed an orange colouration with an absorbance maximum at approximately 500 nm. No such coloured products were generated in the absence of enzyme, or TNT, or NADPH.

Due to the electron-withdrawing nature of the nitro substitutents, the aromatic ring of tri-nitro compounds such as TNT is electron-deficient and is relatively easily reduced to a coloured hydride-Meisenheimer complex (Kaplan and Siedle, 1971) (Figure 5). Such reduction can be accomplished by weak reducing agents such as sodium octahydrotriborate

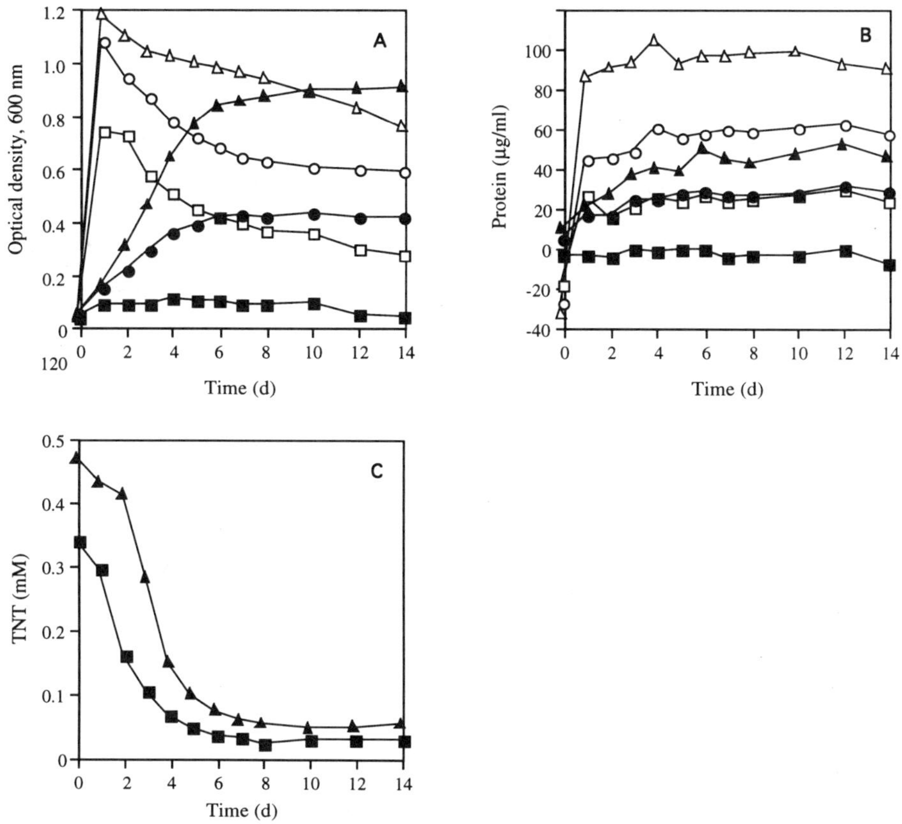

Figure 4. Growth of *Enterobacter cloacae* PB2 with TNT as sole nitrogen source.

(Kaplan and Siedle, 1971). Biological production of the hydride-Meisenheimer complex of TNT has been reported in whole cells of a *Mycobacterium* sp. able to grow with 4-nitrotoluene as sole nitrogen source (Vorbeck *et al*, 1994), and a *Pseudomonas* sp. able to grow with TNT as a sole nitrogen source (Haidour and Ramos, 1996). Isolation of the enzymes responsible has not been reported.

The hydride-Meisenheimer complex of TNT has a characteristic absorption spectrum which does not match that observed in reduction of TNT by PETN reductase. However, experiments in this laboratory suggest that a reducing agent such as sodium borohydride reduces TNT initially to the hydride-Meisenheimer complex, and subsequently to one or more orange products with absorption spectrum similar to that seen with PETN reductase.

Due to the negative charge of the hydride-Meisenheimer complex, analysis by reverse-phase HPLC requires the presence of a counter-ion such as tetrabutylammonium. Reaction mixtures resulting from the reduction of TNT by PETN reductase/NADPH or sodium borohydride were analysed by HPLC using a C18 reverse-phase column and a mobile phase consisting of 10 mM tetrabutylammonium phosphate (pH 7) in 50% (v/v) acetonitrile, 50% (v/v) water. Products were detected at 260 nm (ultraviolet) and 500 nm (visible). Absorbance spectra of each peak were also measured. Chemical reduction of TNT resulted in a peak at 10.0 min, having the characteristic spectrum of the hydride-Meisenheimer complex, and at least

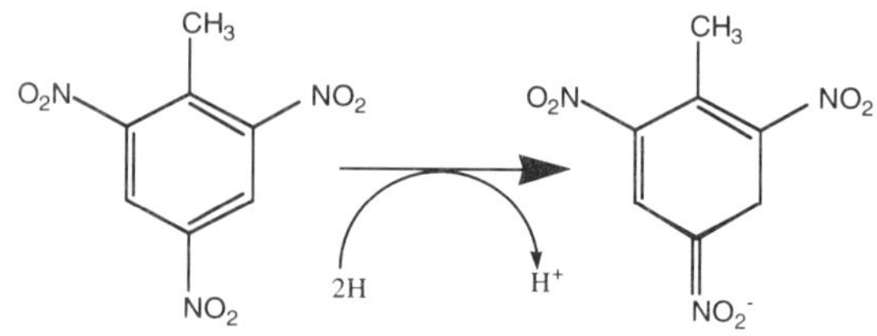

Figure 5. Reduction of TNT to the hydride-Meisenheimer complex.

five other visible peaks, with retention times of 2.7 min (approximately the solvent front), 4.0 min, 4.5 min, 5.1 min and 6.0 min. In enzymic reaction mixtures, two visible peaks appeared, matching the chemical peaks at 4.5 min and 5.1 min in retention times and absorption spectra. Since chemical reduction experiments suggest that the initial product is the hydride-Meisenheimer complex, with further reduction leading to the other products observed, we speculate that PETN reductase initially reduces TNT to the hydride-Meisenheimer complex and then rapidly further reduces it to the observed products. Rapidity of the second reduction step relative to the first would account for our failure to observe the hydride-Meisenheimer complex itself in enzymic reaction mixtures.

The identities of the products derived by further reduction of the hydride-Meisenheimer complex are unknown. Targets for further reduction are the remaining unsaturated bonds and the nitro groups. Since reduction of the aromatic ring should make the nitro groups more stable and the remaining unsaturated bonds less stable, we suspect that the latter are better candidates for reduction by PETN reductase. To determine this, it will be necessary to obtain further data such as mass spectra and ^{1}H-NMR spectra of the reduction products.

Assays of enzymic reaction mixtures using Griess reagents showed that nitrite accumulates. In one reaction mixture left for a period of several days with an enzymic NADPH regeneration system to regenerate NADPH oxidized by PETN reductase, it was found that 1.0 mol of nitrite had been liberated per mol of TNT initially present. The stoichiometry and timing of nitrite release have yet to be determined. Nitrite may be released by slow non-enzymic breakdown of one or more of the reaction products.

Since the final reaction products of TNT reduction by PETN reductase contain less nitrogen than TNT and appear to be water-soluble and non-aromatic, they are likely to be less toxic and less recalcitrant than TNT or nitroreductase products of TNT. Therefore, *E. cloacae* PB2 and recombinant organisms expressing PETN reductase may be useful in the bioremediation of TNT-contaminated soil and water.

ACKNOWLEDGMENTS

This work was supported by the Defence Evaluation and Research Agency and the BBSRC.

REFERENCES

Binks, P. R., French, C. E., Nicklin, S. and Bruce, N. C. (1996). Degradation of pentaerythritol tetranitrate by *Enterobacter cloacae* PB2. Appl. Environ. Microbiol. **62:**1214–1219.

Cleland, W. W. (1970). Steady state enzyme kinetics, p1–65. In P. D. Boyer (ed.), The enzymes, 3rd ed., vol. 2. Academic Press, N.Y.

Fox, K. M. and Karplus, P. A.. (1994). Old yellow enzyme at 2 Å resolution: overall structure, ligand binding, and comparison with related flavoproteins. Structure **2:**1089–1105.

French, C. E. and Bruce, N. C. (1995). Bacterial morphinone reductase is related to old yellow enzyme. Biochem. J. **312:**671–678.

French, C. E., Nicklin S. and Bruce, N. C. (1996). Sequence and properties of pentaerythritol tetranitrate reductase from *Enterobacter cloacae* PB2. J. Bacteriol. **178:**6623–6627.

French, C. E., Nicklin S. and Bruce N. C. (1998). Aerobic degradation of 2,4,6-trinitrotoluene by *Enterobacter cloacae* PB2 and by pentaerythritol tetranitrate reductase. Submitted for publication.

Genetics Computer Group. (1994). Program manual for the Wisconsin package, version 8, September 1994. 575 Science Drive, Madison, WI, USA.

Haïdour, A. and Ramos, J. L. (1996). Identification of products resulting from the biological reduction of 2,4,6-trinitrotoluene, 2,4-dinitrotoluene, and 2,6-dinitrotoluene by *Pseudomonas* sp. Environ. Sci. Technol. **30:**2365–2370.

Kaplan, L. A. and Siedle, A. R. (1971). Studies in boron hydrides IV: stable hydride Meisenheimer adducts. J. Org. Chem. **36:**937–939.

Leatherbarrow, R. J. (1990). GraFit version 2.0, Erithacus Software Ltd., Staines, U.K.

Rosenblatt, D. H., Burrows, E. P., Mitchell W.R. and Parmer, D. L. (1991). Organic Explosives and Related Compounds. In The Handbook of Environmental Chemistry. Edited by O. Hutzinger. Springer-Verlag.

Sambrook, J., Fritsch, E. F. and Maniatis, T. (1989). Molecular cloning: a laboratory manual, 2nd ed. Cold Spring Harbor Laboratory Press, Cold Spring Harbor, N.Y.

Schopfer, L. M. and Massey, V. (1991). Old yellow enzyme, p. 247–269. In S. A. Kuby (ed.), A study of enzymes. CRC Press, Boston.

Vorbeck, C., Lenke, H., Fischer, P., Spain, J. and Knackmuss H.-J. (1998). Initial reductive reactions in aerobic microbial metabolism of 2,4,6-trinitrotoluene. Appl. Environ. Microbiol. **64:**246–252.

CROSS INDUCTION OF 4-NITROBENZOATE AND 4-AMINOBENZOATE DEGRADATION BY *Burkholderia cepacia* STRAIN PB4

Caroline M. Peres, Henry Naveau, and Spiros N. Agathos*

Unit of Bioengineering, Catholic University of Louvain
Place Croix du Sud 2/19, B-1348 Louvain-la-Neuve, Belgium

ABSTRACT

A variety of industrial activities (production of explosives, pesticides, polymers, etc.) have introduced into the environment several nitroaromatic compounds (NACs) which are toxic and recalcitrant to biodegradation by the microorganisms commonly found in nature. The most common reaction undergone by these compounds is the reduction of the nitro group, the aminoaromatic compound (AAC) then produced generally not being further degraded. Due to this ubiquitous nitroreductase activity, AACs are often found together with NACs at contaminated sites or effluents. An efficient bioremediation process should combine treatment of both classes of compounds. *Burkholderia cepacia*, isolated on 4-aminobenzoate as sole source of C, N and energy, was found to be also able to grow on 4-nitrobenzoate. This pure strain has thus the property to mineralize both a NAC and the corresponding AAC. Metabolic pathway studies with *B. cepacia* have shown that it mineralizes 4-nitrobenzoate and 4-aminobenzoate through independent but converging pathways. Curiously, 4-nitrobenzoate degradation was induced by 4-aminobenzoate and vice-versa. The degradation of an equimolar mixture of these two compounds (1 mM each) by washed cells of *B. cepacia* grown on 4-nitrobenzoate, 4-aminobenzoate or glucose was studied in batch cultures. When induced on either 4-nitrobenzoate or 4-aminobenzoate, *B. cepacia* metabolized the inducer faster than the second compound. When grown on glucose, the microorganism mineralized both compounds simultaneously after a short lag. No diauxic growth was observed but rather a normal growth curve. The two pathways of degradation were found to be working simultaneously and such an organism might be appropriate for the bioremediation of a NAC, in this case 4-nitrobenzoate, together with its corresponding AAC, 4-aminobenzoate.

1. INTRODUCTION

As xenobiotic and toxic chemicals released by different industrial and agricultural processes, nitroaromatic compounds need to be considered among classes of compounds

Novel Approaches for Bioremediation of Organic Pollution, edited by
Fass *et al.* Kluwer Academic / Plenum Publishers, New York, 1999.

undesirable in the environment and for which decontamination methods have to be found. Given the electron-attracting property of the nitro moiety, biotic or abiotic reduction of the nitro group into an amino group via the nitroso and hydroxylamino derivatives is a common reaction, while electrophilic attack by oxygenases—enzymes involved in the mineralization of aromatic compounds in aerobic conditions—is hindered. In spite of this, several microorganisms have been isolated which have adapted to use a nitroaromatic compound as sole carbon, nitrogen and energy source by performing an initially oxidative attack on the aromatic ring (Spain, 1995). More recently, microorganisms were discovered which take advantage of the facile reduction and reduce the nitro group as the first reaction of the mineralization pathway, releasing ammonia instead of the nitrites produced by the oxidative attack (Groenewegen *et al.*, 1992; Haigler and Spain, 1993; Nishino and Spain, 1993; Rhys-Williams *et al.*, 1993; Yabannavar and Zylstra, 1995; Michan *et al.*, 1997; Schenzle *et al.*, 1997; Spiess *et al.*, 1998). Contrary to the commonly observed non-specific reduction, it was shown that the reduction accomplished by these newly described bacteria is only partial (Somerville *et al.*, 1995), ending at the hydroxylamino intermediate which is then rearranged in order to be further degraded. In the non-specific reaction, the aminoaromatic compounds formed are usually not further degraded by the microorganism which performed the reduction. As a consequence, at certain locations, the presence of aminoaromatic compounds, sometimes even more toxic than the nitro compound (Couch *et al.*, 1987), makes it necessary to consider a nitroaromatic compound contamination as a dual contamination. For example, analysis of the waste disposal site from a former ammunition plant in Germany revealed the presence of several aminobenzoic acids beside nitroaromatic compounds (Schmidt *et al.*, 1997). Aminoaromatic compounds are normally rather good candidates for aerobic mineralization due to the electron-donating character of the amino group and several bacteria have been described that are mineralizing such chemicals (Field *et al.*, 1995). At sites with a dual contamination, both classes of compounds should be treated. A solution which comes easily to mind is to use in tandem a good nitroreduction-performing microorganism and a microorganism able to mineralize the corresponding aminoaromatic compound, approach which has already been proven (Dickel *et al.*, 1993; Field *et al.*, 1995; Peres *et al.*, 1998). However, it could be less stringent to look after the good survival and activity of a single microorganism (especially with the recent advances made in the construction of probes (Amann *et al.*, 1995)). Therefore, it should be of interest to work with a single pure strain able to mineralize both the nitro- and the aminoaromatic compound. In a study on bacterial metabolism of aminobenzoates, Russ (1991) isolated a *Burkholderia cepacia* strain, named thereafter strain PB4, and proposed that 4-aminobenzoate is degraded by this strain via 4-hydroxybenzoate and protocatechuate, with the intervention of two monooxygenases. Protocatechuate then undergoes an *ortho* cleavage by protocatechuate 3,4-dioxygenase. When searching for other growth substrates, *B. cepacia* strain PB4 was found to be also able to mineralize 4-nitrobenzoate (Russ, 1991). Several other strains have already been described which are able to degrade 4-nitrobenzoate. They mostly proceed via the partial reduction described above to produce 4-hydroxylaminobenzoate rearranged into protocatechuate before ring cleavage (Groenewegen *et al.*, 1992; Yabannavar and Zylstra, 1995; Michan *et al.*, 1997). However, 4-aminobenzoate was postulated as an intermediate of 4-nitrobenzoate degradation by a *Pseudomonas fluorescens* strain (Durham, 1958). 4-Nitrobenzoate is used in industries as a precursor in dye manufacture. Even though 4-aminobenzoate is a precursor of folic acid whose concentration in bacteria can reach around 15 ppm (Stanier *et al.*, 1987), it has been shown to inhibit the same bacteria at concentrations ten times higher (Richards *et al.*, 1995). Although the toxicity and the recalcitrance of 4-nitro- and 4-aminobenzoate is lower than those of certain other nitro- and aminoaromatic compounds, this seldom encountered ability to grow on a nitroaromatic compound and its corresponding aminoaromatic derivative makes *B. cepacia* strain PB4 an interesting bacterium. Our objective was to determine if the strain is able to degrade a nitroaromatic compound—4-nitrobenzoate—simultaneously to the corresponding amino derivative—4-aminobenzoate. Indeed, for the strain to be efficient, assimilation of a compound should not inhibit assimilation of the second compound. As readily assimilated compounds are sometimes also found at nitroaromatic-contaminated sites, we also studied 4-nitrobenzoate degradation by *B. cepacia* strain PB4 in the presence of glucose, succinate and ammonium chloride.

2. MATERIALS AND METHODS

2.1. Microorganisms and Culture Conditions

Strain PB4 was isolated on 4-aminobenzoate as sole source of carbon, nitrogen and energy from a soil contaminated by nitroaromatic compounds, Stadt Allendorf, Hessen, Germany (Russ, 1991). Among other growth substrates, strain PB4 was also able to grow on 4-nitrobenzoate (Russ, 1991). Genus and species were determined using a Biolog GN microplate assay (Hayward, CA, USA). The organism was grown in a nitrogen-free medium (mineral base + phosphate buffer (50 mM, pH 7.4)) (modification of Bruhn *et al.*, 1987 by increasing the concentration of the buffer and reducing the concentration of $CaC_{12}.H_2O$ to 0.5 g/l) supplemented with 1 to 4 mM 4-nitro- or 4-aminobenzoate.

Induced cells were produced by incubating strain PB4 in 400 ml of medium containing 4 mM of the inducer (4-nitrobenzoate or 4-aminobenzoate) as sole source of carbon, nitrogen and energy. Two to four subsequent additions of the inducer were made until a dense suspension ($OD_{540}\approx 1$) was obtained. Cells were harvested two hours after the last addition of 4-nitro- or 4-aminobenzoate. Non-induced cells were grown in a basal mineral salts medium (mineral base + phosphate buffer (65 mM, pH 7.0)) (Stanier *et al.*, 1966) supplemented with a mixture of 13.33 mM glucose and 6.67 mM sodium succinate (molar ratio 2/1) plus ammonium chloride (16 mM), called the BMS-GSA medium. Induced and non-induced cells were harvested by centrifugation and washed twice with phosphate buffer (50 mM, pH 7.4) before being inoculated into the culture medium.

Batch reactors (250 ml-conical flasks, containing 100 ml of medium) were incubated on an orbital incubator (Gallenkamp, Leicester, UK) at 100 rpm and 30°C. All cultures were carried out in duplicate. Non-inoculated flasks served as a control.

2.2. Oxygen Uptake Measurements

A Clark-type oxygen electrode (Yellow Springs Instrument Co., Yellow Springs, OH, USA) was used to measure polarographically the oxygen uptake by concentrated suspensions of induced and non-induced cells of *B. cepacia* strain PB4. The final volume of 1.8 ml contained 55 μM substrate dissolved in dimethylformamide, cells (0.4 to 0.7 mg of protein) and phosphate buffer (0.02 M, pH 7.0) at 19°C. Before addition of substrate, the endogenous oxygen uptake rate was measured.

2.3. Anaerobic Incubation

Anaerobic transformation of 4-nitrobenzoate and 4-aminobenzoate by 4-nitro- and 4-aminobenzoate-induced cells of *B. cepacia* strain PB4 was investigated in 100 ml-serum flask filled with 25 ml of medium and closed tightly by a rubber stopper. Before inoculation of the flask, the atmosphere of the headspace was evacuated and replaced by nitrogen three times.

2.4. Degradation of Equimolar Mixtures of 4-Nitro- and 4-Aminobenzoate

Washed cells of *B. cepacia* strain PB4 induced on 4-nitrobenzoate, 4-aminobenzoate or non-induced were inoculated in 100 ml nitrogen-free medium containing 1 mM of 4-nitrobenzoate and 1 mM of 4-aminobenzoate and were incubated at 30°C and 100 rpm.

2.5. Degradation of 4-Nitrobenzoate in the Presence of Readily Assimilated Substrates

Washed cells of *B. cepacia* strain PB4 grown on BMS-GSA (containing glucose, sodium succinate and ammonium chloride) medium were inoculated in 100 ml BMS-GSA medium supplemented with 1 mM of 4-nitrobenzoate and incubated at 30°C and 100 rpm.

2.6. Analytical Methods

Concentrations of 4-nitrobenzoate, 4-aminobenzoate and other aromatic compounds in the medium were measured by HPLC, on an Alltima C_{18} column (5 mm; 4.6x250 mm; Alltech, Deerfield, IL, USA). An isocratic solvent system consisting of phosphate buffer (25 mM, pH 3.2): acetonitrile 75:25 (v/v) at 1 ml/min was used as the mobile phase. Compounds were monitored at UV A_{230} with a Waters 996 diode array detector (Waters, Milford, MA, USA). Intermediates and products were identified by comparison of retention times and UV spectra with those of authentic standards. An enzymatic method was used to quantify ammonia (Kit Sigma N° 171-UV, St Louis, MO, USA) whereas nitrite release was measured colorimetrically (Smibert and Krieg, 1981). Microorganism concentration was determined through optical density at 540 nm (OD_{540}) or by protein measurement (bicinchoninic acid method (Kit Sigma BCA-1, St Louis, MO, USA)). Liquid chromatography/mass spectrometry (LC/MS) was performed on ethyl acetate-extracted culture medium (after acidification with HCl and saturation with NaCl). Samples were analyzed on a TSP (Thermo Separation Products, San Jose, CA, USA) HPLC coupled to a Finnigan MAT TSQ 7000 (San Jose, CA, USA) triple quadrupole mass spectrometer with an Atmospheric Pressure Chemical Ionization (APCI). Temperatures of the capillary and of the vaporizer chamber were 220°C and 400°C respectively. Pressures were 70 psi for the sheath gas and 10 psi for the auxiliary gas. The corona discharge was 5 μA.

2.7. Chemicals

4-Nitrobenzoate and 4-aminobenzoate were provided by Fluka (Buchs, Switzerland) and their purity was higher than 98%. All other chemicals were of the highest purity commercially available and were obtained from Merck (Darmstadt, Germany), Sigma (St Louis, MO, USA) or Aldrich (Milwaukee, WI, USA).

3. RESULTS

3.1. 4-Nitrobenzoate Metabolism by *Burkholderia cepacia* Strain PB4

Strain PB4, tentatively identified by the Biolog system as *Burkholderia cepacia*, is able to grow on, and thus degrade, 4-nitrobenzoate or 4-aminobenzoate as sole source of

Table 1. Oxygen uptake rate enhancement (+) or lack thereof (-) when whole cells of *B. cepacia* strain PB4 induced on 4-nitrobenzoate, 4-aminobenzoate or non-induced were supplied with some potential intermediates of the degradation pathway of 4-nitro- or 4-aminobenzoate

	Oxygen uptake rate enhancement after growth on		
Substrate	4-Nitrobenzoate	4-Aminobenzoate	Glucose/Succinate
4-Nitrobenzoate	+	+	-
4-Aminobenzoate	+	+	-
4-Hydroxybenzoate	-	+	-
Protocatechuate	+	+	-

carbon, nitrogen and energy. As these compounds were metabolized, ammonia was released and a new, small HPLC peak accumulated. No other intermediates, nor nitrites were detected. MS spectrum of the new peak revealed a compound with a molecular weight of 195 (thus carrying an odd number of nitrogen atoms) and the loss of fragments of 42 and 44, consistent with an acetyl and a carboxyl group respectively. 2-Aminobenzoate and 4-hydroxybenzoate were also able to support growth of *B. cepacia* strain PB4 but not 4-nitrotoluene, 2-nitrobenzoate, 3-nitrobenzoate or 3-aminobenzoate.

In order to obtain indications on 4-nitrobenzoate-degradation pathway by *B. cepacia* strain PB4 (Stanier, 1947), oxygen uptake rate of induced and non-induced cells was measured in the presence of potential intermediates of the 4-nitrobenzoate degradation pathway according to what is described in the literature (Groenewegen *et al.*, 1992; Haigler and Spain, 1993; Rhys-Williams *et al.*, 1993) and also in the presence of the intermediates of the pathway proposed by Russ (1991) for the degradation of 4-aminobenzoate (Table 1). Oxygen uptake rate by 4-nitrobenzoate-induced cells was enhanced by 4-nitrobenzoate, 4-aminobenzoate and protocatechuate but not by 4-hydroxybenzoate. All the compounds identified by Russ (1991) as intermediates in the 4-aminobenzoate-degradation pathway by *B. cepacia* strain PB4 led to an increase of the oxygen uptake rate by 4-aminobenzoate-induced cells. Surprisingly, 4-nitrobenzoate did so too. Oxygen uptake rate by non-induced cells did not increase with any of the substrates tested.

Anaerobic conditions prevent the activity of oxygenase enzymes. Under anaerobic incubation, 4-aminobenzoate was left unchanged whereas 4-nitrobenzoate was transformed mainly into protocatechuate but traces of the compound as described above and of 4-aminobenzoate were also detected.

3.2. Degradation of an Equimolar Mixture of 4-Nitrobenzoate and 4-Aminobenzoate by 4-Nitrobenzoate-Induced, 4-Aminobenzoate-Induced and Non-Induced Cells

Degradation of an equimolar mixture (1 mM each) of 4-nitro- and 4-aminobenzoate was investigated with 4-nitrobenzoate-induced, 4-aminobenzoate-induced and with non-induced cells of *B. cepacia* strain PB4 (Fig. 1).

Whatever the preculturing conditions, 4-nitrobenzoate and 4-aminobenzoate were degraded simultaneously and a mixed-substrate growth (Harder and Dijkhuizen, 1982) was observed. However, as expected, initial degradation rate of the inducer was higher than initial degradation rate of the other compound. Although slower at the beginning, degradation of the second compound accelerated and thus both compounds were ex-

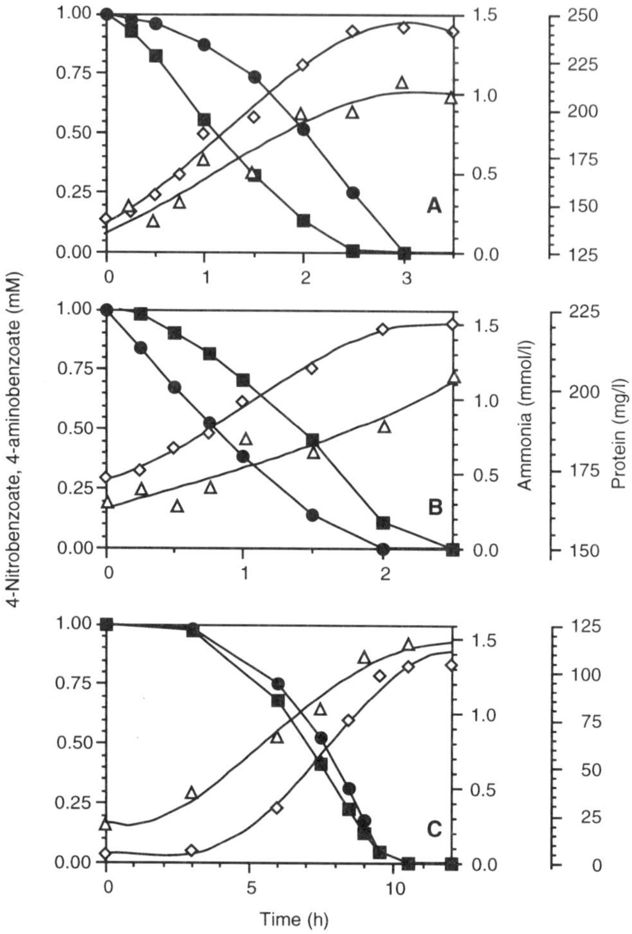

Figure 1. Degradation of equimolar mixtures of 4-nitrobenzoate (■) and 4-aminobenzoate (●) by *B. cepacia* strain PB4 grown on 4-nitrobenzoate (A), 4-aminobenzoate (B), or glucose/succinate/NH_4Cl (C). Protein concentration (◇); ammonia concentration (△). Initial protein content, cells grown on 4-nitrobenzoate: 140 mg/l; cells grown on 4-aminobenzoate: 155 mg/l; cells grown on glucose/succinate/ NH_4Cl: 20 mg/l.

hausted after approximately the same time of incubation. The degradation kinetics of the inducer followed rather closely a logistic curve whereas that of the second compound was more logarithmic (Alexander, 1994). Degradation curves of 4-nitro- and 4-aminobenzoate by non-induced cells matched closely. This means that both the time of onset and the rate of degradation were similar.

Release of 1 to 1.25 mM ammonia accompanied the assimilation of 1 mM 4-nitrobenzoate and 1 mM 4-aminobenzoate and between 40 and 80 mg/l protein were produced.

3.3. Degradation of 4-Nitrobenzoate in the Presence of Readily Assimilated Substrates

The question whether degradation of 4-nitrobenzoate by non-induced *B. cepacia* strain PB4 still takes place when glucose, succinate and ammonium chloride are also present, was subsequently investigated (Fig. 2). 4-Nitrobenzoate transformation started after a lag phase of approximately 1.5 hours, as did growth of the microorganism, and all 4-nitrobenzoate introduced had disappeared in less than 7 hours. Glucose and succinate were

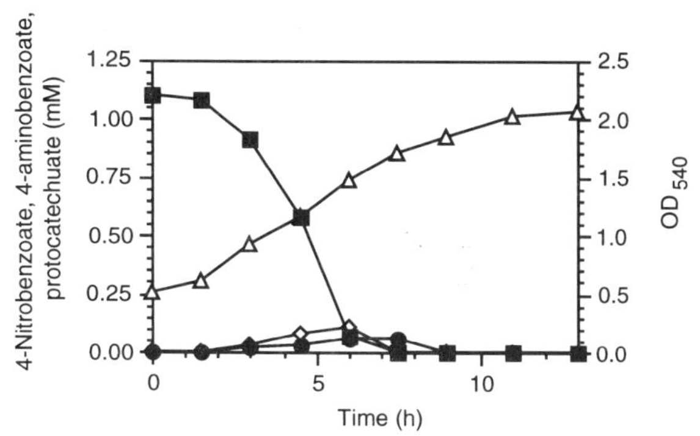

Figure 2. Degradation of 4-nitrobenzoate by non-induced *B. cepacia* strain PB4 in the presence of glucose (13.33 mM), succinate (6.67 mM) and ammonium chloride (16 mM). Symbols: 4-nitrobenzoate (■), 4-aminobenzoate (●), protocatechuate (◇), OD540 (△).

still present at this time (data not shown) and growth continued for approximately 7 more hours. 4-Aminobenzoate and protocatechuate were transiently detected. However, the concentration of 4-aminobenzoate was at any time far lower than stoechiometric to the 4-nitrobenzoate transformed. Except for the metabolite described above (molecular weight: 195) which was accumulating but at a very low concentration—as deduced from the very low surface of the HPLC peak-, no other compounds were detected.

4. DISCUSSION

Burkholderia cepacia strain PB4 was able to use both 4-nitro- and 4-aminobenzoate as sole source of carbon, nitrogen and energy. According to the 'simultaneous adaptation theory' (Stanier, 1947), the fact that 4-nitrobenzoate-induced cells were able to provoke an increase of the oxygen uptake rate when inoculated with 4-aminobenzoate might mean that 4-aminobenzoate is an intermediate of 4-nitrobenzoate degradation pathway. However, since 4-hydroxybenzoate—a proven intermediate in 4-aminobenzoate degradation (Russ, 1991)—did not enhance oxygen uptake rate by 4-nitrobenzoate-induced cells (whereas it did so for 4-aminobenzoate-induced cells) and given that protocatechuate was the metabolite accumulating under anaerobic conditions, it is more likely that the metabolic pathway used by *B. cepacia* strain PB4 for the mineralization of 4-nitrobenzoate is the same as the one previously described for other bacteria (Groenewegen *et al.*, 1992; Haigler and Spain, 1993; Rhys-Williams *et al.*, 1993; Yabannavar and Zylstra, 1995; Michan *et al.*, 1997), i. e. partial reduction of the nitro group to form 4-hydroxylaminobenzoate and lyase-catalyzed rearrangement of the latter into protocatechuate. Although isolated on 4-aminobenzoate and able to degrade 4-nitrobenzoate, *B. cepacia* strain PB4 does not seem to degrade 4-nitrobenzoate via 4-aminobenzoate. Nevertheless, as protocatechuate was cleaved by protocatechuate 3,4-dioxygenase in *B. cepacia* induced on either 4-nitro- and 4-aminobenzoate (Russ, 1991; Lenke, 1997, personal communication), pathways used by *B. cepacia* strain PB4 were probably converging at the intermediate protocatechuate (Fig. 3). Both degradation pathways are inducible since non-induced *B. cepacia* strain PB4 has no activity towards 4-nitro- or 4-aminobenzoate. Although 4-aminobenzoate was not an intermediate in 4-nitrobenzoate degradation (and obviously vice-versa), both compounds were able to induce metabolism of both compounds, thus showing what we could call a cross induction. The consequence of this is that 4-nitro- and 4-aminobenzoate in mixtures were degraded simultaneously irrespective of the inducer, and even if the strain was not induced. The strain showed no preference for one compound

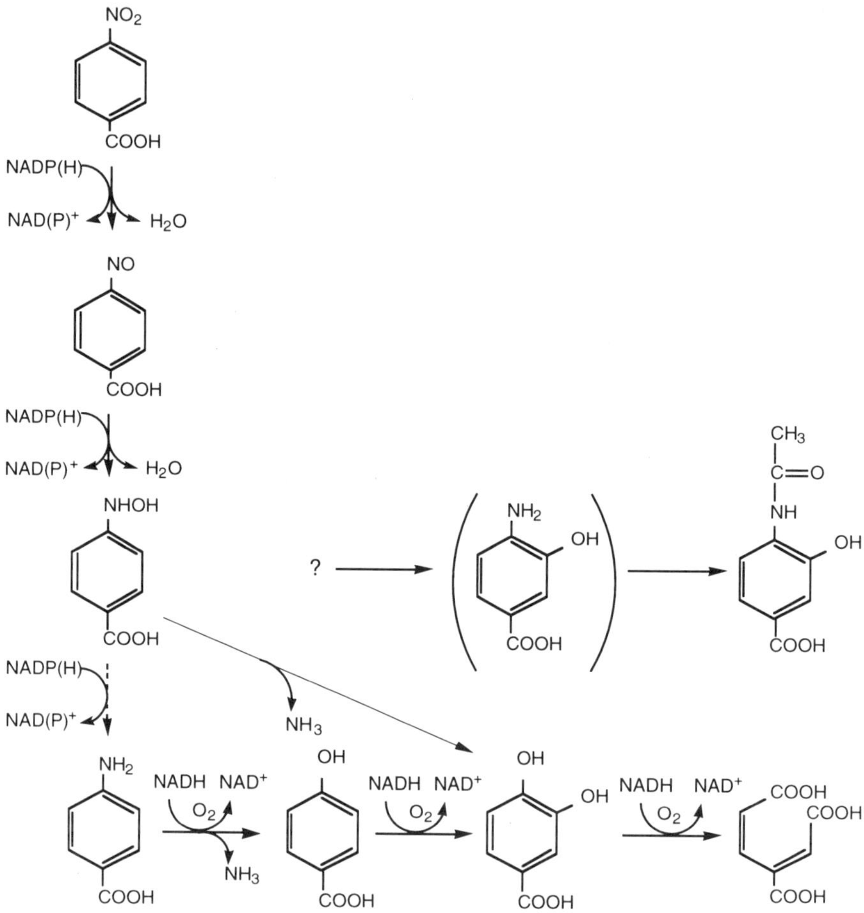

Figure 3. Degradation of 4-aminobenzoate (as determined by Russ (1991)) and of 4-nitrobenzoate (this study) by *Burkholderia cepacia* strain PB4. The dashed line is indicative of the cometabolic pathway.

over the other. However, we should specify that induction of 4-nitrobenzoate transformation by 4-aminobenzoate is stronger than the contrary.

Although both the inducer and the other compound were degraded simultaneously, kinetics of degradation of the inducer seemed logistic whereas those of the second compound were logarithmic. According to Alexander (1994), a logarithmic curve is obtained when concentration of the substrate is much above Ks. In his words, ‘initially, little loss of chemical is evident because the cell density is low, but as the mass of cells becomes large, a rapid loss of chemical is evident’. In our own case, it is more as if initial loss of the second compound is slow but becomes faster when cells get fully induced to it as well. In a logistic model, initial substrate concentration is lower than Ks and the doubling time becomes longer as the substrate is assimilated (Alexander, 1994). In our case, the inducer concentration was not that low but it was apparently small in relation to the ratio (substrate concentration)/(substrate-degrading enzymes concentration), which might explain the observed kinetics.

Also in a *Pseudomonas pickettii* 4-aminobenzoate was able to enhance the oxygen uptake rate of 4-nitrobenzoate-grown cells and, in this strain also, 4-aminobenzoate was

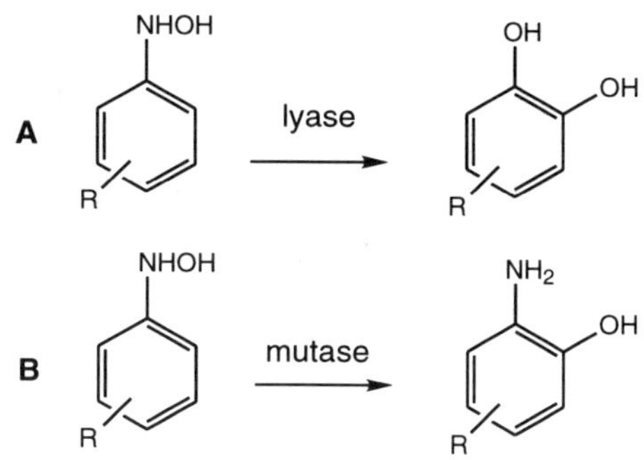

Figure 4. Mechanisms by which an hydroxylaminoaromatic compound can be transformed in order to produce a substrate for subsequent ring cleavage. A: lyase-catalyzed transformation. B: mutase-catalyzed (Bamberger) transformation.

not an intermediate in 4-nitrobenzoate degradation (Yabannavar and Zylstra, 1995). However, P. pickettii did not have the unexpected characteristic exhibited by *B. cepacia* strain PB4, that 4-aminobenzoate induces degradation of 4-nitrobenzoate. The reasons for this cross induction remain unclear. If the monooxygenase from PB4 which is attacking 4-aminobenzoate was also able to transform 4-nitrobenzoate due to a possible non-specificity, it would release nitrites instead of ammonia. But nitrites were not detected. In some cases, an intermediate in a pathway can induce degradation of metabolites upstream or of the parent compound (Stanier, 1947). In our case, it might be that protocatechuate or a later metabolite is able to induce degradation of both 4-nitro- and 4-aminobenzoate. Also, one can reason that *B. cepacia* strain PB4 got adapted in its contaminated site to the frequent occurrence of 4-nitrobenzoate and its corresponding aminoderivative 4-aminobenzoate side by side and that induction on one of the compounds provokes induction of the degradation pathway of both compounds.

The characteristics of the side metabolite produced by *B. cepacia* strain PB4 are those of 3-hydroxy-4-acetamidobenzoate. The latter compound likely derives from 3-hydroxy-4-aminobenzoate whose amino group had been acetylated, a reaction of detoxification of aminoaromatic compounds (Tweedy *et al.*, 1970) that is frequently encountered in the metabolism of nitroaromatic compounds (Gilcrease and Murphy, 1995; Noguera and Freedman, 1996; Schenzle *et al.*, 1997). 4-Acetamidobenzoate, which would be produced by the acetylation of 4-aminobenzoate, was not detected. That might indicate a specific acetylation. Both 4-acetamidobenzoate and 3-hydroxy-4-acetamidobenzoate were produced by a 4-nitrobenzoate-degrading *Pseudomonas fluorescens* (Michan *et al.*, 1997). Similarly, 4-acetamidochlorobenzene and 3-hydroxy-4-acetamidochlorobenzene were detected as metabolites of 4-chloronitrobenzene by the yeast *Rhodosporidium* sp. (Corbett and Corbett, 1981). In the case of *B. cepacia* strain PB4, this hypothetic specificity would be important not to lose 4-aminobenzoate which is a growth substrate. Besides lyase-catalyzed transformation (Fig. 4A) leading to a substrate of ring-cleavage (R = -COOH (Groenewegen *et al.*, 1992; Haigler and Spain, 1993; Rhys-Williams *et al.*, 1993; Yabannavar and Zylstra, 1995; Michan *et al.*, 1997) or R = -OH (Meulenberg *et al.*, 1996)), an hydroxylamino compound was also shown to undergo a mutase-catalyzed rearrangement with production of an aminophenol (Fig. 4B). This mechanism has been shown to lead to the mineralization of nitrobenzene (R = -H (Nishino and Spain, 1993)), 3-nitrophenol (R = -OH (Schenzle *et al.*, 1997)) or 4-nitrotoluene (R = -CH3 (Spiess *et al.*, 1998)). Starting from 4-hydroxylaminobenzoate, it would produce 3-hydroxy-4-aminobenzoate.

B. cepacia strain PB4 was able to degrade 4-nitrobenzoate even when the culture contained exogenous sources of carbon and nitrogen. Usually, growth of a bacterium on exogenous carbon and nitrogen sources produces reducing equivalents which allow the cometabolic reduction of the nitro group into an amino group (Rieger and Knackmuss, 1995). It was thus possible that 4-nitrobenzoate would either be cometabolically reduced into 4-aminobenzoate during growth on glucose, succinate and ammonium chloride, or be mineralized if utilized as sole source of carbon, nitrogen and energy, or both. Since 4-aminobenzoate was detected during 4-nitrobenzoate degradation, this indicates that at least part of the 4-nitrobenzoate was cometabolically reduced. But since 4-aminobenzoate did not accumulate stoechiometrically, we must accept that either the 4-nitrobenzoate- or the 4-aminobenzoate-mineralizing pathway, or even both of them were functioning. The latter assumption is probably the right one considering the cross induction mechanism. As seen on Fig. 1C, with non-induced cells of *B. cepacia* strain PB4, more than 3 hours were needed to induce 4-nitrobenzoate and 4-aminobenzoate degradation pathway. However, the lag phase is expected to be highly dependent on several parameters, such as inoculum size and age, conditions of incubation, etc. It could be that, with the presence of exogenous carbon and nitrogen sources generating a high density of actively growing cells, 1.5 hours were enough to induce 4-nitro- and/or 4-aminobenzoate degradation pathways. It is thus likely that, before the induction of the degradation pathways, 4-nitrobenzoate was transformed into 4-aminobenzoate, and, after, both compounds were simultaneously degraded by the productive pathways. Even when 4-nitrobenzoate was alone in the medium, it could not be excluded that its mineralization proceeded by two parallel pathways, a major one and a minor one caused by a perhaps non-specific nitroreductase, as was reported by Cartwright and Cain (1959).

The versatility of *B. cepacia* strain PB4 (cross induction of 4-nitro- and 4-aminobenzoate degradation pathway, simultaneous degradation of 4-nitro- and 4-aminobenzoate, and degradation of 4-nitrobenzoate in the presence of readily assimilated growth substrates) makes this microorganism a very valuable model system for both mechanistic studies and practical applications of xenobiotic biodegradation.

ACKNOWLEDGMENTS

We wish to thank Anne Pollefeys for technical assistance, Dr Hiltrud Lenke (Fraunhofer Institut für Grenzflächen- und Bioverfahrenstechnik, Stuttgart, Germany) for stimulating discussions and Dr Rainer Russ (Institut für Mikrobiologie, Universität Stuttgart, Germany) for kindly providing us with strain PB4. C.M.P. acknowledges the support of a doctoral fellowship from FRIA (Fund for Training in Research in Industry and Agriculture, Belgium) and S.N.A. is grateful to FNRS (National Fund for Scientific Research, Belgium) for a partial equipment grant.

REFERENCES

Alexander, M. 1994. Kinetics of biodegradation, p. 71–101. In M. Alexander (ed.), Biodegradation and Bioremediation, Academic Press, San Diego, Calif.

Amann, R. I., Ludwig, W., and Schleifer, K.-H. (1995). Phylogenetic identification and in situ detection of individual microbial cells without cultivation. *Microbiol. Rev.* **59**:143–169.

Bruhn, C., Lenke, H., and Knackmuss, H.-J. (1987). Nitrosubstituted aromatic compounds as nitrogen source for bacteria. *Appl. Environ. Microbiol.* **53**:208–210.

Cartwright, N.J., and Cain, R.B. (1959). Bacterial degradation of the nitrobenzoic acids 2. reduction of the nitro group. *Biochem. J.* **73**:305–314.

Corbett, M.D., and Corbett, B.R. (1981). Metabolism of 4-chloronitrobenzene by the yeast Rhodosporidium. *Appl. Environ. Microbiol.* **41**:942–949.

Couch, D.B., Abernethy, D.J., and Allen, P.F. (1987). The effect of biotransformation of 2,4-dinitrotoluene on its mutagenic potential. *Mutagenesis.* **2**:415–418.

Dickel, O., Haug, W., and Knackmuss, H.-J. (1993). Biodegradation of nitrobenzene by a sequential anaerobic-aerobic process. *Biodegradation.* **4**:187–194.

Durham, N.N. (1958). Studies on the metabolism of p-nitrobenzoic acid. *Can. J. Microbiol.* **4**:141–148.

Field, J.A., Stams, A.J., Kato, M., and Schraa, G. (1995). Enhanced biodegradation of aromatic pollutants in cocultures of anaerobic and aerobic bacterial consortia. *Antonie van Leeuwenhoek.* **67**:47–77.

Gilcrease, P.C., and Murphy, V.G. (1995). Bioconversion of 2,4-diamino-6-nitrotoluene to a novel metabolite under anoxic and aerobic conditions. *Appl. Environ. Microbiol.* **61**:4209–4214.

Groenewegen, P.E., Breeuwer, P., van Helvoort, J.M., Langenhof, A.A., de Vries, F.P., and de Bont, J.A. (1992). Novel degradative pathway of 4-nitrobenzoate in Comamonas acidovorans NBA-10. *J. Gen. Microbiol.* **138**:1599–1605.

Haigler, B.E., and Spain, J.C. (1993). Biodegradation of 4-nitrotoluene by Pseudomonas sp. strain 4NT. *Appl. Environ. Microbiol.* **59**:2239–2243.

Harder, W., and Dijkhuizen, L. (1982). Strategies of mixed substrate utilization in microorganisms. *Phil. Trans. R. Soc. London. B* **297**:459–480.

Meulenberg, R., Pepi, M., and de Bont, J.A. (1996). Degradation of 3-nitrophenol by Pseudomonas putida B2 occurs via 1,2,4-benzenetriol. *Biodegradation.* **7**:303–311.

Michan, C., Delgado, A., Haïdour, A., Lucchesi, G., and Ramos, J.L. (1997). In vivo construction of a hybrid pathway for metabolism of 4-nitrotoluene in Pseudomonas fluorescens. *J. Bacteriol.* 179:3036–3038.

Nishino, S.F., and Spain, J.C. (1993). Degradation of nitrobenzene by a Pseudomonas pseudoalcaligenes. *Appl. Environ. Microbiol.* **59**:2520–2525.

Noguera, D.R., and Freedman, D.L. (1996). Reduction and acetylation of 2,4-dinitrotoluene by a Pseudomonas aeruginosa strain. *Appl. Environ. Microbiol.* **62**:2257–2263.

Peres, C.M., Naveau, H., and Agathos, S.N. (1998). Biodegradation of nitrobenzene by its simultaneous reduction into aniline and mineralization of the aniline formed. *Appl. Microbiol. Biotechnol.* **49**:343–349.

Rhys-Williams, W., Taylor, S.C., and Williams, P.A. (1993). A novel pathway for the catabolism of 4-nitrotoluene by Pseudomonas. *J. Gen. Microbiol.* **139**:1967–1972.

Richards, R.M., Xing, D.K., and King, T.P. (1995). Activity of p-aminobenzoic acid compared with other organic acids against selected bacteria. *J. Appl. Bacteriol.* **78**:209–215.

Rieger, P.-G., and Knackmuss, H.-J. 1995. Basic knowledge and perspectives on biodegradation of 2,4,6-trinitrotoluene and related nitroaromatic compounds in contaminated soil, p. 1–18. In J. Spain (ed.), Biodegradation of Nitroaromatic Compounds, Plenum Press, New York.

Russ, R. 1991. Diplomarbeit. Institut für Mikrobiologie der Universität Stuttgart, Stuttgart, Germany.

Schenzle, A., Lenke, H., Fischer, P., Williams, P.A., and Knackmuss, H.-J. (1997). Catabolism of 3-nitrophenol by Ralstonia eutropha JMP 134. *Appl. Environ. Microbiol.* **63**:1421–1427.

Schmidt, T.C., Petersmann, M., Kaminski, L., von Löw, E., and Stork, G. (1997). Analysis of aminobenzoic acids in waste water from a former ammunition plant with HPLC and combined diode array and fluorescence detection. *Fresenius J. Anal. Chem.* **357**:121–126.

Smibert, R.H., and Krieg, N.R. 1981. General Characterization, p. 409–443. In P. Gerhardt, R.G. Murray, R.N. Castilow, E.W. Nester, W.A. Wood, N.R. Krieg, and G.B. Philipps (ed.), Manual of Methods for General Microbiology, American Society for Microbiology, Washington D. C.

Somerville, C. C., Nishino, S. F., and Spain, J. C. (1995). Purification and characterization of nitrobenzene nitroreductase from Pseudomonas pseudoalcaligenes JS45. *J. Bacteriol.* **177**:3837–3842.

Spain, J. C. (1995). Biodegradation of nitroaromatic compounds. *Annu. Rev. Microbiol.* **49**:523–555.

Spiess, T., Desiere, F., Fischer, P., Spain, J.C., Knackmuss, H.-J., and Lenke, H. (1998). A new 4-nitrotoluene degradation pathway in a Mycobacterium strain. *Appl. Environ. Microbiol.* **64**:446–452.

Stanier, R.Y. (1947). Simultaneous adaptation : a new technique for the study of metabolic pathways. *J. Bacteriol.* **54**:339–348.

Stanier, R.Y., Palleroni, N.J., and Doudoroff, M. (1966). The aerobic Pseudomonads : a taxonomic study. *J. Gen. Microbiol.* **43**:159–271.

Stanier, R.Y., Ingraham, J.L., Wheelis, M.L., and Painter, P.R. 1987. In General Microbiology. Macmillan Education LTD, London.

Tweedy, B.G., Loepky, C., and Ross, J. A. (1970). Metobromuron: acetylation of the aniline moiety as a detoxification mechanism. *Science.* **168**:482–483.

Yabannavar, A.V., and Zylstra, G.J. (1995). Cloning and characterization of the genes for p-nitrobenzoate degradation from Pseudomonas pickettii YH105. *Appl. Environ. Microbiol.* **61**:4284–4290.

9

MINERALIZATION OF AZO DYES UNDER INTEGRATED ANAEROBIC/AEROBIC CONDITIONS

N. C. G. Tan,* J. L. Opsteeg, G. Lettinga, and J. A. Field

Wageningen Agricultural University
Department of Agricultural, Environmental and Systems Technology
Sub-department of Environmental Technology
Bomenweg 2, 6703 HD Wageningen, The Netherlands

1. ABSTRACT

Integration of anaerobic and aerobic conditions is a good strategy for the complete mineralization of azo dyes. The success of the strategy relies on anaerobic azo dye reduction occurring inside biofilms exposed to oxygen and subsequent aerobic mineralization of the formed aromatic amines. First, the effect of oxygen on the azo dye reduction by methanogenic granular sludge was studied using Mordant Orange 1 (MO1, Color Index no. 14030) as a model azo dye and ethanol as a cosubstrate. Cosubstrate is required for the formation of anaerobic microniches and supply of reducing equivalents for the reduction. Aromatic amines released from MO1 reduction were not mineralized due to the absence of appropriate aromatic amine degrading bacteria in the methanogenic sludge utilized. Second, the mineralization of 4-phenylazophenol (4-PAP, Color Index no. 11800) was studied under integrated anaerobic/aerobic conditions. Aniline and 4-aminophenol (4-AP) were formed during the anaerobic reduction of 4-PAP. It was examined whether aerobic microorganisms from aerobic sludge and sediment from the Rhine were capable to degrade these aromatic amines. Aniline was readily degraded but 4-AP autoxidized easily under aerobic conditions. No anaerobic degradation could be found for either aromatic amines, indicating the necessity of an aerobic step for removal of these aromatic amines. The bacterial cocultures, anaerobic granular sludge together with the aerobic enrichment culture, were able to degrade the azo dye under integrated anaerobic/aerobic conditions.

* Corresponding author. Telephone (31) 317-484993, Telefax (31) 317-482108, E-mail Nico.Tan@algemeen.mt.wau.nl

Novel Approaches for Bioremediation of Organic Pollution, edited by Fass *et al.* Kluwer Academic / Plenum Publishers, New York, 1999.

2. INTRODUCTION

Dyestuffs are widespread environmental pollutants and these compounds are widely used in textile, printing, leather, pharmaceutical and food industry. The annual dyestuff production world-wide is approximately 7×10^5 tons of which 10–15% are released into the environment (Vaidya and Datye, 1982; Zollinger, 1987). The largest group of dyestuffs are azo dyes, they account for 60–70% of all the textile dyestuffs produced (Carliell *et al.*, 1995).

Azo dyes are often seen as a threat to public health. The azo linkage, the aromatic structure and biotransformation products have shown to be toxic and/or carcinogenic (Chung and Cerniglia, 1992; Chung *et al.*, 1996; Grover *et al.*, 1996). Therefore, the azo dyes are considered as important environmental pollutants and their emission should be minimized. A method to minimize the emission into the environment is via biological degradation.

The initial step in the biotransformation of azo dyes is the cleavage of the azo linkage. This results in a decolorization of azo dyes. In many cases, the biotransformation products are more toxic than the parent compound for mammals (Levine, 1991). Therefore, not only decolorization is required but a complete mineralization must be the final goal of biological treatment.

Under aerobic conditions, azo dyes are very poorly degraded. Due to the strong electron withdrawing character of the azo group an oxygenolitic attack on the aromatic ring or the azo group is almost impossible (Field *et al.*, 1995). Of more than hundred dyes tested, in aerobic biodegradation experiments, only a few were aerobically degraded (Pagga and Brown, 1986; Shaul *et al.*, 1991). In special cases, aerobic bacteria were found that could degrade carboxylated azo dyes after long adaptation and generation times (Kulla *et al.*, 1983). These isolated bacteria posses a oxygen tolerant azo-reductase with narrow substrate range (Zimmermann *et al.*, 1984).

In contrast to the aerobic conditions, azo dyes are readily decolorized under anaerobic conditions. The azo dyes reduction is a ubiquitous anaerobic process if reducing equivalents are available (Carliell *et al.*, 1995; Chung *et al.*, 1978; Nigam *et al.*, 1996; Oxspring *et al.*, 1996; Rafii *et al.*, 1990). Under anaerobic conditions the azo linkage is reduced via four electrons, generating aromatic amines. Azo dyes act as electron acceptors for reduced flavin nucleotides and reduction is enhanced by redox mediators (Gingell and Walker, 1971; Keck *et al.*, 1997; Kudlich *et al.*, 1997).

The aromatic amines are considered to be dead-end metabolites of the anaerobic azo dyes biotransformation. However, some aromatic amines with hydroxyl- and carboxyl-substituents could be degraded under methanogenic conditions (Kuhn and Suflita, 1989; Razo-Flores *et al.*, 1996; Schnell and Schink, 1992). One azo dye, azodisalicylate, was shown to be completely mineralized under anaerobic conditions (Razo-Flores *et al.*, 1997).

In contrast to the anaerobic conditions, aromatic amines are readily degraded under aerobic conditions (Brown and Laboureur, 1983). Many different aromatic amines could be degraded aerobically such as, aniline (Loidl *et al.*, 1990; Lyons *et al.*, 1984), carboxylated aromatic amines (Haller, 1978; Stolz *et al.*, 1992), chlorinated aromatic amines (Loidl *et al.*, 1990; Reber *et al.*, 1979) and sulfonated aromatic amines (Feigel and Knackmuss, 1988; Thurnheer *et al.*, 1988).

Some aromatic amines like phenylenediamines and aminophenols tend to autoxidize under aerobic conditions. This process results in a formation of oligomers and polymers. The initially formed oligomers will not only color the effluent but they are toxic and mu-

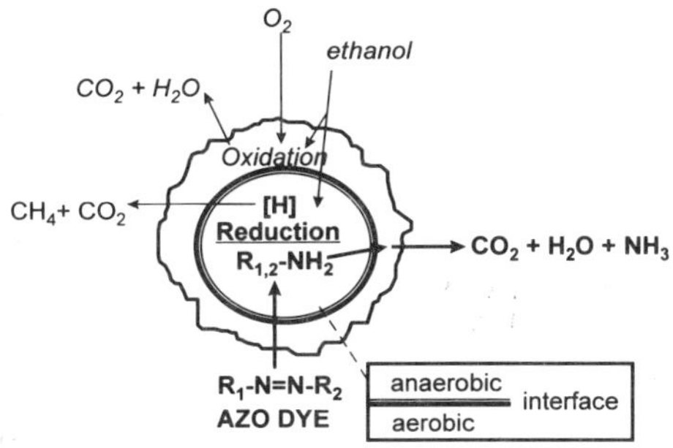

Figure 1. Theoretical representation of the mineralization of an azo dye in a granular sludge particle exposed to oxygen.

tanogenic (Field *et al.*, 1995). A possible way to prevent this autoxidation during the mineralization of azo dyes is an integration of aerobic and anaerobic conditions so that the aromatic amines are only temporarily exposed to low oxygen concentrations. The reduction of the azo dyes and the aerobic mineralization aromatic amines can be performed side by side in the same biofilm. Anaerobic granular sludge can be used as a carrier material for aerobic biofilms and anaerobic microniches (Kato *et al.*, 1993b; Shen and Guiot, 1995; Shen *et al.*, 1996; Zitomer, 1998). Figure 1 shows the theoretical representation of the mineralization of an azo dye within a granular sludge particle.

The first goal was to determine if anaerobic microniches are created in granular sludge exposed to oxygen where azo dyes reduction can occur. The effect of the amount of oxygen on the rate of azo dye reduction was considered, with or without cosubstrate. The second goal was to mineralize an azo dye under the integrated anaerobic/aerobic conditions by including in the coculture aerobic bacteria enrichments which can degrade the aromatic amines formed after azo dye reduction.

3. MATERIALS AND METHODS

3.1. Biomass and Chemicals

Methanogenic granular sludge from a full-scale up-flow anaerobic sludge blanket (UASB) reactor treating wet oxidized industrial effluent of Shell Nederland Chemie (Moerdijk, The Netherlands) and from a full-scale UASB reactor treating effluent of an alcohol producing factory Nedalco (Bergen op Zoom, The Netherlands), were used for the batch experiments. Both granular sludge sources were stored at 4°C and washed and sieved to remove the fine particles before use. Aerobic Rhine sediment was collected along the banks of the river in Wageningen (The Netherlands). The sediment (100 g) was suspended in 500 ml water and sieved (pore 1.5 mm). Before usage, the sediment was stored at 4°C. Activated sludge was collected from a plant treating sewage in Bennekom (The Netherlands). All chemicals were purchased from Merck (Darmstadt, Gemany) or Sigma-Aldrich (Zwijndrecht, The Netherlands) and were of the highest purity commercially available.

3.2. Analyses

The methane, oxygen and the cosubstrates in the serum flasks were determined via gas chromatography (Kato *et al.*, 1993a). The azo dyes and amines were measured spectrophotometrically with a Spectronic 60 spectrophotometer (Milton Roy Analytical Products Division, Oostende, Belgium) at their absorbance maximum, MO1 at 373 nm; 4-PAP at 347 nm; aniline at 230 nm; 4-AP at 230 nm; The formed aromatic amines did not absorb light at the measured wavelength for the azo dye. Liquid samples were centrifuged (7833 x *g*, 10 minutes) and diluted in a 0.10 M phosphate buffer (pH 7.0) solution and measured in a 1 cm 100-QS (Hellma Benelux The Hague, The Netherlands) quartz cuvet.

The azo dyes 4-PAP and the formed aromatic amines aniline and 4-AP were also determined with high performance liquid chromatography (HPLC). The liquid phase, methanol 2% demineralized water and 0.5% acetic acid pH 5.9, was pumped with a flow of 300 $\mu l\ min^{-1}$ (Separations High Precision Pump Model 104, Seperations, Hendrik Ido Ambacht, The Netherlands) first through a Separations GT-103 degasser (Hendrik Ido Ambacht, The Netherlands) and afterwards through a double reverse phase C18 column (Chromosphere C18, Chrompack, Bergen op Zoom, The Netherlands). A gradient program was used for the separation of the aromatic amines and the azo dye. From time 0 minutes to 5 minutes 100 percent acetic acid, afterwards a gradient that finished at 15 minutes with 20 percent of acetic acid and 80 percent of methanol and lasts 5 minutes. At time 25 minutes, 100 percent of acetic acid is reached again and at time 30 minutes, the program stops. Samples (10 μl) were injected with a Marathon autosampler (Separations, Henderik Ido Ambacht, The Netherlands) and the compounds were detected spectrophotometrically with a Spectroflow 783 (Kratos Analytical, Hendrik Ido Ambacht, The Netherlands) at their wavelength of maximum absorbence.

3.3. Assays

3.3.1. Azo Dye Reduction Rate under Influence of Oxygen. The basal medium used in all batch experiments was described previously (Razo-Flores *et al.*, 1997). Total suspended solids (TSS) and volatile suspended solids (VSS) were determined according to AHPA (1985). Azo dye reduction rate experiments were performed in 120 ml and 309 ml serum bottles. These bottles were filled with 25 ml or 64 ml of basal medium, granular sludge (0.4 g VSS l^{-1}), cosubstrate (2.0 g COD l^{-1}) and 50 mg l^{-1} (0.174 mM) MO1 and closed with a butyl rubber septa (Rubber B.V., Hilversum, The Netherlands) and a crimp seal aluminum cap. The headspace of the bottles was flushed with N_2 / CO_2 (70% / 30%) gas for 5 minutes. The same amount of N_2 / CO_2 (70% / 30%) gas was removed and replenished with oxygen gas to arrange an initial headspace oxygen percentage (IHOP). The assays were either performed in duplicate or triplicate. The assay bottles were incubated in a temperature controlled room at 30 ± 2°C on a orbital-motion shaker at 50 strokes min^{-1} (Gerhardt, Bonn, Germany). Methane, oxygen and cosubstrates were measured gas chromatographically and MO1 was measured spectrophotometrically.

3.3.2. Aerobic and Anaerobic Aromatic Amine Degradation Assay. The aerobic aromatic amines degradation was tested via the following assay. Experiments were performed in 120 ml glass serum vials. The bottles contained 22.5 ml and 24.0 ml of basal medium for the aerobic and anaerobic degradation experiments, respectively. The bathes were prepared and treated as described before with 200 mg l^{-1} aromatic amine (aniline or 4-AP). Oxygen and inoculum (aerobic Rhine sediment and aerobic slugde; 10 % v/v) were added for the aerobic degradation experiments. If degradation occurred oxygen and aromatic amines were re-added to enrich aerobic bacteria. Anaerobic (0 IHOP) biodegradation of the aromatic amines were followed in batches with Nedalco granular sludge (1.0 g VSS l^{-1}). Autoclaved batches and no compound containing batches were used as controls. Biodegradation in the anaerobic tests were determined via methane production (Razo-Flores *et al.*, 1996). The aromatic amine concentration was followed spectrophotometrically and oxygen was measured via the gas chromatographic method.

3.3.3. Integrated Anaerobic/Aerobic Degradation of the Azo Dyes. Azo dye degradation experiments were performed in 309 ml serum bottles batches. These bottles were filled with 64 ml of basal medium, seeded with Nedalco granular sludge (1.0 g VSS l^{-1}) from a reactor treating 4-PAP (not described) and the aerobic aromatic amines degrading enrichments (10 % v/v), ethanol (2.0 g COD l^{-1}) and 60 mg l^{-1} (0.30 mM) 4-PAP. The bathes were prepared and treated as described before. Methane, oxygen and ethanol were measured gas chromatographically and the azo dyes and the aromatic amines were measured spectrophotometrically and with HPLC.

4. RESULTS

4.1. Influence of Oxygen on MO1 Reduction

The maximum azo dye reduction rates were calculated using time-courses of the batch experiments. Figure 2 shows a typical time-courses of MO1 exposed to 10 initial headspace oxygen percentage (IHOP) with Nedalco granular sludge and compares the addition of ethanol and no cosubstrates. The reduction of MO1 was mediated biologically since the concentration remained constant in assays with autoclaved sludge (data not shown). Neither the dye was decolorized nor was oxygen consumed if no cosubstrate was added. Addition of ethanol enhanced the azo dye reduction and oxygen consumption rate.

Figure 3 shows the azo dye reduction rate as a function of the IHOP for Nedalco granular sludge. The azo dye reduction was most rapid when ethanol was used as cosubstrate. In the absence of oxygen, this rate was 125 mg MO1 g^{-1} VSS d^{-1} and the rate de-

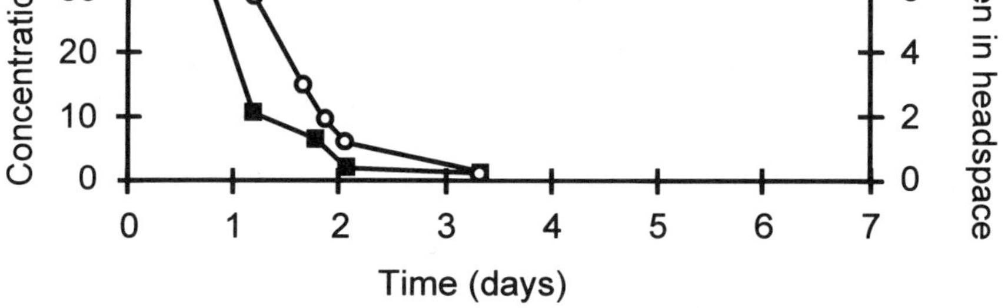

Figure 2. Time courses of MO1 (O) and oxygen (■) at 10 IHOP with ethanol (—) or no cosubstrate (---) and Nedalco granular sludge as biomass source.

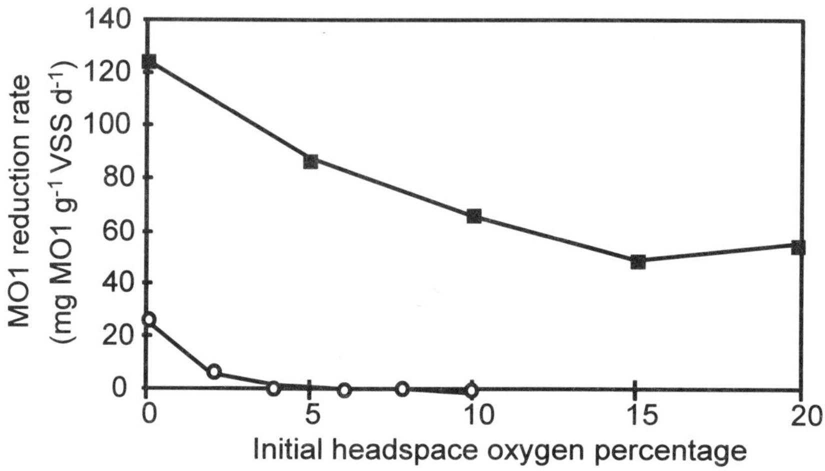

Figure 3. Reduction rates of MO1 in presence of oxygen with Nedalco granular sludge as biomass source with ethanol (■) as cosubstrate or no cosubstrate (O).

creased incrementally with increasing IHOP to 48 mg MO1 g^{-1} VSS d^{-1} at 15 IHOP and remained more less constant thereafter at higher oxygen concentrations. No cosubstrate supported a much lower azo dye reduction rate in the absence of oxygen of approximately 20 mg MO1 g^{-1} VSS d^{-1}. The rate dropped to zero already at 4 IHOP when no cosubstrate was used.

The formed aromatic amines were not degraded aerobically (data not shown) due to the absence of aerobic aromatic amine degrading bacteria in the granular sludge used.

4.2. Mineralization of 4-PAP

Since, no appropriate aerobic bacteria were present in the methanogenic granular sludge to degrade the aromatic amines formed during the reduction of MO1, aerobic enrichments cultures were added to the integrated anaerobic/aerobic biodegradation experiments using 4-PAP as a model azo dye. First, aerobic and anaerobic degradation of the aromatic amines aniline and 4-AP was examined and aerobic aromatic amines degrading bacteria were enriched. Aniline was readily degraded by two different inoculum sources tested under aerobic conditions, aerobic Rhine sediment and activated sludge, respectively. The aerobic degradation of 4-AP was also tested but 4-AP easily reacts with oxygen. The losses of 4-AP was completely accounted for by autoxidation in sterile controls, showing the same degradation rate as the inoculated assay batches. In addition, the medium was colored due to the formation of oligomers and polymers.

The anaerobic degradation of the aromatic amines was also evaluated to investigate whether the aromatic amines could be degraded anaerobically under the integrated anaerobic/aerobic conditions. Anaerobically, 4-AP showed no biodegradation after a period of 100 days Since, aniline is already known to be recalcitrant under methanogenic conditions, no experiments were performed (De *et al.*, 1994; Razo-Flores *et al.*, 1996). Therefore, for complete mineralization of 4-PAP integrated anaerobic/aerobic conditions are required. Firstly, to prevent the autoxidation of 4-AP and secondly, to encourage biodegradation of formed aromatic amines.

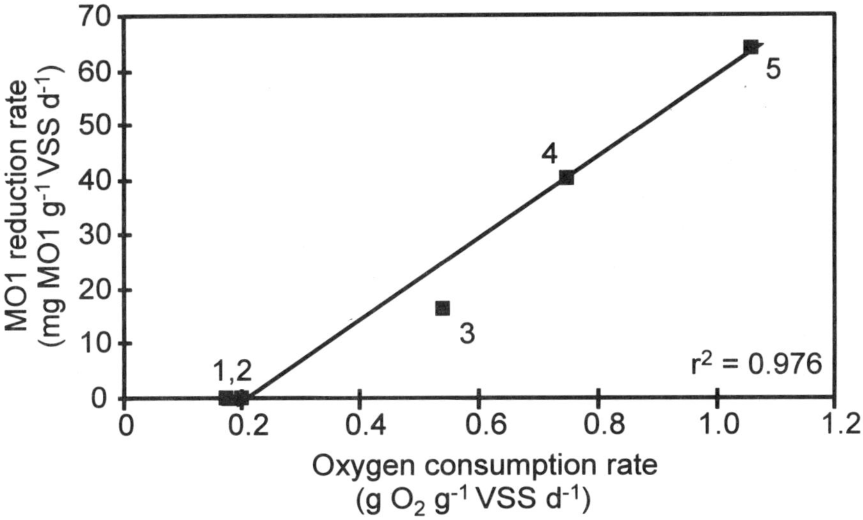

Figure 4. Correlation between oxygen consumption rate and azo dye reduction rate at 10 IHOP (1 = Shell granular sludge, no cosubstrate; 2 = Nedalco granular sludge no cosubstrate; 3 = Nedalco granular sludge, acetate as cosubstrate; 4 = Shell granular sludge, ethanol as cosubstrate; 5 = Nedalco granular sludge, ethanol as cosubstrate).

Figure 4 clearly shows the decolorization of 4-PAP and the degradation of the formed aromatic amines. This figure also shows that the azo dye reduction takes place in presence of oxygen, confirming the results for the influence of oxygen on MO1. The requirement for oxygen to obtain aerobic degradation is also shown in this figure. The formed aromatic amines were not degraded if no oxygen was added (0 IOPH). Due to autoxidation of 4-AP during sample storage prior to HPLC measurements substoichiometric amounts of 4-AP were found. On the other hand, aniline which did not suffer autoxidation during sample storage was recovered in stoichiometric amounts. However, when higher IHOP were applied aniline was quickly degraded by the aerobic enrichments added. The exact mechanism of 4-AP removal (either autoxidation or mineralization) at higher IHOP cannot be differentiated.

5. DISCUSSION

5.1. Influence of Oxygen on MO1 Reduction

The results presented clearly illustrate that azo dye reduction occurs in granular sludge exposed to oxygen. This reduction process only proceeds if cosubstrate is available. The role of the cosubstrate is two-fold. The first role is to provide reducing equivalents via microbial production of reduced flavins for the azo dye reduction (Wuhrmann *et al.*, 1980). By comparing azo dye reduction rates in the assay under completely anaerobic conditions, it was observed that ethanol greatly stimulated the azo dye reduction rate beyond that supported by the meager supply of endogenous substrates in the granular sludge.

The second role of cosubstrate is to create anaerobic microniches (Figure 1). This phenomenon occurs if the cosubstrate is aerobically respired and diffusion of oxygen is limited due to the resistance of the biofilm. In actively respiring aerobic biofilms, the

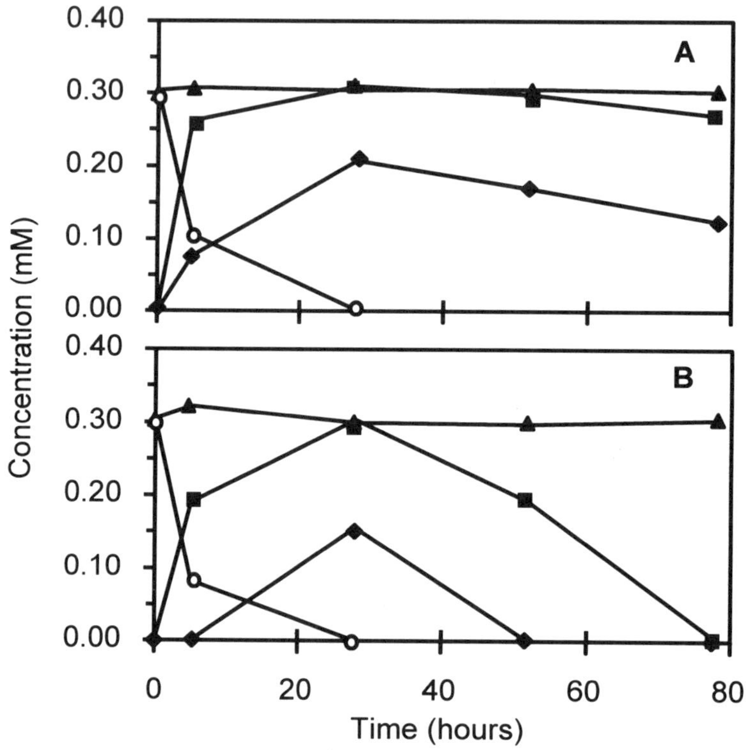

Figure 5. Integrated degradation of 4-PAP (○) and its mineralization products aniline (■) and 4-AP (◆) and 4-PAP in the sterile control (▲), with 0 IHOP (A), and 20 IHOP (B).

depth of oxygen penetration is limited to a few hundred μm (Costerton *et al.*, 1994; Hooijmans *et al.*, 1990; Muller *et al.*, 1994; Nielsen *et al.*, 1990). Cosubstrate is preferably metabolized aerobically by facultative aerobes naturally present in granular sludge resulting in consumption of oxygen. The rate of oxygen consumption was found to be highly correlated with the rate of azo dye reduction in various assays exposed to a given concentration of oxygen (Figure 5). This correlation can be explained on basis of oxygen penetration into biofilm as a function of the oxygen respiration rates (Figure 6). It is well established in the literature that the depth of oxygen penetration into a biofilm is lowered by increasing oxygen consumption rates (Kudlich *et al.*, 1996; Kurosawa and Tanaka, 1990). Consequently, it can be expected that the volume of anaerobic microniches which can support azo dye reduction would increase with greater oxygen consuming activity of the biofilm.

5.2. Mineralization of 4-PAP

The results clearly show the degradation of azo dye 4-PAP under the integrated anaerobic/aerobic conditions. In the literature only one other example was described for complete mineralization azo dyes under these integrated conditions (Kudlich *et al.*, 1996). First, 4-PAP was reduced and the released aromatic amines, aniline and 4-AP, were aerobically degraded in the same batch system.

Due to the autoxidation reactions of 4-AP with oxygen it did not became clear whether this compound was mineralized or autoxidized under the integrated condition. However, the aerobic degradation experiments described here and described in literature

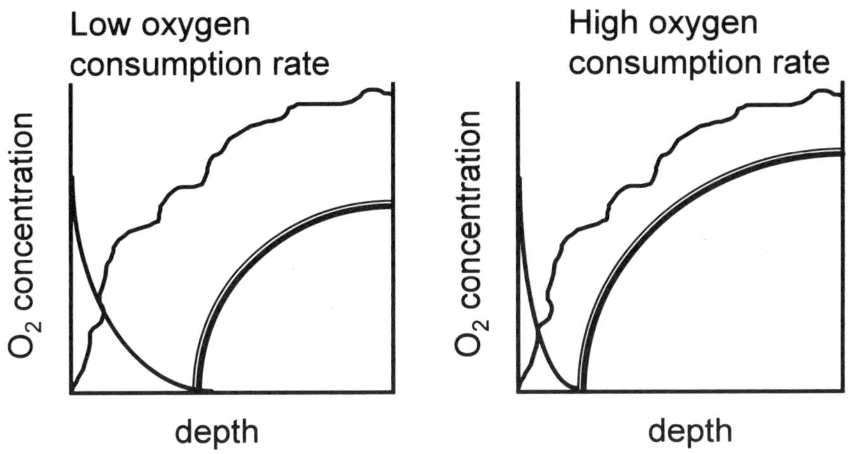

Figure 6. Theoretical representation of the influence of oxygen penetration depth on the formation of anaerobic microniches.

for 4-AP show that this compound is easily autoxidized (Haller, 1978). In general hydroxyl substituted aromatic amines tend to autoxidize easily (Field *et al.*, 1995).

In the anaerobic (0 IHOP, Figure 4a) experiment no aniline degradation was found under methanogenic conditions (De *et al.*, 1994; Razo-Flores *et al.*, 1996). However, aerobic degradation experiments and the integrated anaerobic/aerobic experiment described here show that the aniline removal is easily achieved. The results indicate the necessity of an oxygenolitic attack for the complete mineralization of aniline.

6. CONCLUSIONS

The results confirm the possibility of azo dye reduction under integrated aerobic/anaerobic conditions. Anaerobic microniches were formed in granular sludge exposed to oxygen and the azo dye reduction could take place if the appropriate cosubstrate was present. Additionally, the results demonstrate the applicability of integrated anaerobic/aerobic treatment systems with granular sludge as a biomass source for the degradation of azo dyes that require both anaerobic and aerobic phases for complete mineralization. However, appropriate aerobic microorganisms must be present that can degrade the anaerobically formed aromatic amines products and the aerobic microorganisms have to be incorporated in the aerobic biofilm.

REFERENCES

AHPA. 1985. Standard methods for examination of water and wastewater. American Public Health Association, Washington D.C.

Brown, D., and Laboureur, P. (1983). The aerobic biodegradability of primary aromatic amines. *Chemosphere* **12**:405–414.

Carliell, C.M., Barclay, S.J., Naidoo, N., Buckley, C.A., Mulholland, D.A., and Senior, E. (1995). Microbial decolourisation of a reactive azo dye under anaerobic conditions. *Water SA* **21**:61–69.

Chung, K.T., and Cerniglia, C.E. (1992). Mutagenicity of azo dyes: Structure-activity relationships. *Mutation Res* **277**:201–220.

Chung, K.T., Fulk, G.E., and Egan, M. (1978). Reduction of azo dyes by intestinal anaerobes. *Appl Environ Microbiol* **35**:558–562.

Chung, K.T., Murdock, C.A., Zhou, Y.G., Stevens, S.E., Li, Y.S., Wei, C.I., Fernanda, S.Y., and Chou, M.W. (1996). Effects of the nitro group on the mutagenicity and toxicity of some benzamines. *Environmental and Molecular Mutagenesis* **27**:67–74.

Costerton, J.W., Lewandowski, Z., De Beer, D., Caldwell, D., Korber, D., and James, G. (1994). Biofilms, the customized microniche. *J Bacteriol* **176**:2137–2142.

De, M.A., O'Connor, O.A., and Kosson, D.S. (1994). Metabolism of aniline under different anaerobic electron-accepting and nutritional conditions. *Environ Toxicol Chem* **13**:233–239.

Feigel, B.J., and Knackmuss, H.J. (1988). Bacterial catabolism of sulfanilic acid via catechol-4-sulfonic acid. *FEMS Microbiol Lett* **55**:113–118.

Field, J.A., Stams, A.J.M., Kato, M., and Schraa, G. (1995). Enhanced biodegradation of aromatic pollutant in coculture of anaerobic and aerobic bacterial consortia. *Antonie van Leeuwenhoek* **67**:47–77.

Gingell, R., and Walker, R. (1971). Mechanisms of azo reduction by *Streptococcus faecalis*II. The role of soluble flavins. *Xenobiotica* **1**:231–239.

Grover, I.S., Kaur, A., and Mahajan, R.K. (1996). Mutagenicity of some dye effluents. *National Academy Science Letters India* **19**:149–158.

Haller, H.D. (1978). Degradation of mono-substituted benzoates and phenols by wastewater. *J Water Pollut Control Fed* **12**:2771–2777.

Hooijmans, C.M., Briasco, C.A., Huang, J., Geraats, B.G.M., Barbotin, J.N., Thomas, D., and Luyben, K.C.A.M. (1990). Measurement of oxygen concentration gradients in gel-immobilized recombinant *Escherichia coli*. *Appl Microbiol Biotechnol* **33**:611–618.

Kato, M.T., Field, J.A., and Lettinga, G. (1993a). High tolerance of methanogens in granular sludge to oxygen. *Biotechnol Bioeng* **42**:1360–1366.

Kato, M.T., Field, J.A., and Lettinga, G. (1993b). Methanogenesis in granular sludge exposed to oxygen. *FEMS Microbiol Lett* **114**:317–323.

Keck, A., Klein, J., Kudlich, M., Stolz, A., Knackmuss, H.J., and Mattes, R. (1997). Reduction of azo dyes by redox mediators originating in the naphthalenesulfonic acid degradation pathway of *Sphingomonas* sp. Strain BN6. *Appl Environ Microbiol* **63**:3684–3690.

Kudlich, M., Bishop, P.L., Knackmuss, H.J., and Stolz, A. (1996). Simultaneous anaerobic and aerobic degradation of the sulfonated azo dye Mordant Yellow 3 by immobilized cells from a naphthalenesulfonate-degrading mixed culture. *Appl Microbiol Biotechnol* **46**:597–603.

Kudlich, M., Keck, A., Klein, J., and Stolz, A. (1997). Localization of the enzyme system involved in anaerobic reduction of azo dyes by *Sphingomonas* sp. Strain BN6 and effect of artificial redox mediators on the rate of azo dye reduction. *Appl Environ Microbiol* **63**:3691–3694.

Kuhn, E.P., and Suflita, J.M. (1989). Anaerobic biodegradation of nitrogen-substituted and sulfonated benzene aquifer contaminants. *Hazardous Waste and Hazardous Materials* **6**:121–134.

Kulla, H.G., Klausener, F.K., Meyer, U., Ludeke, B., and Leisinger, T. (1983). Interference of aromatic sulfo groups in the microbial degradation of the azo dyes Orange I and Orange II. *Arch Microbiol* **135**:1–7.

Kurosawa, H., and Tanaka, H. (1990). Advances in immobilized cell culture: development of co-immobilsed mixed culture system of aerobic and anaerobic micro-organisms. *Process Biochem* **25**:189–196.

Levine, W.G. (1991). Metabolism of azo dyes: implications for detoxcification and activation. *Drug Metabol Rev* **23**:253–309.

Loidl, M., Hinteregger, C., Ditzelmuller, G., Ferschl, A., and Streichsbier, F. (1990). Degradation of aniline and monochlorinated anilines by soil-born Pseudomonas acidovorans strains. *Arch Microbiol* **155**:56–61.

Lyons, C.D., Katz, S., and Bartha, R. (1984). Mechanisms and pathways of aniline elimination from aquatic environments. *Appl Environ Microbiol* **48**:491–496.

Muller, W., Winnefeld, A., Kohls, O., Scheper, T., Zimelka, W., and Baumgart, H. (1994). Real and pseudo oxygen gradients in Ca-alginate beads monitored during polarographic PO-2-measurements using Pt-needle microelectrodes. *Biotechnol Bioeng* **44**:617–625.

Nielsen, L.P., Christensen, P.B., Revsbech, N.P., and Sorensen, J. (1990). Denitrification and oxygen respiration in biofilms studied with a microsensor for nitrous oxide and oxygen. *Microbial Ecology* **19**:63–72.

Nigam, P., McMullan, G., Banat, I.M., and Marchant, R. (1996). Decolourisation of effluent from the textile industry by a microbial consortium. *Biotechnol Lett* **18**:117–120.

Oxspring, D.A., McMullan, G., Smyth, W.F., and Marchant, R. (1996). Decolourisation and metabolism of the reactive textile dye, Remazol Black B, by an immobilized microbial consortium. *Biotechnol Lett* **18**:527–530.

Pagga, U., and Brown, D. (1986). The degradation of dyestuffs: Part II. Behaviour of dyestuffs in aerobic biodegradation tests. *Chemosphere* **15**:479–491.

Rafii, F., Franklin, W., and Cerniglia, C.E. (1990). Azoreductase activity of anaerobic bacteria isolated from human intestinal microflora. *Appl Environ Microbiol* **56**:2146–2151.

Razo-Flores, E., Donlon, B.A., Field, J.A., and Lettinga, G. (1996). Biodegradability of N-substituted aromatics and alkylphenols under methanogenic conditions using granular sludge. *Water Sci Technol* **33**:47–57.

Razo-Flores, E., Luijten, M., Donlon, B.A., Lettinga, G., and Field, J.A. (1997). Complete biodegradation of the azo dye azodisalicylate under anaerobic conditions. *Environ Sci Technol* **31**:2098–2103.

Reber, H., Helm, V., and Karanth, N.G.K. (1979). Comparitive studies on the metabolism of aniline and chloroaniline by *Psuedomonas multivorans* strain an 1. *Eur J Appl Microbiol Biotechnol* **7**:181–189.

Schnell, S., and Schink, B. (1992). Anaerobic degradation of 3-aminobenzoate by a newly isolated sulfate reducer and a methanogenic enrichment culture. *Arch Microbiol* **158**:328–334.

Shaul, G.M., Holdsworth, T.J., Demmpsey, C.R., and Dostal, K.A. (1991). Fate of water soluble azo dyes in the activated sludge process. *Chemosphere* **22**:107–119.

Shen, C.F., and Guiot, S.R. (1995). Long term impact of dissolved O2 on the activity of anaerobic granules. *Biotechnol Bioeng* **49**:611–620.

Shen, C.F., Miguez, C.B., Borque, D., Groleau, D., and Guiot, S.R. (1996). Methanothroph and methanogen coupling in granular biofilm under O2-limited conditions. *Biotechnol Lett* **18**:495–500.

Stolz, A., Nortemann, B., and Knackmuss, H.J. (1992). Bacterial metabolism of 5-aminosalicylic acid: Initial ring cleavage. *Biochem J* **282**:675–680.

Thurnheer, T., Alasdiar, M., and Leisinger, T. (1988). Co-culture of defined bacteria to degrade seven sulfonated aromatic compounds efficiency, rates and phenotypic variations. *Appl Microbiol Biotechnol* **29**:605–609.

Vaidya, A.A., and Datye, K.V. (1982). Environmental pollution during chemical processing of synthetic fibers. *Colourage* **14**:3–10.

Wuhrmann, K., Menscher, K., and Kappeler, T. (1980). Investigation on rate - determining factors in the microbial reduction of azo dyes. *Eur J Appl Microbiol Biotechnol* **9**:325–338.

Zimmermann, T., Gasser, F., Kulla, H., and Leisinger, T. (1984). Comparison of two bacterial azoreductases acquired during adaptation to growth on azo dyes. *Arch Microbiol* **138**:37–43.

Zitomer, D.H. (1998). Stoichiometry of combined aerobic and methanogenic COD transformation. *Wat Res* **32**:669–676.

Zollinger, H. 1987. Color chemistry - Syntheses, properties and applications of organic dyes pigments. VCH, New York, NY.

10

MICROBIAL ENZYMES IN BIODEGRADATION

Lawrence P. Wackett*

Department of Biochemistry, Molecular Biology, and Biopysics
Biological Process Technology Institute, and Center for Biodegradation Research and Informatics
University of Minnesota, St. Paul, Minnesota, 55108

1. INTRODUCTION

Biodegradation of organic pollutants cleanses the Earth much like the mammalian immune system cleanses the body of foreign viruses and bacteria. Both systems have evolved to handle millions of different foreign agents. More than 10 million organic compounds are known on Earth and many of those are likely to be biodegradable.

How does nature biodegrade such a vast range of compounds? First, hundreds of thousands, if not millions, of compounds are formed naturally by biosynthesis or diagenesis. The latter process has generated fossil fuel hydrocarbons and heterocyclic compounds (Blumer, 1976). Second, life has likely existed on Earth for at least 3.6 billion years (Lazcano & Miller, 1996), during which time evolution has trained life to catabolize many carbon sources. Consider that early life forms were thought to be bathed in a soup of organic compounds formed by prebiotic chemical reactions and these provided some of the first energy sources for metabolism. Single celled life forms, Eubacteria and Archae in the modern world, have continued to feed on the "soup" around them. Their evolution over 3.6 billion years has generated enormous phylogenetic diversity which is matched by enormous metabolic versatility in the prokaryotic world (Pace, 1997).

But while prokaryotes have recycled organic matter on Earth over eons, organic chemists in the last century have fashioned millions of new organic structures, posing new challenges to microbial metabolism. Yet, microbes have apparently responded. There are numerous reports of industrial chemicals initially evading microbial catabolism, as evidenced by their persistence in the environment, only to be later documented to be biodegradable. Examples include polychlorinated biphenyls (Brown, et.al, 1987), tetrachloroethene (Maymo-Gatell, et.al, 1997), and atrazine (Cook, 1987); all compounds that in early studies were poorly, if at all, cleared from the environment. All of these com-

* Phone: 612-625-3785; fax: 612-625-1700; e-mail: wackett@biosci.cbs.umn.edu

Novel Approaches for Bioremediation of Organic Pollution, edited by
Fass *et al.* Kluwer Academic / Plenum Publishers, New York, 1999.

pounds are, more recently, well-documented to be metabolized by bacteria and are readily cleared in some environments. Do these observations reflect the evolution of new enzymes to handle new compounds or have pre-existing enzymes proliferated due to selective pressure posed by the organic pollutant?

This paper addresses the evolution of biodegradation in two ways. First, it describes an ongoing effort to systematically organize biodegradation information with sufficient rigor to allow the prediction of biodegradation pathways. An understanding of organismal evolution required a systematically organized phylogenetic system and a thorough understanding of biodegradative functions will require an analogous organization. Second, this paper discusses recent experimentally-derived insights into the molecular basis of microbial evolution for the biodegradation of one commercially important compound, atrazine.

2. SYSTEMATIC ORGANIZATION OF BIODEGRADATION INFORMATION

We must begin this topic by realizing the extent of our ignorance of microbial biodegradation. As stated previously, there are over 10 million organic compounds. At most, 10,000 compounds, or 0.1% of the total, have ever been studied in any manner with respect to biodegradation. And only on the order of 100 compounds have been studied in detail for biodegradation, in which the pathway, enzymes and genes are well-characterized.

With such a lack of knowledge, how can we answer fundamental questions on the evolution of biodegradative functions and applications questions as to how newly-synthesized chemicals might be biodegraded? To begin to address these questions, we contributed the University of Minnesota Biocatalysis/Biodegradation Database (Table 1) to the World Wide Web beginning in February 1995. The UM-BBD contains information on bacterial catabolic pathways, enzymes and genes for the biodegradation of principally industrial organic chemicals (Ellis & Wackett, 1997). This represents an attempt to systematically organize biodegradation information and make it readily available. The UM-BBD has proven useful and used, as evidenced by an increase in accesses from 83,385 in 1995 to a pace that will bring over one million accesses in 1998.

The UM-BBD is one of many Web databases used by microbial biochemists and biotechnologists (Table 1). How do we use these enormous information resources, particularly in light of the greater than fifty bacterial genome sequencing projects finished or underway? The key is that information must be organized systematically to produce new knowledge. In the context of biodegradation, a true knowledge of the subject would allow one to predict the biodegradation pathway(s) for any experimentally untested compound.

Predicting biodegradation requires one to deduce the biochemical fate of a compound based on general chemical and biological principles. The chemistry to be considered is the reactivity of the compound; the biology to be considered is the known microbiological transformations of similar molecular fragments. Applying biological and chemical knowledge effectively necessitates compartmentalizing a molecule into manageable fragments. An analogy will illustrate how we are beginning to accomplish this for predicting the biodegradation pathway(s) of experimentally untested compounds; that is, most of the greater than ten million known organic compounds.

We are using the concept of the Periodic Table of the Elements to guide our systematic organization of biodegradation reactions. In biodegradation, the "elements" are best defined as organic functional groups of which a representative set are shown in Figure 1. Forty functional groups are being considered initially. They can exist singly or in combi-

Table 1. Selected web databases useful in biodegradation research

Chemical compounds and toxicology	
Klotho: Biological Compounds Declarative Database at Washington University	http://www.ibc.wustl.edu/klotho/
Cambridge Scientific ChemFinder Chemical Search Service	http://chemfinder.camsoft.com/
Extension Toxicology Network (EXTOXNET) at Oregon State University	http://ace.ace.orst.edu/info/extoxnet/
Metabolism	
The University of Minnesota Biocatalysis/ Biodegradation Database (UM-BBD)	http://www.labmed.umn.edu/umbbd/
KEGG: Kyoto Encyclopedia of Genes and Genomes	http://www.genome.ad.jp/kegg/kegg.html
EcoCyc: Encyclopedia of E. coli Genes and Metabolism	http://ecocyc.PangeaSystems.com/ecocyc/ecocyc.html
WIT at Michigan State University	http://www.cme.msu.edu/WIT/
Enzymes and genes	
LIGAND Chemical Database for Enzyme Reaction at Kyoto University	http://www.genome.ad.jp/dbget_bin/www_bfind?ligand
ExPASy ENZYME Database at the University of Geneva	http://expasy.ch/sprot/enzyme.html
NCBI GenBank	http://www.ncbi.nlm.nih.gov/
Microorganisms	
Bacterial Nomenclature Up-To-Date from the Deutsche Sammlung von Mikroorganismen (DSM)	http://www.gbf-braunschweig.de/DSMZ/bactnom/bactname.htm
Word Federation for Culture Collections (WFCC) World Data Centre for Microorganisms (WDCM)	http://wdcm.nig.ac.jp/
Microbe Zoo at Michigan State University	http://commtechlab.msu.edu/CTLProjects/dlc-me/zoo/

nation to make millions of organic compounds and, in this way, they resemble the chemical elements. Many of the organic functional groups in Figure 1 are known to be transformed enzymatically. For example, a nitrile, or -C≡N functional group, can undergo addition of water to yield an amide. The amide C-N bond, in turn, can undergo addition of water with C-N bond cleavage. Oxygen esters and thioesters also undergo similar hydrolytic reactions. Such organic functional groups that undergo facile hydrolysis undergo these reactions during biodegradation.

Other organic functional groups may be subjected to more than one biochemical transformation. For example, benzene is oxidized by many aerobic soil bacteria to produce benzene *cis*-dihydrodiol, via a dioxygenase, or phenol, via a monooxygenase enzyme. Benzene ring compounds are now known to undergo anaerobic biodegradation via carboxylation reactions, thioesterification with coenzyme A and benzene ring reduction and cleavage.

It is a very reasonable exercise to compile information on the reactions undergone by all 40 organic functional groups under consideration. With an average of two reactions per group, this is only eighty enzymatic transformations in total. Benzene, cyanide and ethylene are examples of important industrial chemicals for which biodegradation pathways can be readily predicted.

The next step up in complexity is to consider all combinations of two functional groups. This would be eight hundred reactions if all were considered. However, about half can be excluded as being irrelevant to biodegradation, for example acyl chlorides, -C(O)-Cl, that undergo very rapid spontaneous hydrolysis yielding a carboxylic acid and chloride anion. Of the four hundred combinations remaining, one must consider, on the

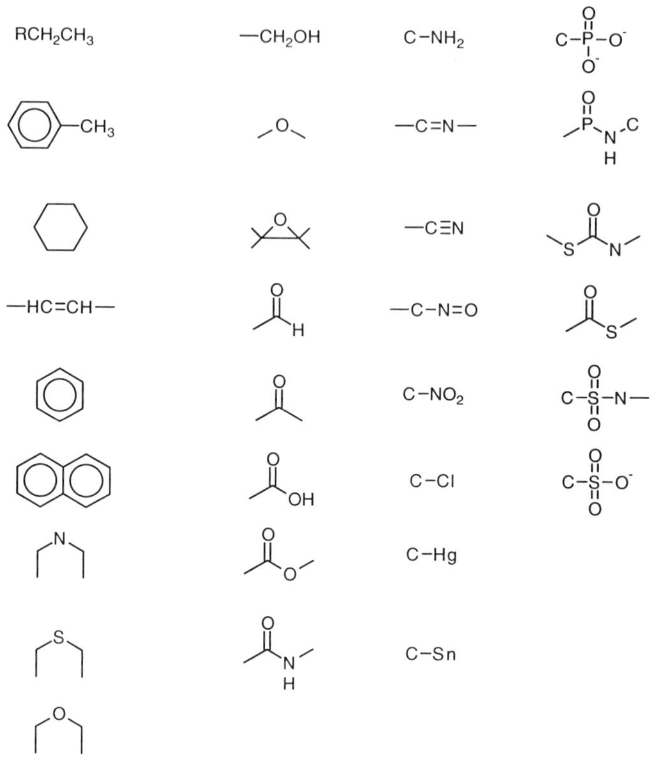

Figure 1. The "elements" of biodegradation as organic functional groups.

average, two reactions by which those functions groups may be cleaved from each other or about eight hundred reactions, a reasonable number. Of course, the individual functional groups may also be transformed and this has already been considered.

The main new consideration is to decide on the priority of reaction. Figure 2 illustrates this point. With benzonitrile, or cyanobenzene, the cyano group has priority as two facile water addition reactions generate benzoic acid that can be readily metabolized aerobically or anaerobically. One can envision initial biodegradation reactions on the benzene ring, but those liberating cyanide may well prove toxic to the bacterium.

Organic compounds containing more than two combinations of functional groups are more challenging and may necessitate a heuristic approach. However, it should be realized that many important industrial chemicals are comprised of two or less functional group combinations. And predicting biodegradation for more complex compounds can build on the extensive empirical biodegradation information currently entered in the UM-BBD.

We are currently investigating both expert and self-learning systems to encode the biochemical data and ultimately produce a predictive computer software. The software will be made freely available via the World Wide Web. Moveover, the interactive nature of the Web will facilitate our compilation of expert knowledge. It is currently planned to convene a panel of biodegradation and computer metabolic logic experts to compile comprehensive and accurate knowledge that is necessary to construct the predictive software. Making our findings accessible to the international scientific community via the Web will lead to further iterative processes, increasing and improving on the expert knowledge that is compiled.

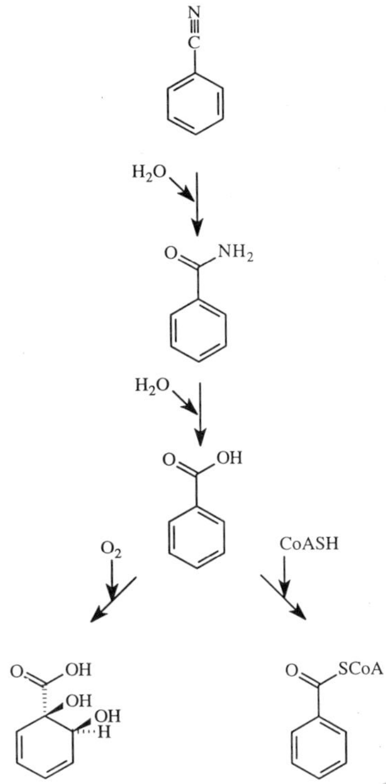

Figure 2. Priority of functional group attack observed with benzonitrile.

3. EVOLUTION OF AN ATRAZINE CATABOLIC PATHWAY

While a holistic approach to biodegradation can be immensely useful, an in-depth analysis of a single catabolic pathway can bring fresh insights into the evolution of new biodegradative functions. As indicated previously, only a small fraction of organic compounds will ever be investigated for their biodegradation. The selection of compounds to be studied will be skewed toward those of greatest economic and environmental significance. The herbicide atrazine is a major agricultural chemical product used in the United States, Europe and Israel. Moreover, over one billion pounds of atrazine have been applied globally, it is only moderately biodegradable, and it is sometimes found in ground and surface waters. These issues have spurred research on atrazine biodegradation.

There have been many reports of atrazine biodegradation in soils, mixed bacterial cultures, and, most recently, isolated bacteria. However, the molecular genetic basis of atrazine catabolism to carbon dioxide, ammonia and chloride was largely undefined before 1993. In fact, most reports stated that atrazine was catabolized largely via oxidative dealkylation reactions that removed either or both of the ethyl and isopropyl side chains (Cook, 1987; Erickson & Lee, 1989). These metabolites were often observed in soils and thus oxidative dealkylation was thought of as a dead-end metabolism. The genes encoding bacterial oxidative dealkylation reactions, *thcBCD*, and the respective cytochrome P450 monooxygenase enzyme system they encode, are described by Shao and Behke (1996). A similar bacterial system catalyzing enzymatic dealkylation is described by Nagy and coworkers (1995).

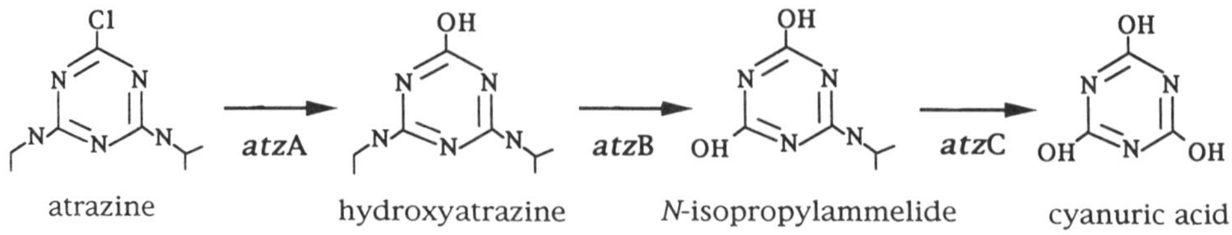

Figure 3. Pathway of atrazine catabolism to cyanuric acid by *Pseudomonas* sp. ADP.

Most recently, a number of bacteria were described that rapidly catabolize atrazine with liberation of ammonia and carbon dioxide derived from the triazine ring carbon atoms (Mandelbaum, et.al, 1995; Radosevich, et.al, 1995). Some of those strains were shown to produce hydroxyatrazine as an intermediate during atrazine metabolism. A *Rhizobium* sp. stoichiometrically metabolizes atrazine to hydroxyatrazine (Bouquard, et.al, 1997). These, and other studies, convincingly refute the often-stated idea that hydroxyatrazine in the environment is derived exclusively via abiotic hydrolytic cleavage of the carbon-chloride bond (Armstrong & Chesters, 1968; Erickson & Lee, 1989).

Further studies identified the gene (*atzA*) and enzyme (AtzA) responsible for the hydrolytic dechlorination of atrazine by *Pseudomonas* sp. ADP (deSouza, et.al., 1996). *Escherichia coli* clones containing *atzA* stoichiometrically convert atrazine to hydroxyatrazine. An *E. coli* clone containing a somewhat larger DNA fragment from *Pseudomonas* sp. ADP produces hydroxyatrazine transiently, which is subsequently converted to N-isopropylammelide (deSouza, et.al, 1995). The *atzB* gene is responsible for this transformation and encodes an enzyme catalyzing a hydrolytic deamidation reaction (Boundy-Mills, et. al, 1997). The *atzC* gene was subsequently cloned and shown to encode another hydrolytic deamidase that transforms N-isopropylammelide to cyanuric acid (Sadowsky, et.al, 1998).

Elucidation of the first three steps in atrazine catabolism by *Pseudomonas* sp. ADP (Figure 3) revealed the metabolic and evolutionary logic underlying efficient atrazine catabolism. Three consecutive enzymatic hydrolysis reactions transform atrazine, a compound few bacteria can mineralize, to cyanuric acid, a compound many soil bacteria can mineralize. Cyanuric acid, in its preferred tautomeric form, is a cyclic amide and thus likely susceptible to hydrolysis by many non-specific bacterial amidases. This is supported by the studies of Cook and coworkers (Cook, et.al, 1985; Cook, 1987) and Eaton & Karns (1991), who collectively demonstrated bacteria, enzymes and genes that function to transform cyanuric acid to biuret, urea and, ultimately, carbon dioxide and ammonia. Common soil bacteria with a cyanuric acid pathway, that acquire enzymes isofunctional to AtzA, B and C, will be able to mineralize atrazine and use it as a sole nitrogen source. But what is the origin of the *atzA,B* and *C* genes and might they be acquired concurrently?

In *Pseudomonas* sp. ADP, the *atzABC* genes reside on a 96 Kb broad host range plasmid that we have denoted pADP-1 (deSouza, et.al, 1998c). Similar sized plasmids have been found in other atrazine-catabolizing bacteria (Topp, 1997), but the relationship of those to pADP-1 are presently unknown. Plasmid pADP-1 will replicate in *E. coli*. Moreover, DNA that is highly homologous to the *Pseudomonas* sp. ADP *atzABC* genes are present in six recently isolated atrazine catabolizing bacteria belonging to disparate genera: *Ralstonia, Alcaligenes, Agrobacterium* and *Clavibacter* (deSouza, et al., 1998b). It remains to be established if the genes are plasmid-encoded in these bacteria. We anticipate that different arrangements of atrazine genes exist in different organisms. A *Rhizobium* sp.

contains a functional atzA gene only (Bouquard, et.al, 1997; deSouza, unpublished), thus accumulating hydroxyatrazine from atrazine. A *Clavibacter* sp. contains *atzA, B* and *C* homologs, yet it catalyzes only two enzymatic reactions in transforming atrazine to N-ethylammelide. *Pseudomonas* sp. CN, that was isolated from a stable consortium containing the atrazine-metabolizing *Clavibacter*, has an *atzC* gene and metabolizes N-ethylammelide via cyanuric acid as an intermediate (de Souza et al., 1998a). Consistent with these data, we've detected DNA sequences homologous to known insertion elements in regions flanking the *atzA* and *C* genes in *Pseudomonas* sp. ADP (de Souza et al., 1996; de Souza et al., unpublished). These data suggest potential mechanisms for variant assemblages of atrazine catabolism genes in single bacteria or in different species in consortia where collective metabolism of atrazine leads to its mineralization.

But where did the *atzA, B* and *C* genes come from to confer atrazine mineralizing ability on various bacterial strains and consortia? We can now begin to answer this question. The sequences of the proteins atrazine chlorohydrolase (AtzA), hydroxyatrazine N-ethylaminohydrolase (AtzB), and N-isopropylammelide N-isopropylaminohydrolase (AtzC), respectively, were derived from the corresponding gene sequences. Pairwise comparisons of these sequences did not immediately reveal a relationship; percent amino acid sequence identities were only on the order of 20%. It was revealing, however, to compare the sequences to conserved motifs of different hydrolytic enzyme families (Sadowsky, et. al, 1998). Significant sequences identity was observed in comparing AtzA, B or C with the conserved sequence motif discovered in a large amidohydrolase protein superfamily (Holm & Sanders, 1997). For the structurally-defined superfamily members, the conserved region includes amino acids involved in providing ligands to a functionally important transition metal. This enzyme class includes bacterial adenine deaminase, adenosine deaminase and cytosine deaminase. Thus, there is a potential mechanistic connection between metalloenzymes that catalyze purine or pyrimidine hydrolytic deamination reactions and the atrazine enzymes AtzABC that catalyze hydrolytic deamination or dechlorination reactions, also with nitrogen heterocyclic substrates.

The extensive sequence divergence of AtzA, B and C indicate that one did not give rise to another recently. Rather, AtzA, B and C all arose from members of an ancient amidohydrolase superfamily and have presumably reunited on pADP-1 under the selective pressure imposed by the presence of greater than one billion pounds of atrazine in the environment.

4. CONCLUDING REMARKS ON CATABOLIC ENZYME EVOLUTION AND GENETIC ENGINEERING

These studies highlight how nature recruits enzyme *mechanisms* encoded by existing genomes for transforming new compounds to funnel the major carbon fragment(s) into existing metabolic pathways. With atrazine, a metal-dependent hydrolytic mechanism was used to accomplish both dechlorination and deamination to generate a readily metabolizable intermediate cyanuric acid. More generally, the metapathway map of the University of Minnesota Biocatalysis/Biodegradation Database (Table 1) illustrates the economy displayed by naturally evolved catabolic pathways. That is, selective pressure operates on bacteria to minimize the number of enzymatic steps, and the energy input required, to funnel new compounds into pre-existing metabolism. The widespread atrazine pathway discussed here removes three *s*-triazine ring substituents in three consecutive enzyme catalyzed reactions. Each reaction is hydrolytic and thus does not consume ATP or reducing equivalents, unlike the NAD(P)H-dependent cytochrome P450-dependent atrazine dealkylation reaction demonstrated in *Rhodococcous* (Behki & Khan,1986; Nagy, et. al, 1985).

These lessons learned from nature can be used to enhance the use of genetic engineering to biodegrade pollutants. Those pollutants which are poorly metabolized by naturally-occurring microorganisms can increasingly be targeted for biodegradation by creative metabolic engineering. Computer simulation of biodegradative pathways, as described in this paper, will be a prerequisite to designing efficient new microorganisms by genetic engineering. The pathway from the laboratory to the field will, increasingly in the future, start *in silico*.

ACKNOWLEDGMENTS

The research described here is the work of many excellent coworkers who have made biodegradation research both enlightening and enjoyable. The research has been supported by grants from the National Science Foundation, USDA/BARD and Novartis Crop Protection.

REFERENCES

Armstrong, D.E., and G. Chesters. 1968. Adsorption catalyzed chemical hydrolysis of atrazine. *Enviorn. Sci. Technol.* 2:683–689.

Behki, R. M. and S. U. Khan. 1986. Degradation of atrazine by *Pseudomonas*: N- dealkylation and dehalogenation of atrazine and its metabolites. *J. Agric. Food Chem.* 34:746–749.

Blumer, M. (1976) Polycyclic aromatic compounds in nature. *Sci. Amer.*, p. 35 - 45.

Boundy-Mills, K., M. L. de Souza, R. M. Mandelbaum, L. P. Wackett, and M. J. Sadowsky. 1997. The *atz*B gene of *Pseudomonas* sp. strain ADP encodes the second enzyme of a novel atrazine degradation pathway. *Appl. Environ. Microbiol.* 63: 916–923.

Bouquard, C., J. Ouazzani, J.-C. Prome, Y. Michel-Briand, and P. Plesiat. 1997. Dechlorination of atrazine by a *Rhizobium* sp. isolate. *Appl. Environ. Microbiol.* 63: 862- 866.

Brown, J.F., D.L. Bedard, M.J. Brennan, J.C. Carnahan, H. Feng and R.E. Wagner. 1987. Polychlorinated biphenyl dechlorination in aquatic sediments. *Science* 236: 709- 712.

Cook, A. M. 1987. Biodegradation of *s*-triazine xenobiotics. *FEMS Microbiology Rev.* 46:93–116.

Cook, A. M., P. Bellstein, H. Grossenbacher and R. Hutter. 1985. Ring cleavage and degradative pathway of cyanuric acid in bacteria. *Biochem. J.* 231:25–30.

Cook, A. M. and R. Hutter. 1984. Deethylsimazine: Bacterial dechlorination, deamination, and complete degradation. *J. Agric. Food. Chem.* 32:581–585.

DeSouza, M.L., D. Newcombe, Crowley, D.E, S. Alvey, M.J. Sadowsky, and L.P. Wackett (1998) Molecular basis of a bacterial consortium: Interspecies catabolism of atrazine. *Appl. Environ. Microbiol.* 64:178–184.

DeSouza, M. L., M. J. Sadowsky, J. Seffernick, B. Martinez, and L. P. Wackett (1998) The *atzABC* genes encoding atrazine catabolism are widespread and highly conserved *J. Bacteriol.* 180:1951–1954.

DeSouza, M. L., M. J. Sadowsky and L. P. Wackett. 1996. Atrazine chlorohydrolase from *Pseudomonas sp.* Strain ADP: Gene sequence, enzyme purification, and protein characterization. *J. Bacteriol.* 178:4894–4900.

De Souza, M. L., L. P. Wackett, K. L. Boundy-Mills, R. T. Mandelbaum and M. J. Sadowsky. 1995. Cloning, characterization, and expression of a gene region from *Pseudomonas sp.* strain ADP involved in the dechlorination of atrazine. *Appl Environ Microbiol.* 61:3373–3378.

DeSouza, M.L. , L.P. Wackett, and M.J. Sadowsky (1998) The genes encoding atrazine catabolism are located on a self-transmissable plasmid in *Pseudomonas* sp. ADP. *Appl. Environ. Microbiol.* 64: 2323–2326.

Eaton, R. W. and J. S. Karns. 1991. Cloning and comparison of the DNA encoding ammelide aminohydrolase and cyanuric acid amidohydrolase from three s-triazinedegrading bacterial strains. *J. Bacteriol.* 173:1363–1366.

Ellis, L.B.M. and L.P. Wackett (1997) Microbial biocatalysis and biodegradation informatics. *Nature Biotechnology* 15:1406.

Erickson, L. E. and K. H. Lee. 1989. Degradation of atrazine and related *s*-triazines. *Critical Rev. in Environ. Control.* 19:1–14.

Holm, L. and Sander, C. 1997. An evolutionary treasure: Unification of a broad set of amidohydrolases related to urease. *Proteins* 28:72–82.

Lazcano, A. & S.L. Miller (1996) The origin and early evolution of life: Prebiotic chemistry, the pre-RNA world,and time. *Cell* 85:793–8

Mandelbaum, R.T., D.A. Allan and L.P. Wackett. 1995. Isolation and characterization of a *Pseudomonas* sp. that mineralizes the *s*-triazine herbicide atrazine. *Appl. Environ. Microbiol.* 61:1451–1457.

Maymo-Gatell X, Chien Y, Gossett JM, Zinder SH (1997) Isolation of a bacterium that reductively dechlorinates tetrachloroethene to ethene. *Science* 276:1568–1571

Mulbry, W. W. 1994. Purification and characterization of an inducible s-triazine hydrolase from *Rhodococcus corallinus* NRRL B-15444R. *Appl. Environ. Microbiol.* 60:613–618.

Nagy, I., Schoofs, G., Compernolle, F., Proost, J., Vanderleyden, J. and de Mot, R. 1995. Degradation of the thiocarbamate herbicide EPTC (S- ethyldiisopropyl- carbamothioate) and biosaftening by *Rhodococcus* sp. strain NI86/21 involve an inducible cytochrome P450 system and aldehyde dehydrogenase. *J. Bacteriol.* 177:676–687.

Pace, W.R. (1997) A molecular view of microbial diversity and the biosphere. *Science* 276:734–740.

Radosevich, M., S. J. Traina, Y. Hao and O. H. Tuovinen. 1995. Degradation and mineralization of atrazine by a soil bacterial isolate. *Appl. Environ. Microbiol.* 61:297- 302.

Sadowsky, M.J., M.L. DeSouza, Z. Tong, and L.P. Wackett (1998) AtzC is a member of the amidohydrolase protein superfamily and is homologous to other atrazine-metabolizing enzymes. *J. Bacteriol.* 180:152–158.

Shao, Z. Q. and R. Behki. 1996. Characterization of the expression of the *thcB* gene, coding for a pesticide-degrading cytochrome P-450 in *Rhodococcus* strains. *Appl. Environ. Microbiol.* 62:403–407.

Topp, E., Tessier, L., and Lewis, M. 1997. Characterization of atrazine-degrading bacteria isolated from agricultural soils. p. 252. *Abstr. 97th Ann. Meet. Am. Soc. Microbiol.* Washington, DC.

11

ENGINEERING HYDROLASES FOR THE CONVERSION OF HALOGENATED ALIPHATIC HYDROCARBONS AND EPOXIDES

Dick B. Janssen, Gerrit J. Poelarends, and Rick Rink

Biochemical Laboratory
Groningen Biomolecular Sciences and Biotechnology Institute
University of Groningen
Nijenborgh 4, 9747 AG Groningen, The Netherlands

1. INTRODUCTION

Halogenated compounds are important environmental chemicals which may have adverse ecotoxicological effects and act as carcinogens in man. Information about microbial detoxification reactions towards these compounds has mainly become available from studies on bacterial cultures that use them as a growth substrate. Our work has focused on the enzymes involved in the bacterial utilization of halogenated aliphatic compounds. Cleavage of the carbon-halogen bond obviously is a critical step in the metabolism of these compounds. This cleavage reduces or eliminates toxicity. It often is a difficult step, however, because the carbon-halogen bond occurs only in a restricted range of naturally occurring chemicals and only few organisms produce specific dehalogenating enzymes that recognize and act on carbon-halogen bonds in synthetic compounds. In this paper, we discuss the properties of some enzymes that carry out this reaction. We also discuss some protein engineering studies aimed at expanding the catabolic activities of dehalogenases and at understanding the underlying catalytic and kinetic mechanisms.

2. ENVIRONMENTAL SIGNIFICANCE OF HALOGENATED ALIPHATIC COMPOUNDS

Industrially synthesized low molecular weight halogenated aliphatic compounds are applied for a number of different purposes. Some of these applications require bulk production which may be accompanied by the formation of side products in considerable quantities. Both losses of the desirable products during synthesis, transport, or usage, and

Novel Approaches for Bioremediation of Organic Pollution, edited by
Fass *et al.* Kluwer Academic / Plenum Publishers, New York, 1999.

the emission of side products may cause pollution of air, water and soil (for examples, see van Agteren *et al.*, 1998).

The largest halogenated compound from a production point of view is 1,2-dichloroethane. More than 10.10^6 tonnes are produced annually worldwide. Most of it is used for making vinylchloride and PVC, but it can also be used as the starting material for other chemicals and as a solvent. Chloroacetic acid is also used as a starting material for further chemical synthesis, e.g. in the production of carboxymethylcellulose. Epichlorohydrin is used for the production of resins, and is also synthesized in large amounts. The latter two compounds can also be used as intermediates in the production of fine chemicals, for which it may be preferred to use the pure stereoisomers.

Chlorinated solvents and halogenated aerosol propellants are also bulk chemicals, with production volumes of over 1.10^6 tonnes/year for several components. Methylchloride, dichloromethane, chlorofluorocarbons, trichloroethylene and tetrachloroethylene are the most important examples. It has been suggested that the annual production of compounds which are mainly used as solvents is about equal to their loss to the environment plus the amount that ends up in chemical wastes (Leisinger, 1983). Thus, large amounts of chlorinated volatile organics must be removed by proper treatment methods to protect the environment.

Another group of bulk halogenated chemicals are used as agrochemicals for crop protection. Examples are 1,2-dibromoethane and 1,3-dichloropropylene, which have been widely used for soil treatment in citrus fields and potato crops, respectively. The soil fumigant 1,2-dibromoethane may be persistent in groundwater and subsurface soils for decades. Dichloropropylene is chemically unstable, but commercial preparations have been contaminated with 1,2-dichloropropane, which is very recalcitrant and sorbs to soil.

For these low-molecular weight compounds, there is a direct relationship between their biodegradability and the possibility to develop biological treatment systems. Degradation is much less limited by sorption to soil than with high molecular hydrophobic compounds such as polychlorinated biphenyls and polycyclic aromatics. Furthermore, halogenated aliphatics are relatively water soluble and can migrate with groundwater flows, thus threatening groundwater quality. These factors make it highly desirable to develop treatment strategies and make biological treatment an attractive option if a suitable microbial activity can be obtained.

3. UTILIZATION OF HALOGENATED ALIPHATICS AS GROWTH SUBSTRATES

Microorganisms that grow on halogenated aliphatics have been isolated from soil, water, and activated sludge. Compounds that can be used as growth substrate include methylchloride, dichloromethane (Leisinger, 1996), chlorinated and brominated acetic acids, 2-chloropropionic acid, 2-chloroethanol, simple 1-halo-n-alkanes, several α/ω-dihalo-n-alkanes (Janssen *et al.*, 1985, 1994; Scholtz *et al.*, 1988), and a number of chloropropanols (for a recent overview, see van Agteren *et al.*, 1998). With the chloroethenes, only a bacterial culture that grows on vinylchloride has been isolated (Hartmans *et al.*, 1991). Generation times vary from about 1 h for chloroacetic acid and 2-chloroethanol-utilizing *Pseudomonas* strains to about 40 h for a vinylchloride-degrading *Mycobacterium*.

With most bacterial cultures that utilize simple halogenated aliphatics for growth, the first catabolic step in the degradation pathway is cleavage of the carbon-halogen bond. A number of these reactions are included in the routes shown in Fig. 1 (Janssen *et al.*,

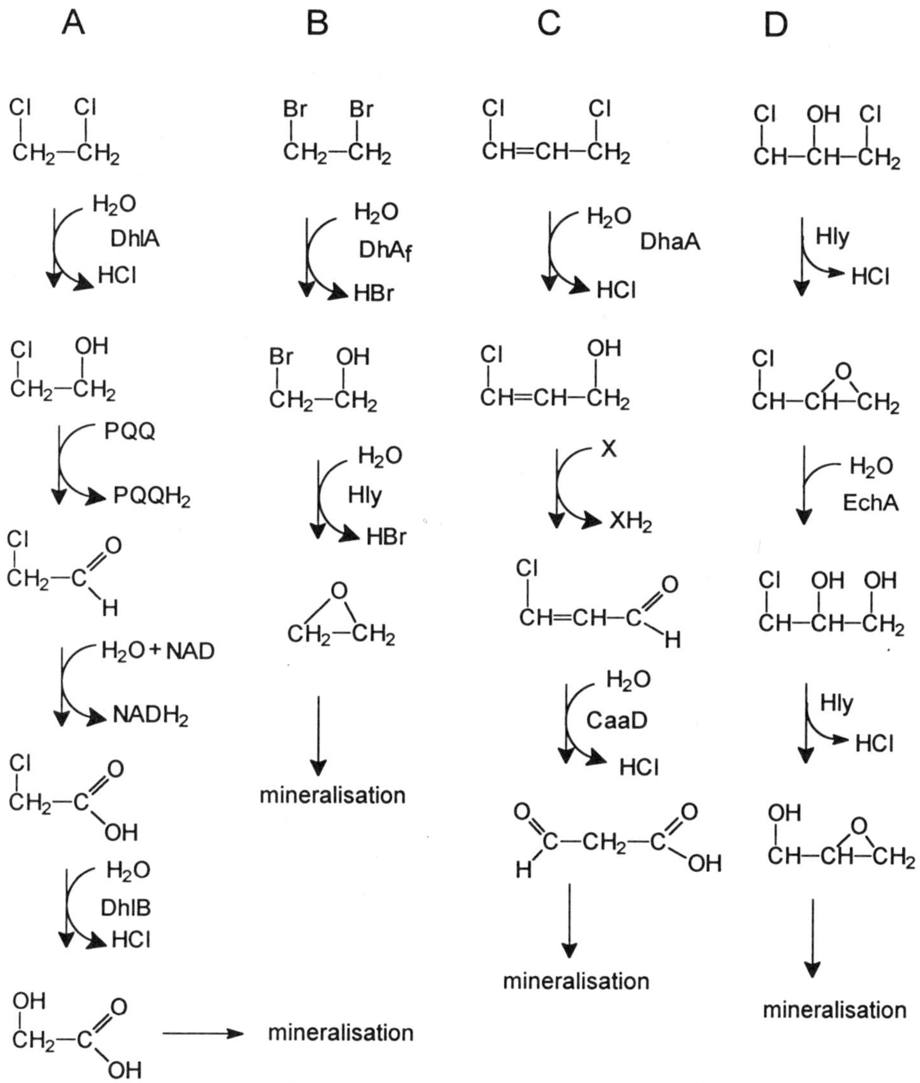

Figure 1. Catabolic route for 4 major halogenated aliphatics: A, 1,2-dichloroethane, the precursor of industrial vinylchloride synthesis and also widely used as solvent; B, 1,2-dibromoethane, a soil fumigant and antiknock agent in fuels; C, 1,3-dichloropropylene, a soil fumigant for potato crops; D, epichlorohydrin, an intermediate in the synthesis of resins.

1994; Fetzner and Lingens, 1994). The substrate range of the enzymes involved and the properties of the intermediates greatly influence the biodegradability of individual haloaliphatics. Thus, a careful analysis of the dehalogenation reactions allows an understanding of the factors causing recalcitrance and the potential of overcoming recalcitrance by evolution of new enzyme activities. If a certain enzyme in a pathway is missing, this may result in inertness of the starting compound, accumulation of an intermediate, or serious toxic effects on the active organism if the intermediate is a reactive chemical.

The role of toxic intermediates can be well illustrated by comparing the catabolic pathways of 1,2-dichloroethane and 1,2-dibromoethane (Fig. 1). The former compound is dehalogenated to 2-chloroethanol, which is oxidized in two steps to chloroacetic acid. Chloroacetate can subsequently be rapidly dehalogenated to glycolic acid (Janssen *et al.*, 1985). It was expected that 1,2-dibromoethane would be degradable according to the same route, since both dehalogenases involved in 1,2-dichloroethane conversion have a high activity with the brominated analogs. However, enrichment of dibromoethane-degrading organisms or adaptation of dichloroethane degraders to the brominated analogs proved extremely difficult due to toxicity of bromoacetaldehyde, which would be an intermediate if 1,2-dibromoethane metabolism occurred along this pathway (van der Ploeg *et al.*, 1996).

After prolonged adaptation, we have recently isolated a culture that grows on 1,2-dibromoethane from a mixed culture supplied by Dr. L. Freitas dos Santos (Freitas dos Santos *et al.*, 1996). The gram-positive organism obtained circumvents the formation of toxic bromoacetaldehyde by converting 2-bromoethanol to ethylene oxide, rather than by oxidizing it. This dehalogenase carries out a lyase reaction which is a net intramolecular substitution leading to elimination of HBr. In this way, toxicity of a reactive intermediate is prevented by using another catabolic route (Fig. 1). A high haloalcohol dehalogenase activity is crucial in this pathway, since it captures 2-bromoethanol and channels it in the right pathway. Similar haloalcohol dehalogenases are involved in the degradation of epichlorohydrin and chloropropanols (Fig. 1).

Another chlorinated aliphatic of which the catabolic pathway was recently investigated is the soil fumigant 1,3-dichloropropylene. Degradation again starts with conversion to 3-chloroallylalcohol, both for the *cis* and *trans* isomers, which were used as a mixture (Poelarends *et al.*, 1998). Chloroallylalcohols can be oxidized to chloroacrylic acids, which undergo dehalogenation by up to date poorly characterized chloroacrylic acid dehalogenases (van der Waarde *et al.*, 1992; van Hylckama Vlieg *et al.*, 1992). Most organisms tested produce two of such enzymes, one selective for the *cis* isomer, the other for *trans*-3-chloroacrylic acid. We have recently cloned the gene encoding the *trans*-3-chloroacrylic acid dehalogenase of *Pseudomonas cichorii* 170. The sequence is currently analyzed in order to attempt to understand the mechanism of this enzyme from sequence similarities. The reaction catalyzed by chloroacrylic acid dehalogenases is of special interest in view of the fact that the enzymes act on a halogen atom that is bound to an unsaturated carbon atom. This reaction type would be of great value if it were also found for other compounds containing vinylic halogens.

4. HALOALKANE DEHALOGENASES: MECHANISM, DIVERSITY, AND DISTRIBUTION

Haloalkane dehalogenases catalyze the hydrolytic removal of halide from haloalkanes, producing alcohols. All haloalkane dehalogenases analyzed in detail up to now were found to belong to the so called α/β-hydrolase fold enzymes (Ollis *et al.*, 1992). Because of the sequence similarity to haloalkane dehalogenase of *X. autotrophicus*, for which the X-ray structure is known, it can be concluded that these proteins share a similar overall structure and catalytic mechanism. The enzymes do not require oxygen or cofactors for their catalytic activity. Instead, catalysis proceeds by attack of a carboxylate oxygen of a conserved aspartate residue on the carbon atom to which the halogen is bound.

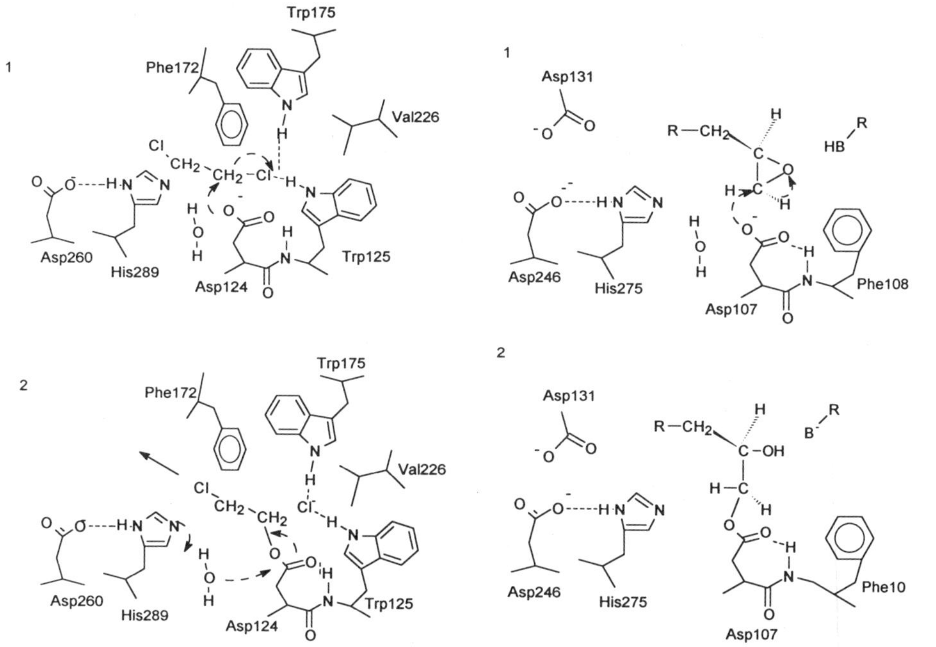

Figure 2. The catalytic mechanisms of haloalkane dehalogenase (left) and epichlorohydrin hydrolase (right). Both enzymes use an aspartic acid carboxylate as the nucleophile, and form a covalent intermediate (Step 1). This is subsequently hydrolyzed by a water molecule that is activated by the histidine which is part of the catalytic triad (Step 2).

The covalent alkyl-enzyme intermediate is subsequently hydrolyzed by water (Fig. 2, Verschueren *et al.*, 1993; Pries *et al.*, 1994a). The last and slowest step in the catalytic cycle is the release of the halide ion (Schanstra and Janssen, 1996).

At least three other dehalogenases with sequence similarity to the *X. autotrophicus* enzyme have been discovered. They hydrolyze fluoroacetate (DehH1), 1-chlorobutane (DhaA) and tetrachlorocyclohexadiene (LinB), which is an intermediate in hexachlorocyclohexane degradation. We have recently investigated the haloalkane dehalogenases in a series of 1,2-dichloroethane, 1-chlorobutane, and 1,3-dichloropropene utilizing bacteria. It appeared that all 1,2-dichloroethane degrading enzymes are identical, including the occurrence of short sequence repeats in the cap domain, which were proposed to be involved in recent adaptation of these enzymes to synthetically produced 1,2-dichloroethane (Pries *et al.*, 1994b). The gene was detected in organisms belonging to 3 different genera: *Xanthobacter, Ancylobacter,* and *Pseudomonas*.

A second group of mutually identical dehalogenases was detected in 1-chlorobutane degrading organisms. The corresponding gene was first sequenced in *Rhodococcus* strains (Kulakova *et al.*, 1997), and we detected the same sequence in a 1,3-dichloropropene degrading *Pseudomonas* and in 6 independently isolated gram-positive organisms from Japan, the US, Switzerland, the UK, and the Netherlands. These observations suggest that horizontal transfer plays an important role in the distribution of haloalkane dehalogenase genes (Poelarends *et al.*, in press).

A role for horizontal transmission is also suggested by the fact that the haloalkane dehalogenase genes in *Rhodococcus* and *Xanthobacter* are harbored on transmissible plasmids. It is well possible that the massive use of 1,3-dichloropropene in agriculture has contributed to the spread of a haloalkane dehalogenase gene that allows degradation of this compound. Accelerated degradation and ineffectivity of the nematocide in some soils has indeed been observed (Verhagen *et al.*, 1995).

5. KINETICS AND ENGINEERING OF HALOALKANE DEHALOGENASE

Using protein engineering studies and kinetic experiments, much insight has been obtained in the kinetics of haloalkane dehalogenase and the potential of improving the properties of the enzyme (Schanstra *et al.*, 1996b; 1997). From a practical point of view, different engineering goals can be envisaged, such as improving the stability of the enzyme, enhancing the catalytic rate with substrates such as 1,2-dibromoethane, 1,3-dichloropropylene, or 1,2-dichloroethane, and expanding the substrate range to include recalcitrant compounds like 1,2-dichloropropane and 1,1,2-trichloroethane. The latter two compounds are substrate analogs. 1,2-Dichloropropane has only a single methyl group in addition to 1,2-dichloroethane, but surprisingly it is a very poor substrate which is converted at less than 1% of the rate found with 1,2-dichloroethane (5 mM concentrations). The K_m for 1,2-dichloropropane is very high, which is likely to be due to poor binding and/or slow cleavage of the carbon-chlorine bond. 1,1,2-Trichloroethane is a low affinity competitive inhibitor of haloalkane dehalogenase (K_i= 27 mM). Both chemicals were found to be persistent in enrichment experiments carried out in our laboratory.

Improving enzyme stability and increasing rates with known substrates do not seem of major importance for application of organisms harboring haloalkane dehalogenase genes for waste- or groundwater treatment. Environmental biotechnology processes preferably make use of whole cells rather than isolated enzymes, for the simple reasons that the cell provides a protective environment, renews itself and thereby multiplies the catalyst. Using intact cells also reduces the number of process steps, since cell cultivation does not have to take place separately and enzyme isolation is unnecessary.

For the best substrates 1,2-dichloroethane and 1,2-dibromoethane, cleavage of the carbon-halogen bond appeared not to be the slowest step in the catalytic cycle (Schanstra and Janssen, 1996; Schanstra *et al.*, 1996a). Instead, we have found that release of the halide ion is slowest and occurs with a rate which is only slightly higher than the k_{cat} (Fig. 3). Stopped-flow fluorescence measurements have shown that there are two parallel kinetic

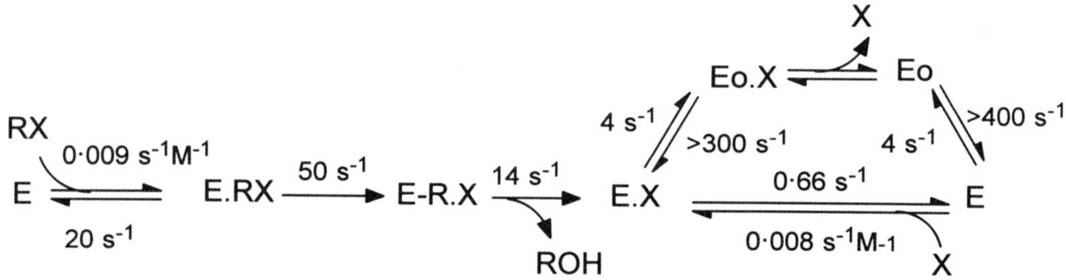

Figure 3. Kinetic mechanism of haloalkane dehalogenase. With 1,2-dichloroethane, halide release is the slowest step. It mainly occurs via the upper route, which starts with a slow conformational change, leading to an 'open' enzyme form from which the halide ion can rapidly dissociate.

Table 1. Mutants of haloalkane dehalogenase with altered specificity and activity

Isolation	Phenotype	Position in structure	Kinetics
in vivo selection, spontaneous mutants	Better with 1-chlorohexane	Various in cap domain	Faster with chlorohexane, slower with 1,2-DCE
SDM: Trp125Phe, Trp125Tyr, Trp175Tyr	Slower with DCE, good with DBE	Halogen & halide binding	Slower C-X cleavage
SDM: Phe172Trp	Poor with DCE, faster with DBE	T-interaction with Trp175 wall of cavity	Slower C-X cleavage, faster halide release, faster conf. change
SDM: Val226Ala	Same with DCE, faster with DBE	Stabilizes Trp175	Slower C-X cleavage, faster halide release, faster conf. change
SDM: Pro168Ser, Pro168Thr; spontaneous: Pro168Ser	Faster with DBE, same with DCE, better with 1-chlorohexane	Stabilizes cap domain	Slower C-X cleavage, faster halide release, faster conf. change

Abbreviations: SDM, site-directed mutagenesis; DCE, 1,2-dichloroethane; DBE, 1,2-dibromoethane.

mechanisms for halide release. One (the upper route in Fig. 3) starts with a conformational change, and leads to an enzyme form from which the halide ion can rapidly dissociate. After this, the enzyme reverts to its original conformation. This route is of major importance during the normal catalytic cycle. The other (lower) route, which is of minor importance, also starts with a conformational change in case of bromide release, but this leads to a collision complex with enzyme in the normal state. For 1,2-dichloroethane, it is even simpler, since halide release is a slow unimolecular step (Fig. 3).

Several mutants with increased activity towards 1,2-dichloroethane or 1,2-dibromoethane have been constructed (Schanstra *et al.*, 1996b; 1997; Krooshof *et al.*, 1997). In agreement with the rate of halide release being rate limiting, these mutants show an increased rate of the conformational change (Table 1). The properties of the mutants indicate that there is an inverse correlation between the rate of the conformational change required for halide release (according to the upper route, Fig. 3) and the rate of carbon-halogen bond cleavage.

Haloalkane dehalogenase is better evolutionary optimized for 1,2-dichloroethane than for 1,2-dibromoethane, although the latter is a better substrate. For 1,2-dibromoethane, the rate of carbon-bromine cleavage is much higher than the rate of halide release, making it possible to improve the enzyme by making mutants with faster halide release. For 1,2-dichloroethane, the rates of carbon-chlorine bond cleavage and chloride export are well balanced and the catalytic performance cannot be increased by accelerating halide release since then the rate of carbon-chlorine bond cleavage becomes too low (Schanstra *et al.*, 1997). In a general sense, one can conclude that it is feasible to improve the dehalogenase activity for substrates for which there is not a good balance between halide-release and carbon-halogen bond cleavage, or for compounds which are poorly bound in the active site. Spontaneous mutants with increased activity towards 1-chlorohexane, which does not fit into the cavity as it is observed in the X-ray structure of the wild-type enzyme, have indeed been obtained (Pries *et al.*, 1994b).

Further engineering studies on haloalkane dehalogenase are ongoing in our laboratory. Although the enzyme has a modest activity for 1,2-dichloroethane (the k_{cat} is about 4 sec^{-1}), improving the enzyme for this compound is not a primary goal since the activity is apparently high enough to realize a full scale application (Stucki *et al.*, 1995). It would be much more desirable to obtain improved enzyme variants for substrates which are really resistant to hydrolytic dechlorination.

6. HALOALKANOIC ACID DEHALOGENASES

Several 2-haloalkanoic acid dehalogenases have been described in the literature. These enzymes used to be classified on basis of their substrate range and electrophoretic mobility, but recent molecular studies and sequence analysis have shown that they can be divided in different mechanistic and structural classes. Two of these are well defined.

First, there is a large group of enzymes that catalyze carbon-halogen bond cleavage by covalent catalysis involving the carboxylate group of an aspartate residue that is close to the N-terminus of the protein (Fig. 4). The structure of the *Pseudomonas* YL and *X.*

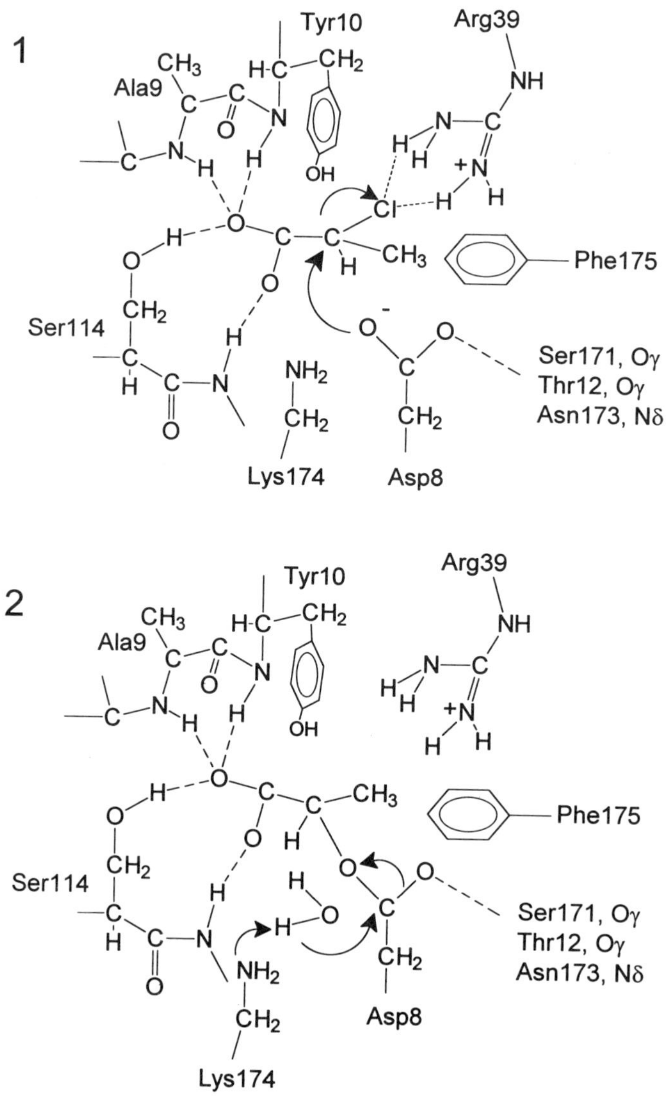

Figure 4. Catalytic mechanism of 2-haloalkanoic acid dehalogenases. A covalent intermediate is formed by attack of an aspartate carboxylate oxygen on C-2 of the substrate, leading to displacement of the halogen. The intermediate is hydrolyzed by water.

autotrophicus GJ10 enzymes which both belong to this class have been solved (Hisano *et al.*, 1996; Ridder *et al.*, 1997; Li *et al.*, 1998). The catalytic mechanism is slightly similar to that of haloalkane dehalogenase, but the groups involved are very different. Leaving group stabilization is mediated by an arginine sidechain rather than by hydrogens bound to indole nitrogens of tryptophans. Furthermore, in halocarboxylic acid dehalogenases the water molecule that cleaves the covalent intermediate is activated by a lysine sidechain amine group, not by a histidine as in haloalkane dehalogenase. This partly explains the higher pH optimum of these dehalogenases.

A haloalkanoic acid dehalogenase of a completely different class is the fluoroacetate dehalogenating enzyme of *Moraxella* (Kawasaki *et al.*, 1992). Based on sequence analysis, this protein is more similar to the α/β-hydrolase fold enzymes to which also haloalkane dehalogenases belong. Other haloacid dehalogenases, including the enzymes that are stereoselective with D-2-chloropropionic acid, are similar to neither of these groups of enzymes and thus may have still other mechanisms (Nardi-Dei *et al.*, 1997).

Whereas we have found that identical haloalkane dehalogenases occur in at least 10 independently isolated organisms from different locations and from different genera, it appeared that there is a very large diversity of 2-haloalkanoic acid dehalogenases. At least 9 different but related genes encoding members of the *Pseudomonas* YL class of L-2-chloropropionic acid dehalogenases have been obtained from organisms growing on 1,2-dichlorethane, chlorobenzoate, chloroacetate and 2-chloropropionate. Thus, it appears that the diversity of the haloacid dehalogenases is much larger, suggesting that they are much older on an evolutionary timescale. Several 2-haloalkanoic acids occur naturally, and they also may be formed as intermediates during the degradation of other halogenated compounds. In contrast, the haloalkane dehalogenases may have become distributed in response to environmental contamination by synthetic compounds, with only a few ancestral genes serving as a starting point.

7. BIOCATALYTIC POTENTIAL OF ENZYMES INVOLVED IN THE DEGRADATION OF HALOALIPHATICS

Thousands of synthetic chemicals are produced industrially. A recent extensive survey of the biodegradation of such chemicals (van Agteren *et al.*, 1998) reveals that at least 50% of the compounds which have been investigated may serve as a growth substrate for some specific microbial culture. Only in a very few cases, biodegradation pathways use identical enzymes for different compounds. In fact, it is much more common that a number of different routes can be used for the same chemical. This indicates that there is a huge natural diversity of enzymes that can recognize and transform synthetic chemicals. Such enzymes have great potential in the development of biocatalytic routes for synthesis of fine chemicals, especially if they are enantioselective or regioselective.

At least three such enzymes can be found in the catabolic routes shown in Fig. 1. First, the 2-haloalkanoic acid dehalogenase of strain GJ10 is selective for the L-enantiomer of 2-chloropropionic acid, similar to what has been found for other enzymes of this class. These enzymes may be used for the preparation of enantiopure D-lactate and D-2-chloropropionic acid. Second, haloalcohol dehalogenases involved in chloropropanol degradation may be used for the kinetic resolution of 1,2-dichloro-3-propanol to produce epichlorohydrin (Nakamura *et al.*, 1992; Assis *et al.*, 1998). The enzymes from the epichlorohydrin-degrading organism *Agrobacterium radiobacter* AD1 and *Lorynebacterium* N1074 (Yu *et al.*, 1994) appear to be enantioselective for the (S) enantiomer of 1,2-

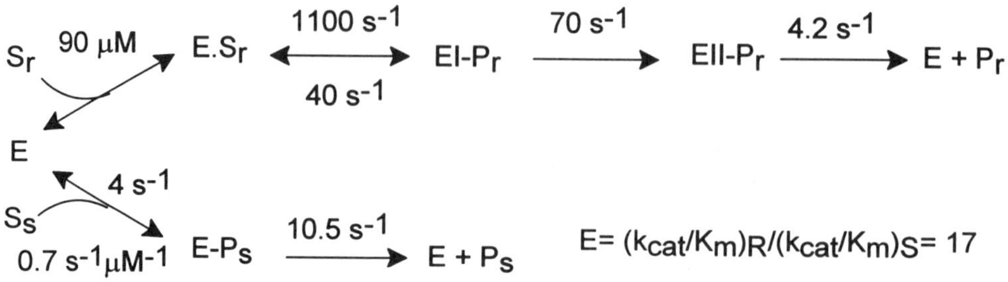

Figure 5. Kinetic mechanism of epoxide hydrolase with (R)- and (S)-styrene oxide. The rate and equilibrium constants explain the observed enantioselectivity. An unimolecular step that 'traps' the (R)-styrene oxide intermediate is an important factor for the preferential conversion of the (R)-enantiomer.

dichloro-3-propanol. Third, we have recently found that the epoxide hydrolase of the same organism can be used for the kinetic resolution of various styrene epoxides with different substituents on the aromatic ring (Lutje Spelberg *et al.*, 1998).

This epoxide hydrolase has a catalytic mechanism that is very similar to that of haloalkane dehalogenase (Rink *et al.*, 1997). On basis of sequence similarity and site-directed mutagenesis experiments, it can be concluded that catalysis also proceeds by nucleophilic attack of an aspartic acid carboxylate on the oxirane ring. The hydroxylalkyl-enzyme intermediate is subsequently hydrolyzed by water, which is activated by a conserved histidine residue (Fig. 2; Lacourciere & Armstrong, 1994; Arand *et al.*, 1994; Rink *et al.*, 1997). The observed enantioselectivity with styrene epoxide could be understood by solving the kinetic mechanism for different enantiomers using stopped-flow fluorescence and rapid-quench methods (Fig. 5). The most remarkable features of the kinetic mechanism are a very rapid formation of the covalent intermediate, the reversibility of this first step, the role of a unimolecular conformational change in the kinetic mechanism of the (R)-enantiomer, and the rate-limiting hydrolysis of the covalent intermediate. This detailed kinetic mechanism serves as a basis for the rational design and analysis of mutants with altered enantioselectivity or modified substrate range.

Since useful biocatalysis can often be detected in organisms that degrade xenobiotics and in view of the large numbers of catabolic pathways and enzymes that have not yet been analyzed, it can be concluded that there still is an enormous reservoir of potentially useful industrial biocatalysts that remain to be exploited.

REFERENCES

Arand, M., Grant, D.F., Beetham, J.K., Friedberg, T., Oesch, F., and Hammock, B.D. (1994). Sequence similarity of mammalian epoxide hydrolases to the bacterial haloalkane dehalogenase and other related proteins. Implication for the potential catalytic mechanism of enzymatic epoxide hydrolysis. *FEBS Lett.* **338**:251–256.

Assis, H.M.S., Bull, A.T., and Hardman, D.J. (1998). Synthesis of chiral epihalohydrins using haloalcohol dehalogenase A from *Arthrobacter erithii* H10a. *Enz. Microb. Technol.* **22**:545–551.

Fetzner, S., and Lingens, F. (1994). Bacterial dehalogenases: biochemistry, genetics, and biotechnological applications. *Microbiol. Rev.* **58**:641–685.

Freitas dos Santos, L.M., Leak, D.J., and Livingston, A.G. (1996). Enrichment of mixed cultures capable of aerobic degradation of 1,2-dibromoethane. *Appl. Environ. Microbiol.* **62**:4675–4677.

Hartmans, S., and de Bont, J.A.M. (1991). Aerobic vinyl chloride metabolism in *Mycobacterium aurum*. *Appl. Environ. Microbiol.* **58**:1220–1226.

Hisano, T., Hata, Y., Fujii, T., Liu, J.Q., Kurihara, T., Esaki, N., and Soda, K.J. (1996). Crystal structure of L-2-haloacid dehalogenase from *Pseudomonas* sp. YL. An α/β hydrolase structure that is different from the α/β hydrolase fold. *J. Biol. Chem.* **271**:20322–20330.

Janssen, D.B., Scheper, A., Dijkhuizen, L., and Witholt, B. (1985). Degradation of halogenated aliphatic compounds by *Xanthobacter autotrophicus* GJ10. *Appl. Environ. Microbiol.* **49**:673–677.

Janssen, D.B., Pries, F. and Van der Ploeg, J.R. (1994). Genetics and biochemistry of dehalogenating enzymes. *Ann. Rev. Microbiol.* **48**:163–191.

Kawasaki, H., Tsuda, K., Matsushita, I., and Tonomura, K.J. (1992). Lack of homology between two haloacetate dehalogenase genes encoded on a plasmid from *Moraxella* sp. strain B. *J. Gen. Microbiol.* **138**:1317–1323.

Krooshof, G.H., Kwant, E.M., Damborsky, J., Koca, J., and Janssen, D.B. (1997). Repositioning the catalytic triad aspartic acid of haloalkane dehalogenase: effects on stability, kinetics, and structure. *Biochemistry* **36**:9571–9580.

Kulakova, A.N., Larkin, M.J., and Kulakov, L.A. (1997). The plasmid-located haloalkane dehalogenase gene from *Rhodococcus rhodochrous* NCIMB 13064. *Microbiology* **143**:109–115.

Lacourciere, G.M., and Armstrong, R.N. (1994). Microsomal and soluble epoxide hydrolases are members of the same family of C-X bond hydrolase enzymes. *Chem. Res. Toxicol.* **7**:121–124.

Leisinger, T. (1983). Microorganisms and xenobiotic compounds. *Experientia* **39**:1183–1191.

Leisinger, T. (1996). Biodegradation of chlorinated aliphatic compounds. *Curr. Opin. Biotechnol.* **7**:295–300.

Li, Y.F., Hata, Y., Fujii, T., Hisano, T., Nishihara, M., Kurihara, T., and Esaki, N. (1998). Crystal structures of reaction intermediates of L-2-haloacid dehalogenase and implications for the reaction mechanism. *J. Biol. Chem.* **273**:15035–15044.

Lutje Spelberg, J.H., Rink, R., Kellogg, R.M., and Janssen, D.B. (1998). Enantioselectivity of a recombinant epoxide hydrolase from *Agrobacterium radiobacter*. *Tetrah. Assym.* **9**: 459–466.

Nakamura, T., Nagasawa, T., Yu, F., Watanabe, I., and Yamada, H. (1992). Resolution and some properties of enzymes involved in enantioselective transformation of 1,3-dichloro-2-propanol to (*R*)-3-chloro-1,2-propanediol by *Corynebacterium* strain N-1074. *J. Bacteriol.* **174**:7613–7619.

Nardi-Dei, V., Kurihara, T., Park, C., Esaki, N., and Soda, K. (1997). Bacterial DL-2-haloacid dehalogenase from *Pseudomonas* sp. strain 113: gene cloning and structural comparison with D- and L-2-haloacid dehalogenases. *J. Bacteriol.* **179**:4232–4238.

Ollis, D.L., Cheah, E., Cygler, M., Dijkstra, B.W., Frolow, F., Franken, S.M., Haral, M., Remington, S.J., Silman, I., Schrag, J., Sussman, J.L., Verschueren, K.H.G., and Goldman, A. (1992). The α/β-hydrolase fold. *Protein Eng.* **5**:197–211.

Poelarends, G.J., Wilkens, M., Larkin, M.J., van Elsas, J.D., and Janssen, D.B. (1998). Degradation of 1,3-dichloropropene by *Pseudomonas cichorii* 170. *Appl. Environ. Microbiol.*, in press.

Pries, F., Kingma, J., Pentenga, M., Van Pouderoyen, G., Jeronimus-Stratingh, C.M., Bruins, A.P., and Janssen, D.B. (1994a). Site-directed mutagenesis and oxygen isotope incorporation studies of the nucleophilic aspartate of haloalkane dehalogenase. *Biochemistry* **33**:1242–1247.

Pries, F., van den Wijngaard, A.J., Bos, R., Pentenga, M., and Janssen, D.B. (1994b). The role of spontaneous cap domain mutations in haloalkane dehalogenase specificity and evolution. *J. Biol. Chem.* **269**:17490–17494.

Ridder, I.S., Rozeboom, H.J., Kalk, K.H., Janssen, D.B., and Dijkstra, B.W. (1997). Three-dimensional structure of L-2-haloacid dehalogenase from *Xanthobacter autotrophicus* GJ10 complexed with the substrate-analogue formate. *J. Biol. Chem.* **272**:33015–33022.

Rink, R., Fennema, M., Smids, M., Dehmel, U., and Janssen, D.B. (1997). Primary structure and catalytic mechanism of the epoxide hydrolase from *Agrobacterium radiobacter* AD1. *J. Biol. Chem.* **272**:14650–14657.

Schanstra, J.P., and Janssen D.B. (1996). Kinetics of halide release of haloalkane dehalogenase: evidence for a slow conformational change. *Biochemistry* **35**:5624–5632.

Schanstra, J.P., Kingma, J., and Janssen, D.B. (1996a). Specificity and kinetics of haloakane dehalogenase. *J. Biol. Chem.* **271**:14747–14753.

Schanstra, J.P., Ridder, I.S., Heimeriks, G.J., Rink, R., Poelarends, G.J., Kalk, K.H., Dijkstra, B.W., and Janssen, D.B. (1996b). Kinetic characterization and X-ray structure of a mutant of haloalkane dehalogenase with higher catalytic activity and modified substrate range. *Biochemistry* **35**:13186–13195.

Schanstra, J.P., Ridder, A., Kingma, J., and Janssen, D.B. (1997). Influence of mutations of Val226 on the catalytic rate of haloalkane dehalogenase. *Protein Engin.* **10**:53–61.

Scholtz, R., Messi, F., Leisinger, T., and Cook, A.M. (1988). Three dehalogenases and physiological restraints in the biodegradation of haloalkanes by *Arthrobacter* sp. strain HA1. *Appl. Environ. Micobiol.* **54**:3034–3048.

Stucki, G., and Thüer, M. (1995). Experiences of a large-scale application of 1,2-dichloroethane degrading microorganisms for groundwater treatment. *Environ. Sci. Technol.* **29**:2339–2345.

Van Agteren, M.H., Keuning, S., and Janssen, D.B. (1998). Handbook on biodegradation and biological treatment of hazardous organic compounds. Kluwer Academic Publishers, Dordrecht, The Netherlands.

Van der Ploeg, J.R., Kingma, J. De Vries, E.J., Van der Ven, J.G.M., and Janssen, D.B. (1996). Adaptation of *Pseudomonas* sp. GJ1 to 2-bromoethanol caused by overexpression of an NAD-dependent aldehyde dehydrogenase with low affinity for halogenated aldehydes. *Arch. Microbiol.* **165**:258–276.

Van der Waarde, J.J., Kok, R., and Janssen, D.B. (1992). Degradation of 2-chloroallylalcohol by a *Pseudomonas* sp. *Appl. Environ. Microbiol.* **59**:528–535.

Van Hylckama Vlieg, J.E.T., and Janssen, D.B. (1992). Bacterial degradation of 3-chloroacrylic acid and characterization of *cis*- and *trans*-specific dehalogenases. *Biodegradation* **2**:139–150.

Verhagen, C., Smit, E., Janssen, D.B., and Van Elsas, J.D. (1995). Bacterial dichloropropene degradation in soil; screening of soils and involvement of plasmids carrying the *dhlA* gene. *Soil Biol. Biochem.* **27**:1547–1557.

Verschueren, K.H.G., Seljée, F., Rozeboom, H.J., Kalk, K.H., and Dijkstra, B.W. (1993). Crystallographic analysis of the catalytic mechanism of haloalkane dehalogenase. *Nature* **363**:693–698.

Yu, F., Nakamura, T., Mizunashi, W., and Watanabe, I. (1994). Cloning of two halohydrin hydrogen-halide-lyase genes of *Corynebacterium* sp. strain N-1074 and structural comparison of the genes and gene products. *Biosci. Biotechnol. Biochem.* **58**:1451–1457.

12

FERREDOXIN-MEDIATED REACTIVATION OF CATECHOL DIOXYGENASE IMPROVES AROMATIC RING CLEAVAGE IN PSEUDOMONADS

Y. Jouanneau,[1] J. Armengaud,[2] N. Hugo,[1] C. Meyer,[1] and K. N. Timmis[2]

[1]CEA-Grenoble et CNRS UMR 314, DBMS/BBSI
F-38054 Grenoble, France
[2]GBF, Division of Microbiology
D-38124, Braunschweig, Germany

1. INTRODUCTION

Cleavage of the aromatic ring is a critical step in the biodegradation of aromatic hydrocarbons and is generally catalyzed by dioxygenases. These enzymes incorporate each atom of molecular oxygen in diol substrates carrying two hydroxyl groups on two adjacent carbons of the aromatic ring, and eventually cause the opening of the ring. They are classified in two families depending on the mode of scission of the aromatic substrates. Intradiol dioxygenases cleave the aromatic ring between the two hydroxyls groups, whereas extradiol enzymes cleave the ring in a position adjacent to the diol. The two groups of enzymes share no amino acid sequence similarities, and are also structurally distinct in that the former contain a non-heme ferric ion at the catalytic site, and the latter contain a ferrous ion (Harayama and Rekik, 1989). The catechol 2,3-dioxygenase (XylE) of the TOL pathway in *Pseudomonas putida* mt2 is a typical extradiol dioxygenase which utilizes catechol and methyl-substituted catechols as substrates. Like many extradiol dioxygenases, it is a rather labile enzyme which tends to inactivate during catalytic turnover, or upon exposure to air oxygen. Inactivation is caused by oxidation of the ferrous iron atom at the enzyme active site. In the presence of certain substrates such as 4-methylcatechol and chlorocatechols, the enzyme rapidly loses activity (Bartels et al., 1984; Cerdan et al., 1995).

A natural mechanism that prevents irreversible inactivation of catechol 2,3-dioxygenase has been discovered in *P. putida* (Polissi and Harayama, 1993). This mechanism relies on a functional *xylT* gene and allows bacterial growth on *p*-xylene and *p*-toluate as

Novel Approaches for Bioremediation of Organic Pollution, edited by
Fass *et al.* Kluwer Academic / Plenum Publishers, New York, 1999.

carbon sources. It was found that 4-methylcatechol, a metabolite formed upon degradation of *p*-xylene, inactivated XylE in *xylT* mutants, while the enzyme remained active in wild-type *P. putida*. Recently, we have investigated the XylT-dependent mechanism of catechol 2,3-dioxygenase protection *in vitro* (Hugo et al., 1998). In the first part of this report, we summarize the properties of the isolated XylT protein, a novel [2Fe-2S] ferredoxin which specifically reduces the ferric ion of inactive XylE and thereby reactivates the enzyme. Secondly, we present an inventory of XylT analogous sequences available in the data bases and make comparisons based on their predicted properties as well as on the biochemical properties of selected XylT analogues that we have purified. Finally, the physiological function of some of the XylT analogues has been tested *in vivo* by complementation analysis of a *xylT*-deleted mutant of *P. putida*.

2. A SPECIFIC [2FE-2S] FERREDOXIN (XylT) MEDIATES REACTIVATION OF CATECHOL 2,3-DIOXYGENASE (XylE)

2.1. Purification and Molecular Properties of XylT

The predicted amino acid sequence of the *xylT* gene product contains six cysteines, four of which are arranged in a motif typically found in certain [2Fe-2S] ferredoxins, including plant ferredoxins (Harayama et al., 1991). These Cys residues generally serve as ligands for the Fe-S cluster. In order to facilitate the isolation of the XylT product, the *xylT* gene was cloned in an expression vector and overexpressed in *E. coli*. The recombinant XylT protein was isolated and purified as a pink fraction exhibiting UV-visible absorption maxima at 336, 416 and 456 nm. It is composed of a single monomeric polypeptide with a molecular size of 12,034 Da, which binds one [2Fe-2S] cluster (Table 1). Evidence for the presence of one such cluster in purified XylT is based on UV-visible absorption and EPR spectroscopy, as well as on chemical determination of the protein iron content (Hugo et al., 1998). Redox titration of the cluster gave a mid-point potential of -373 mV, similar to that reported for plant-type ferredoxins.

In addition, XylT shows peculiar properties that makes it stand apart from most other [2Fe-2S] ferredoxins. As predicted from the amino acid sequence, it is a basic protein with an isoelectric point of 8.27. This feature, which is unique among all known ferredoxins, may be of critical importance for the XylT-XylE molecular interaction (see below). Moreover, XylT appeared rather unstable and became denatured upon exposure to oxygen or upon incubation at moderate temperature. For example, the half-lives of the protein were 69 min in air at 25°C, and 70 min at 37°C under anaerobic conditions. The XylT instability might reflect a relatively high degree of exposure of the [2F-2S] cluster to the solvent. In accordance, we observed that the EPR signal of the reduced cluster was

Table 1. Molecular properties of XylT

Parameter	Value
Molecular mass (Da)	12,034
Isoelectric point (theoretical)	8.27
Prosthetic group	[2Fe-2S]
Mid point redox potential (mV)	-373 ± 6
Light absorption	
λ_{max} (nm)	278, 336, 416, 456
ε_{416} ($mM^{-1}.cm^{-1}$)	9.52

markedly affected by the addition of glycerol in the XylT protein solution (Hugo et al., 1998).

2.2. XylT-Dependent Reactivation of XylE *in Vitro*

Catechol 2,3-dioxygenase becomes inactivated when incubated with substituted catechols, including 3- and 4-methylcatechol (Cerdan et al., 1994). Inactivation is primarily due to oxidation of the ferrous ion present at the catalytic site of the enzyme. The fact that XylT shows properties common to all ferredoxins, suggested that it might be involved in a reduction reaction of the XylE iron atom. To test this hypothesis, purified inactive XylE enzyme was incubated in the presence of XylT in reaction vials also containing 5-deazaflavin as a light-driven reductant generating system. Inactivated XylE was obtained *in vitro* by treating the enzyme with 4-methyl catechol. Upon subsequent dialysis and incubation with the 5-deazaflavin alone, slow and little reactivation of XylE was observed (Fig. 1). In contrast, addition of various amounts of XylT to the inactivated enzyme promoted a rapid recovery of the XylE activity. The kinetics and extent of enzyme reactivation were found to be dependent upon the concentration of XylT added in the reactivation assays (Fig. 1).

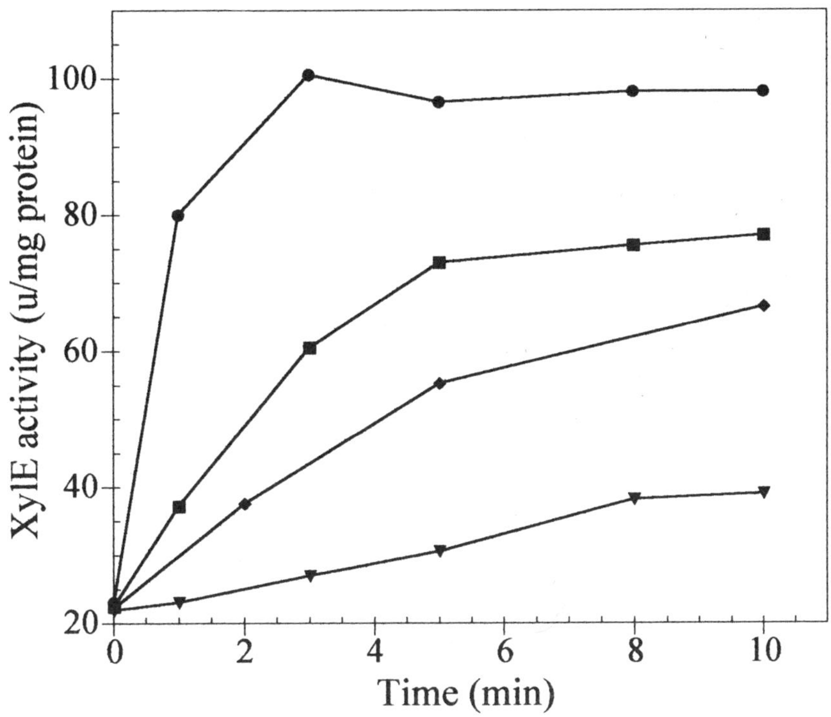

Figure 1. Kinetics of XylT-dependent XylE reactivation. Samples of inactivated XylE were incubated in anaerobic vials in a total of 200 μl of reactivation mixture. Vials contained 50 nM XylE and the following concentration of XylT: 5 nM (diamonds), 50 nM (squares), or 250 nM (circles). A control vial contained no XyT (triangles). At time zero, vials were exposed to light. XylE activity was assayed on 5 to 50 μl samples anaerobically withdrawn from the vials at the times indicated. Reproduced from Hugo et al., 1998 with permission.

Control experiments run with heat-denatured XylT or in the absence of reductant generating system gave essentially no reactivation. Other ferredoxins of different sources were tested in replacement of XylT, none of which was able to induce a significant reactivation of XylE. For example, spinach ferredoxin was ineffective even when added in 10-fold excess over the XylE molar concentration. Since spinach ferredoxin has a mid-point redox potential similar to that of XylT, its observed lack of reactivity might reflect poor interaction with XylE due to charge incompatibility. Indeed, spinach ferredoxin, like the vast majority of ferredoxins, is an acidic protein with an overall negative charge, whereas XylT has a net positive charge.

The molecular interaction between XylT and XylE is currently being investigated in our laboratory using cross-linking experiments. Preliminary results indicate that a covalent bond is specifically formed between a basic residue of XylT and an acidic residue of XylE. Peptidic analyses will be performed to map the regions which come into contact when the two partner proteins associate. Taken together, our results indicate that XylT mediates the reactivation of XylE in a highly specific reaction which only requires a source of reductant (Hugo et al., 1998).

2.3. XylE Reactivation Occurs through a XylT-Mediated Reduction of the Enzyme Active Site

The results described above provided evidence but did not prove that XylT reactivates XylE through reduction of the enzyme active site. Other possible explanations were considered, including transfer of ferrous ions from the XylT cluster to the XylE active site, or removal of a bound metabolite from the enzyme active site. The redox state of the XylE iron atom was then monitored by EPR spectroscopy in the inactivated and reactivated states. While the active catechol 2,3 dioxygenase is EPR silent, the 4-methylcatechol-inactivated enzyme displayed a signal with a S=5/2 ground state typical for a ferric iron. This signal disappeared upon subsequent reactivation of XylE with a catalytic amount of XylT, suggesting that reduction of the enzyme active site accompanied reactivation. This interpretation was further confirmed by monitoring the Fe(II)-NO EPR signal of XylE-nitrosyl complexes. The Fe(II)-NO signal was found to increase upon XylE reactivation (Hugo et al., 1998). Hence, all our biochemical and spectroscopic data are consistent with a XylT-dependent reductive reactivation of catechol 2,3 dioxygenase, according to the mechanism depicted in Fig. 2. It is a rare example of an enzyme reactivation that is directly mediated by a ferredoxin.

3. PROPERTIES OF XylT ANALOGUES FOUND IN VARIOUS BIODEGRADATION PATHWAYS

3.1. General Properties of the XylT Analogues

In recent years, several genes homologous to *xylT* have been identified in various bacterial species. A search in the data bases led us to establish a list of genes the products of which resemble XylT at the amino acid sequence level. The source and predicted properties of these XylT homologues are reported in Table 2. All the selected genes are associated with catabolic genes involved in the degradation of aromatic hydrocarbons, and most of them have been found in *Pseudomonas* species. Moreover, the *xylT*-like genes are generally located in the immediate vicinity of a gene encoding a catechol dioxygenase, sug-

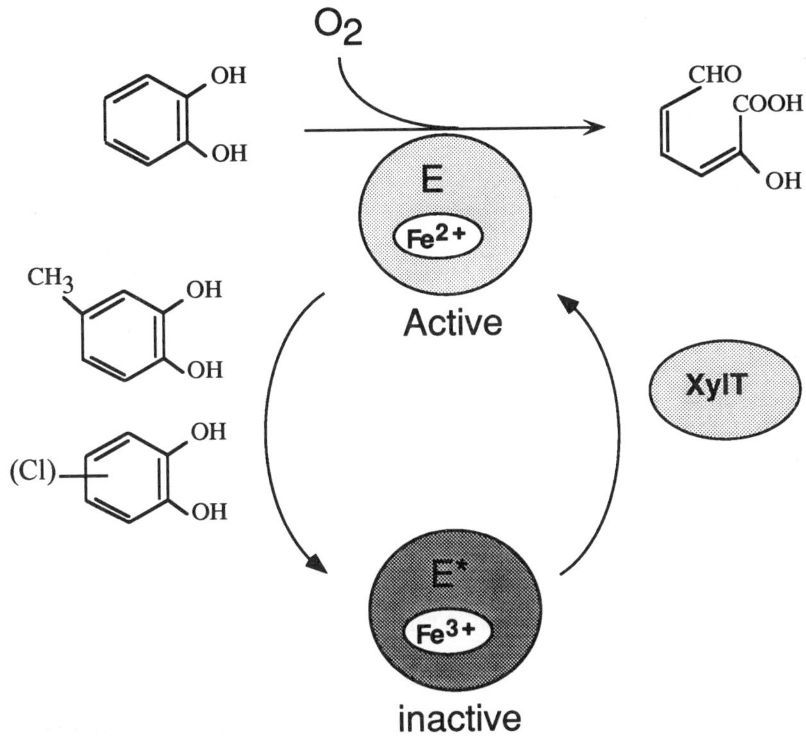

Figure 2. Proposed mechanism of catechol 2,3-dioxygenase reactivation by XylT. Catechol 2,3-dioxygenase (XylE) becomes inactivated upon incubation with certain substrates such as 4-methylcatechol or chlorocatechols. Inactivation results primarily from an oxidation of the active site iron. XylT mediates a specific reduction of Fe(III) at the XylE catalytic site, thereby restoring an active enzyme.

gesting that they may have a function analogous to that of XylT in their relevant host strain. However, a *xylT*-like gene from *Sphingomonas yanoikuyae* has been reported to map close to *xylH* in a position remote from *xylE* (Zylstra and Kim, 1997). Genes homologous to *xylT* were found in the catabolic pathways of BTEX (*xylT, tbuW*), naphthalene and salicylate (*nahT*), phenol and substituted phenols (*phhQ, phlG, dmpQ*), aniline (*atdS*) and nitrobenzene (*cdoT*). The XylT analogues have predicted sizes between 11 and 13 kDa,

Table 2. Source and predicted properties of XylT analogues

Gene	Strain	Catabolic pathway	Polypeptide size Da	Isoelectric point	Sequence identities %	Accession number
xylT	*P. putida* mt2	Tol/Xylene	12,034	8.27	100	M64747
nahT	*P. putida* G7	Naphthalene	11,899	7.47	47	J05317
nahT	*P. putida* KF715	Salicylate	12,504	7.75	53	S78585
phhQ	*P. putida* P35X	(met)-phenol	10,953	7.86	57	X79063
phlG	*P. putida* H	Phenol	10,966	8.11	58	X80765
dmpQ	*Pseudomonas* CF600	$(met)_2$-phenol	12,336	8.40	65	X60835
atdS	*Acinetobacter* YAA	Aniline	12,032	7.45	47	AB008831
tbuW	*P. picketti*	Tol/Xylene	13,023	12.2	24	U20258
cdoT	*Comamonas* JS765	Nitrobenzene	12,983	8.77	29	U93090

and share around 50 % sequence similarity or over with XylT, except for TbuW (24 %) and CdoT (29 %). Remarkably, all the analogues display a basic theoretical isoelectric point, suggesting that this feature is important for function..

3.2. Comparison of Amino Acid Sequences

The alignment of the 9 selected XylT-like sequences reveals that 19 residues are strictly conserved (Fig. 3). These invariable residues include four cysteines, which have the typical spacing observed in plant-type [2Fe-2S] ferredoxins, and which certainly serve as ligands for the cluster. Most of the other conserved amino acids are clustered around the putative cysteine ligands and are either hydrophobic or glycine residues. They likely have a structural function in facilitating proper folding of the protein or in stabilizing the folded polypeptide around the cluster. Similar patterns of conservation were observed in plant-type ferredoxins (Tsukihara et al., 1990). Other relatively conserved residues include two additional cysteines in positions 19 and 36 (numbering refers to the XylT sequence). The TbuW sequence is remarkable in that it contains no cysteine in either of these positions. In addition, four proteins (PhhQ, PhlG, DmpQ and NahT from *P. putida* KF715) have an extra pair of cysteines in positions 59 and 64. The significance and possible roles of these cysteine residues remain to be assessed.

Since all the XylT-like proteins show a basic character, and since this property seems to be important for the interaction with catechol dioxygenase (at least in the case of XylT), we inspected the sequences looking for conserved lysine or arginine residues. Although none of these basic amino acids appear to be absolutely invariant, a lysine in position 50 and an arginine in position 52 are found in 8 out of the 9 sequences. In XylT, a conservative replacement of the lysine by an arginine is observed in position 50. Besides, an arginine in position 42 and a pair of basic residues at positions 34 and 35 also appear to be partially conserved in the sequences considered.

Phylogenetic comparisons of the nine XylT-like sequences indicate that TbuW and CdoT are relatively distant from the group of the other ferredoxins (Fig. 4). This was already suggested from sequence alignment (Fig. 3) and percentages of similarity (Table 2). TbuW and CdoT have around 10 extra residues compared to XylT, which appear as short sequence insertions in variable regions of the protein found at the N- and C-termini. A comparison extended to 22 sequences representative of [2Fe-2S] ferredoxins revealed that XylT-like proteins form a distinct phylogenetic group (data not shown). However, this result could be obtained on condition that the more divergent sequences of the CdoT and TbuW proteins were excluded from the XylT-like subgroup.

3.3. Purification and Biochemical Properties of Selected XylT Analogues

Recently, we have purified three of the XylT analogues, using the same strategy as described for the isolation of XylT. The three proteins, namely PhhQ, DmpQ and NahT (from *P. putida* KF715), were characterized as [2Fe-2S] ferredoxins similar to XylT with respect to their biochemical and spectroscopic properties. They showed however, signicant differences in stability and redox properties. The PhhQ protein for instance, appeared to be about 10-fold more stable than XylT when exposed to air oxygen, and displayed a midpoint redox potential approximately 100 mV higher.

From a functional point of view, it was of interest to determine whether these purified XylT analogues would react with and reactivate catechol dioxygenase *in vitro*. A

```
NahT-KF715    M-----PTVF EITVRPDGES FVCQPQQSVL RAMEAQN-KH CLPVGCRGGG CGLCKVRVLT GDYE-CGRMS CKHVP    68
NahT-G7       M-----SEVF EITVQPGGER FVCQPQQSAL HAMETQG-KR CLPVGCRGGG CGLCKVRVLA GDYE-SGRVS CKHLP    68
PhlG-H        S-----SPPF QVHETNSGQS FTCRPDQSVL RAMEEQG-KR CVPVGCRGGG CGLCKVRVLS GDYQ-CGRMS CSQVP    68
PhhQ-P35X     S-----SPPF QVHETNSGQS FTCRPDQSVL RAMEEQG-KR CVPVGCRGGG CGLCKVRVLS GDYQ-CGRMS CSQVP    68
AtdS-YAA      M-----GESY QITEQCSGQR FPCKAGQSVL KAMEQQG-LE CAPVGCRGGG CGLCKVTVRE GDYE-CGKMS RVHAP    68
DmpQ-CF600    N-----RAGY EIRETVSGQT FRCLPDQSVL SAMEQQGKKR CVPVGCRGGG CGLCKVRVLS GTYQ-CHKMS CNHVP    69
XylT-MT2      N-----SAGY EVFEVLSGQS FRCAEGQSVL RAMEAQG-KR CIPVGCRGGG CGLCRVRVLS GAYR-SGRMS RGHVP    68
TbuW-PK01     MLIPDPRWTY AWNRPRS-TT HARPPKAS-L TGMLRLG-RK GIPVGCVNGG CGVCKVRVLD GSTR-LGRRQ PCPRQ    71
CdoT-JS765    MSLFDSRPKF AVHVAQTDET FPCAGNESLL TGMVRLG-RK GIPVGCVNGG CGVCKVRIVE GQIKALGPIS RAHVT    74

NahT-KF715    VEAREQGYAL ACRLFARS-- DLCI-ERYSK PCYENTV--D PQQREKVTS     112
NahT-G7       VEAREQGYAL ACRLFARS-- DLCI-ERYSK PCSESTV--D QQQRE----     108
PhlG-H        PEAAQQGLAL ACQLYPRA-- DLYI-ESLRQ ------V--R SNP------     100
PhhQ-P35X     PEAGEQGLAL ACQLYPRA-- DLYI-ESLRQ ------V--R SNP------     100
AtdS-YAA      PEALAQGEVL ACRIYPLS-- DLII-ECRPR QSAAGLA--N EYTTTKAMR     112
DmpQ-CF600    PEAAKQGLAL ACQLFPQT-- DLNI-ECLRR QGPGDHN--N KNQQE-VSS     112
XylT-MT2      AKAAAEGLAL ACQVFPQT-- DLTI-EYFRH VGGNKPD--N MNYEE-VTS     111
TbuW-PK01     RRRRSAGLTL ACREAPLTAV RLAVLGKFEK PFSRASA-FA GAQSDTKPQ     119
CdoT-JS765    LDEENQGYTL ACRVAPQTPV NLEVAGKLSK PFSKGRAESA TASPSIQQQ     123
```

Figure 3. The sequences of the nine ferredoxins described in the text were compiled using the GeneWorks software (version 2.5N) from IntelliGenetics and the multiple alignment analysis was performed using the Phylips Package Programs. The sources and references for each sequence are as follows: NahT from *P. putida* sp. KF715 (Lee et al., 1995), NahT from *P. putida* PpG7 (You et al., 1991), PhhQ from *P. putida* P35X (Ng et al., 1994), PhlG from *P. putida* strain H (Herrmann et al., 1995), DmpQ from *Pseudomonas* sp. CF600 (Shingler et al., 1992), AtdS from the plasmid pYA1 of *Acinetobacter* sp YAA (Takeo et al., 1998), XylT from the TOL plasmid of *P. putida* mt-2 (Harayama et al., 1991), TbuW from *P. pickettii* PK01 (Kukor and Olsen, 1996) and CdoT from *Comamonas* sp. JS765 (Parales et al., 1997). Accession numbers are indicated in Table 2.

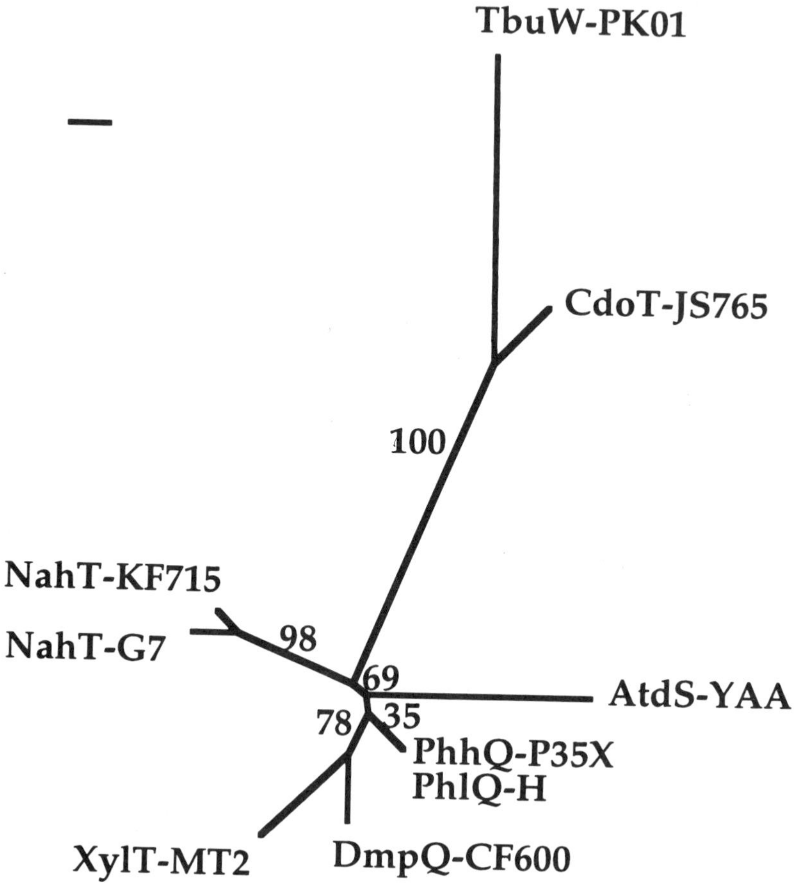

Figure 4. Phylogenetic relationships among XylT-like sequences. The sequence were compared using the same software as mentioned in the legend of Fig. 3. The phylogenetic unrooted tree has been drawn using TreeView. The horizontal bar indicates the percent divergence (distance). The numbers on the branches refer to the confidence (%) estimated by bootstrap analysis (100 replications). The proteins are labelled in the figure by trivial abbreviations. Their accession codes in the SwissProt or GenBank databases and their sources are reported in Table 2.

preparation of 4-methylcatechol-inactivated XylE was used as a test enzyme in reactivation assays as defined above. The three XylT analogues proved to be capable of reactivating XylE on a comparable time scale. Differences were observed however between the ferredoxins when the extent of reactivation was plotted versus the concentration of XylT-like protein. Based on this criterion, the NahT protein appeared to be as efficient as XylT, whereas PhhQ and DmpQ promoted slower reactivation. Hence, our results provide strong evidence that the XylT analogues do catalyse reactivation of catechol dioxygenase through a mechanism similar to that proposed for XylT.

We also studied the interaction of XylE with the purified XylT analogues by cross-linking experiments. A specific cross-linked product was identified in each case, the size of which corresponds to a complex between one XylE subunit and the relevant ferredoxin. An effect of ionic strength on the dissociation of the XylE-ferredoxin complexes was also obsereved, which appeared to vary depending on the XylT analogue. In the case of PhhQ, ionic strength has a more dramatic effect on complex dissociation than for XylT, whereas the contrary is true in the case of DmpQ. These observations certainly reflect differences between the XylT-like proteins in their affinity for the XylE enzyme. The relative affinities of the four ferredoxins of interest for the catechol dioxygenase are currently being investigated by measurements of the variation of the intrinsic fluorescence of XylE upon association of the two partner proteins.

4. COMPLEMENTATION OF A *XylT* NULL MUTANT BY XylT ANALOGUES

In order to assess the physiological role of XylT-like proteins *in vivo*, the XylT protein was replaced by each of the three proteins described above in the natural host *P. putida*. For this purpose, a *xylT*-null mutant of *P. putida* (Polissi and Harayama, 1993) was used as a recipient strain in genetic complementation experiments. The *xylT, phhQ, dmpQ* and *nahT* genes were individually cloned in a plasmid suitable for expression in *Pseudomonas*, and the recombinant genes were then introduced in the *xylT* mutant. The recombinant strains were allowed to grow on a minimal medium containing *p*-toluate as sole carbon source. While the control *xylT* mutant failed to grow under such conditions, all the complemented strains except that carrying the *nahT* gene, were capable of utilizing *p*-toluate as carbon source. The growth rates of the complemented strains were found to be comparable to that of the wild-type strain. In addition, when 4-methylcatechol was supplemented to growing cultures, a transient inhibition of XylE was observed followed by a rapid recovery of enzyme activity, indicating that the expressed ferredoxins reactivated the enzyme *in vivo*. These results provide unequivocal evidence that at least two of the XylT analogues, PhhQ and DmpQ, can readily replace XylT in *P. putida*. The apparent absence of complementation by the plasmid-borne copy of *nahT* is currently being examined. One simple explanation would be that NahT was not synthesized in the *xylT* mutant, due to instability or an unidentified default of expression.

5. CONCLUSION

A novel subgroup of [2Fe-2S] ferredoxins with distinct biochemichal properties has been identified in certain bacteria which degrade aromatic hydrocarbons. These ferredoxins are genetically and functionally associated with ring-cleavage extradiol dioxygenases. Presumably, the role of the XylT-like proteins is to reactivate the intrinsically labile extradiol dioxygenases by reducing the iron atom present at the enzyme active site, as demonstrated for the XylT protein from *P. putida* mt2. Because ring-cleavage extradiol dioxygenases catalyze a critical step in the catabolic pathway of various aromatic pollutants, the mechanism of regeneration of an active catalyst described in this study is probably beneficial to strains containing a XylT-like protein. From the data reported here, strategies may be envisioned to engineer microorganisms with greater degradation efficiency towards aromatic pollutants.

REFERENCES

Bartels, I., Knackmuss, H. J., and Reineke, W. (1984). Suicide inactivation of catechol 2,3-dioxygenase from *Pseudomonas putida* mt-2 by 3-Halocatechols. Appl. Environ. Microbiol. **47**, 500–505.

Cerdan, P., Rekik, M., and Harayama, S. (1995). Substrate specificity differences between two catechol 2,3-dioxygenases encoded by the TOL and NAH plasmids from *Pseudomonas putida*. Eur. J. Biochem. **229**, 113–118.

Cerdan, P., Wasserfallen, A., Rekik, M., Timmis, K. N., and Harayama, S. (1994). Substrate specificity of catechol 2,3-dioxygenase encoded by TOL plasmid pWW0 of *Pseudomonas putida* and its relationship to cell growth. J. Bacteriol. **176**, 6074–6081.

Harayama, S., Polissi, A., and Rekik, M. (1991). Divergent evolution of chloroplast-type ferredoxins. FEBS Lett. **285**, 85–88.

Harayama, S., and Rekik, M. (1989). Bacterial aromatic ring-cleavage enzymes are classified into two different gene families. J. Biol. Chem. **264**, 15326–15333.

Herrmann, H., Muller, C., Schmidt, I., Mahnke, J., Petruschka, L., and Hahnke, K. (1995). Localization and organization of phenol degradation genes of Pseudomonas putida strain H. Mol Gen Genet **247**, 240–246.

Hugo, N., Armengaud, J., Gaillard, J., Timmis, K. N., and Jouanneau, Y. (1998). A Novel [2Fe-2S] Ferredoxin from *Pseudomonas putida* mt2 Promotes the Reductive Reactivation of Catechol 2,3-Dioxygenase. J Biol Chem **273**, 9622–9629.

Kukor, J. J., and Olsen, R. H. (1996). Catechol 2,3-dioxygenases functional in oxygen-limited (hypoxic) environments. Appl. Environ. Microbiol. **62**, 1728–1740.

Lee, J., Min, K. R., Kim, Y. C., Kim, C. K., Lim, J. Y., Yoon, H., Min, K. H., Lee, K. S., and Kim, Y. (1995). Cloning of salicylate hydroxylase gene and catechol 2,3-dioxygenase gene and sequencing of an intergenic sequence between the two genes of Pseudomonas putida KF715. Biochem Biophys Res Commun **211**, 382–388.

Ng, L., C., Shingler, V., Sze, C. C., and Poh, C. L. (1994). Cloning and sequences of the first eight genes of the chromosomally encoded (methyl)phenol degradation pathway from *Pseudomonas putida* P35X. Gene **151**, 29–36.

Parales, R. E., Ontl, T. A., and Gibson, D. T. (1997). Cloning and sequence analysis of a catechol 2,3-dioxygenase gene from the nitrobenzene-degrading strain Comamonas sp JS765. J Ind Microbiol Biotechnol **19**, 385–391.

Polissi, A., and Harayama, S. (1993). *In vivo* reactivation of catechol 2,3-dioxygenase mediated by a chloroplast-type ferredoxin: a bacterial strategy to expand the substrate specificity of aromatic degradative pathways. EMBO J. **12**, 3339–3347.

Shingler, V., Powlowski, J., and Marklund, U. (1992). Nucleotide sequence and functionnal analysis of the complete phenol/3,4-dimethylphenol catabolic pathway of *Pseudomonas* sp. strain CF600. J. Bacteriol. **174**, 711–724.

Takeo, M., Fujii, T., Takenaka, K., and Maeda, Y. (1998). Cloning and sequencing of a gene cluster for the *meta*-cleavage pathway of the aniline degradation in *Acinetobacter* sp. strain YAA. J. Ferment. Bioeng., in press.

Tsukihara, T., Fukuyama, K., Mizushima, M., Harioka, T., Kusunoki, M., Katsube, Y., Hase, T., and Matsubara, H. (1990). Structure of the [2Fe-2S] Ferredoxin I from the blue-green *Alga Aphanothece* sacrum at 2.2 A resolution. J. Mol. Biol. **216**, 399–410.

You, I. S., Ghosa, D., and Gunsalus, I. C. (1991). Nucleotide sequence analysis of the *Pseudomonas putida* PpG7 salicylate hydrolase gene (*nahG*) and its 3'-flanking region. Biochemistry **30**, 1635–1641.

Zylstra, G. J., and Kim, E. (1997). Aromatic hydrocarbon degradation by *Sphingomonas yanoikuyae* B1. J. Ind. Microbiol. Biotechnol. **19**, 408–414.

13

DIRECTED EVOLUTION OF A BACTERIAL PESTICIDES DEGRADING ENZYME

Y. Flashner,[1] I. Mendelson,[1] A. Ordentlich,[1] H. Grosfeld,[1] M. Leitner,[1] A. Balan,[2] Y. Segall,[2] D. Barak,[2] and B. Velan[1]

[1]Department of Biochemistry and Molecular Biology
[2]Department of Synthetic Chemistry
Israel Institute for Biological Research, Ness-Ziona, Israel

1. ABSTRACT

The wide use of pesticides is a threat to the public health. An attractive tool for bioremediation of pesticides pollution is the bacterial pesticides degrading enzyme, phosphotriesterase (OPH), which is able to hydrolyze a wide range of organophosphate compounds. Yet, engineering of a more potent hydrolytic machinery into this enzyme could yield an advantageous bioremediation tool.

We undertook the *in-vitro* evolution strategy in order to improve the hydrolyzing power of phosphotriesterase towards certain organophosphates. The *opd* gene, coding for the phosphotriesterase enzyme, was amplified from *Flavobacterium* by PCR, and introduced into *pmalE* gene fusion system, expressed in *E. coli*, to allow simple procedure for purification of the recombinant mature native (and future mutant) enzyme. Random mutagenesis of the *opd* gene was carried out by error-prone PCR, and controlled low level mutagenesis was adopted. Mutagenesis products served for generation of an expression library, using a modified pUC18 as an expression vector. A screening system was developed to allow monitoring of enzymatic kinetics with whole expressing cells, and to isolate phosphotriesterase mutants with altered enzymatic properties.

2. INTRODUCTION

The pollution of the environment with man-made organic compounds causes a major human health problem. This is mainly due to their persistence, toxicity, or transformation into hazardous metabolites. Although microorganisms may play a crucial role in bioremediation of these compounds several obstacles should be resolved. These include slow

Novel Approaches for Bioremediation of Organic Pollution, edited by
Fass *et al.* Kluwer Academic / Plenum Publishers, New York, 1999.

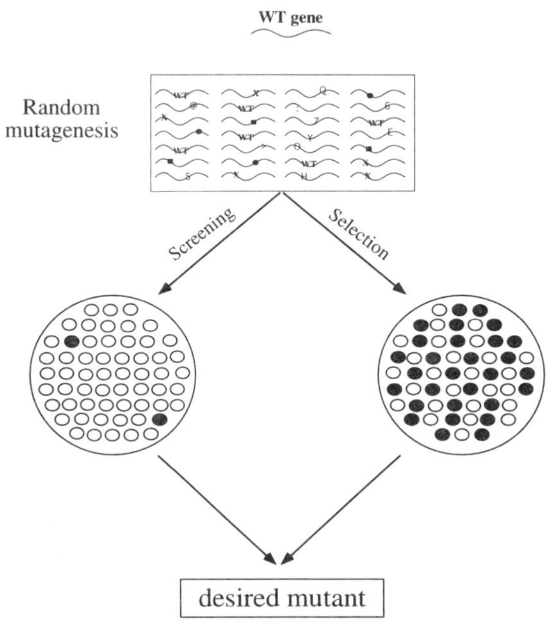

Figure 1. *In-vitro* evolution scheme.

biodegradation kinetics, especially in unfavorable physicochemical conditions, low availability of nutrients, and limited accessibility of substrates (Timmis *et al.*, 1994). The low biodegradability could be caused not only by the environmental conditions, but also by the limited intrinsic degradative power of the bacterial enzymes towards the specific substrates. The limited enzymatic efficiency may be elevated by long-term exposure of microorganisms to the substrates. This evolutionary process is expected to be inefficient towards natural chemical compounds that are not essential for bacterial growth, and even less efficient towards xenobiotics, compounds alien to nature.

Directed enzyme evolution has emerged in the last years as a powerful tool for engineering enzymes with desired properties, as an alternative to the rational approach for protein engineering (Kuchner and Arnold, 1997). The first step in directed enzyme evolution is random mutagenesis of the target gene, and creation of a library of mutated genes. This step is followed by a screening or selection processes, aimed at the isolation of improved mutant enzymes. The selected genes are subjected to further cycles of mutagenesis and screening/selection in order to accumulate beneficial mutations (Fig 1). This strategy has been applied successfully, in recent years, towards several bacterial enzymes (Kuchner and Arnold, 1997).

In this study the pesticide degrading enzyme, phosphotriesterase (OPH), from *Flavobacterium*, is the target enzyme for directed evolution. The need for a convenient and safe system for pesticides detoxification is evident considering the extensive use of these compounds and their accumulation in food products and water supplies. The phosphotriesterase enzyme, encoded by the *opd* gene, is unique in its ability to hydrolyze a wide range of pesticides and nerve agents, cleaving P-O, P-F, P-S as well as P-CN bonds (Dumas *et al.*, 1989). The enzymatic hydrolysis rates towards these substrates range from very high (paraoxon) to very low (P-S bond cleavage) (Dumas *et al.*, 1989; Dumas *et al.*, 1990).

Here we describe the establishment of the various tools needed for carrying out the *in-vitro* evolution aimed at increasing OPH reactivity towards methyl-parathion. This work is a part of a project aiming at engineering efficient pesticides degrading bacteria to be used in the field. All the *opd* gene manipulations have been carried out in *E. coli.* The optimal mutant *opd* gene will then be obviously inserted under a suitable regulation/expression system in soil bacteria, to be functional in the field.

3. RESULTS AND DISCUSSION

3.1. Cloning of the *opd* Gene and Purification of the Recombinant OPH Enzyme

Both *Pseudomonas diminuta MG* and *Flavobacterium* (ATCC#27551) were found to express phosphotriesterase activity; these activities were attributed to identical *opd* genes, present in the two bacterial strains (Harper *et al.*, 1988). The coding sequence encompasses codons for 365 amino acids, 29 of which serve as a signal. The *opd* gene has been cloned into a wide variety of host systems, including the bacteria *E. coli* and *Streptomyces lividans* as well as into insect tissue cultures (Mulbry and Karns, 1989; Steiert *et al.*, 1989; Dave *et al.*, 1994). All these systems express either the mature or the signal containing protein. The established enzyme purification method includes several steps of ion-exchange and gel filtration chromatography (Dumas *et al.*, 1989).

Here, we adopted the protein-fusion expression system for generation and purification of recombinant OPH enzyme. This strategy was undertaken in view of the need for an efficient and simple procedure of enzyme purification when analysis of several mutant enzymes would be required following mutant library screening. The *opd* gene (without the signal sequences) was amplified by PCR from *Flavobacterium* ATCC#27551, and inserted into the pMAL-cII expression vector, producing a fused *malE-opd* gene downstream to the *tac* promoter (Fig 2). Following induction of the fused gene expression by addition of IPTG a band of the expected Maltose-Binding-Protein-OPH fused protein could be observed (Fig. 2, first two lanes). The collected supernatant, containing the fused protein, was subjected to maltose affinity chromatography and Xa-factor protease cleavage to release the native OPH followed by Q-sepharose column to resolve the pure OPH. This procedure resulted in a highly purified OPH preparation (Fig. 2, last lane).

The purified enzyme was tested for its hydrolyzing activity towards several substrates containing P-O bond: paraoxon, parathion, and methyl-parathion. The respective specific activities of the recombinant enzyme manifested the expected spectrum of activities, where hydrolysis of paraoxon is the most efficient, and that of methyl parathion is the least efficient (not shown) (Dumas *et al.*, 1989; Dumas *et al.*, 1990).

3.2. Establishment of Random Mutagenesis Conditions for *opd*

The first step in carrying out the *in-vitro* evolution procedure (Fig. 1) is to introduce random mutations within the gene of interest, and to generate a library of mutated genes.

Random mutagenesis should meet two requirements: 1. No preference for a specific type or site of mutation. 2. Low rate mutagenesis of about 1–3 mutations per gene (Moore

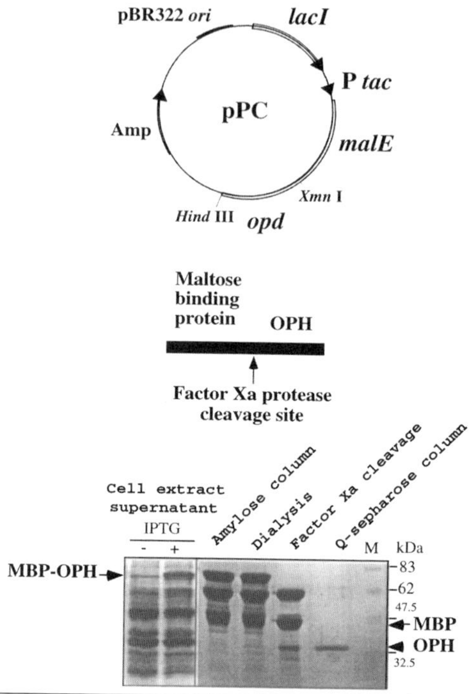

Step	Total protein		Activity		Sp. activity	Purification
	mg	%	units *	%	(units/mg)	(fold)
Lysis sup.	197	100	420	100	0.1	1
Amylose col.	5.2	2.6	406	97	78	780
Q-Sepharose.col.	0.14	0.07	112	27	812	8120

$mO.D_{405}/min$ *

Figure 2. Purification of recombinant OPH through expression of a fused *malE-opd* gene. The upper part shows schematically the plasmid expression system, and the fused protein generated. The protein gel and the table summarize the purification process.

and Arnold, 1997), to allow progress of the *in-vitro* evolution process in small increments. Several approaches have been developed for random mutagenesis of DNA sequences (Cadwell and Joyce, 1992; Stemmer, 1994). We chose the error prone PCR strategy mainly since it allows better control of the mutagenesis rate.

We have utilized several protocols for error prone PCR to evaluate the resulting level of mutagenesis. The evaluation process was based on the estimation that random mutagenesis leading to a single nucleotide change would generate stop codons within the *opd* gene at a rate of about 5% (corresponding to the fraction number of stop codons among the total number of codons), and thereby yielding the same percentage of mutants with truncated OPH proteins. Hence, the evaluation of the mutagenesis conditions consisted of the following steps:

1. Ligation of mutagenized *opd* gene PCR fragments to an expression vector and introduction into *E. coli* cells to generate a library of OPH mutants producers.

Level of mutagenesis:

Type of mutant enzyme	number	%
Active	18	21
Inactive	68	79

* Protocol 1: 10% DMSO, 10mM βME, 6.25mM MgCl2
5U Taq pol., 30 cycles

A. Gene presence - (PCR)

B. OPH production - (western blot)

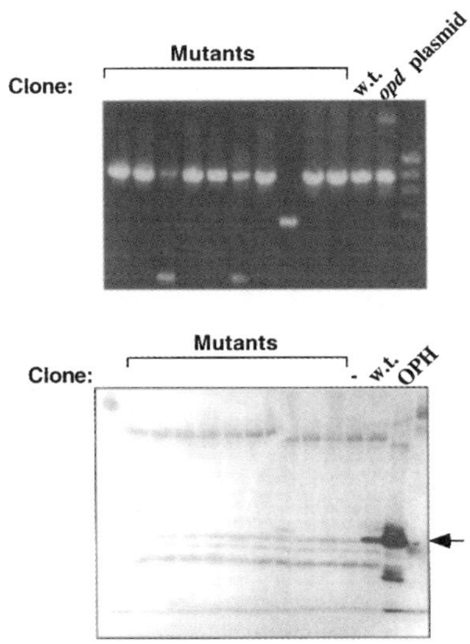

Figure 3. Evaluation of *opd* gene error-prone-PCR mutagenesis conditions: protocol 1. Following carrying out PCR according to protocol 1 (see details), the *opd* gene fragment was ligated to an expression vector (see text), and introduced into *E. coli* HB101 cells. 100 clones were grown to early logarithmic phase in microtiter plates, induced for OPH production by addition of IPTG (1mM) and $CoCl_2$ (50μM). Parathion hydrolysis activity was monitored (reading at 405nm) following cell lysis by addition of lysozyme (0.25mg/ml) and polymyxin B (10μM). Ten inactive mutants were taken for analysis of *opd* gene presence (A), and for OPH production assay (B).

2. Examination of 50–100 library clones for OPH activity (hydrolysis of parathion).
3. The inactive clones were subjected to PCR analysis to verify "true ligation" (the presence of the *opd* gene), and to western blot analysis to determine production of full-length OPH polypeptide.

Several PCR protocols were examined. The two PCR protocols presented here consisted both of 30 cycles, but differed in the buffer composition and duration of PCR steps. Protocol 1 buffer constituents included DMSO, β-mercaptoethanol, and high concentration of $MgCl_2$, (Fig 3), whereas in protocol 2 DMSO and β-mercaptoethanol were missing, and $MgCl_2$ concentration was low (Fig. 4).

Level of mutagenesis:

Type of mutant enzyme	Protocol 1 %	Protocol 2 %
Active	21	78
Inactive	79	22

* Protocol 2: 1.5mM $MgCl_2$, 5U Taq pol., 30 cycles

OPH protein production (western blot):

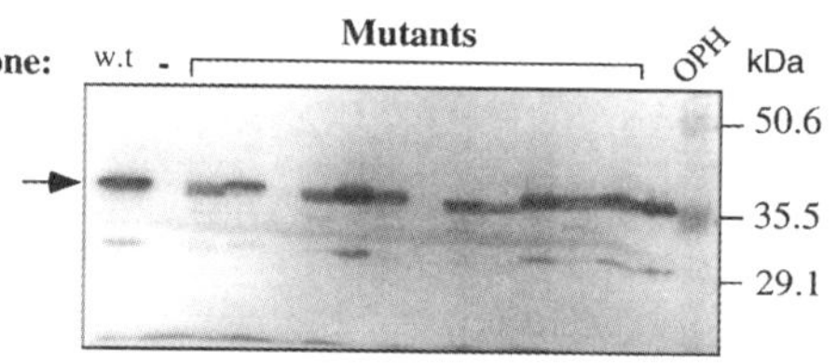

Figure 4. Evaluation of error-prone-PCR conditions: protocol 2. PCR was carried out following protocol 2 (see details). Evaluation procedure was done as described in Fig 3.

Evaluation of the rate of mutagenesis following protocol 1 (Fig. 3) reveals that these conditions yield a high percentage of inactive mutants (79%). Whereas PCR analysis shows that ligation of the PCR *opd* gene fragment was efficient (Fig 3A), none of the 10 clones analyzed by western blot exhibited production of full length OPH protein (Fig 3B). These results indicate that protocol 1 for error-prone PCR yields a rate of mutagenesis which is too high for practical use. Indeed, when three mutants were submitted to "sample sequencing", four mutations were identified (including stop codons within 400bp sequenced), suggesting the presence of about 10 mutations per gene.

The second protocol for error-prone-PCR (protocol 2, Fig. 4) proved to be more suitable for directed evolution. The percentage of inactive clones (21%) was different from that exhibited by the first protocol (79% inactive, Fig 3). Western blot analysis of 12 clones indicated that 10 clones produced full length OPH protein (Fig 4). This suggests that mutagenesis rate was low as indeed confirmed by sample sequencing of few mutants. Protocol 2 was adopted for generation of *opd* mutant library.

3.3. Screening of Library of *E. coli* Cells Harboring Random Mutagenized *opd* Genes

Generation and screening of a library of cells expressing mutant OPH enzymes requires a suitable expression vector (to gain a soluble enzyme), and appropriate conditions for cell growth and for OPH assay (to allow mass-screening).

To generate a vector which allows a moderate level of expression to prevent possible formation of inclusion bodies, we constructed a modified pUC18 expression vector, in which the *lacZ* coding sequences downstream to the *lac* promoter were deleted. The *opd*

Activity of titrated OPH producing cells

Culture Dilution	mOD_{405} min^{-1} Parathion		mOD_{405} min^{-1} Methyl-Parathion	
Source		48		7
1:2		25		3
1:4		12		1

Figure 5. Evaluation of screening conditions. Cells expressing wild-type OPH were grown to mid-log, and then were subjected to two-fold dilutions. The source and the diluted cultures were placed in a microtiter plate, and then cell lysis followed by monitoring parathion hydrolysis were carried out (see also legend to Fig 3).

coding sequences (without the signal sequence) were positioned immediately under the *lac* promoter. Expression assays of *E. coli* cells harboring this recombinant *opd* showed a moderate level of OPH production, and most of the OPH produced was found to be soluble (not shown).

Since the *opd* genes in the mutant library are expressed intracellularly it was necessary to adopt suitable ways to expose the enzyme to the substrates. Several methods for cell lysis were tested for their efficiency and reproducibility. The adopted treatment includes addition of lysozyme (0.25mg/ml) and polymyxin B (10μM) (Hancock and Wong, 1994). The combination of this treatment with screening for OPH activity was evaluated for its ability to allow detection of two-fold difference in enzymatic activity. This last point is important since the improvement in the enzyme capacity through directed evolution process is expected to proceed in small increments. Fig 5 shows that under the screen conditions it was possible to detect about two-fold difference in OPH activity.

A library of 1100 clones harboring random mutagenized *opd* genes was screened. Mutant library cells were grown to early logarithmic phase in microtiter plates, and OPH production was induced by addition of 1mM IPTG in the presence of 50μM $CoCl_2$. Following lysis with lysozyme and polymyxin B the kinetics of methyl-parathion hydrolysis by the mutant OPHs was monitored. Five clones showed apparent "improved" hydrolyzing power towards methyl-parathion (Fig. 6), and their higher activity was evaluated again in three independent assays. Two of these clones were submitted to sequencing of the *opd* gene, and in both a single amino acid substitution was found. Hence, the first cycle of mutagenesis and screening yielded *opd* mutants which could serve as candidate templates in a second round of mutagenesis and screening. It has been already noted, that directed evolution is a multistep process, and usually 3–4 cycles of mutagenesis and screening are required for gaining an order of magnitude increase in the target activity (Arnold and Moore, 1997).

3.4. Selection for Clones Expressing Mutated *opd*

An alternative way to obtain mutants with altered specificities is to subject the mutant library population to a selection force for cells producing "improved" enzymes. This strategy has, of course, the clear benefits of not requiring the tedious process of mass-screening. The problem is, that in many cases, including that of OPH, there is no trivial way for such selection, since the target enzymatic activity does not confer any advantage on the bacterial host.

To overcome this we designed a synthetic substrate, which links an essential factor for bacterial growth to the "core" of the target substrate. This synthetic substrate should meet the following requirements:

a. Solubility and stability in aqueous solution.
b. Cleaveability at a reasonable rate by the target enzyme, to release the essential factor.
c. Ability to penetrate into the bacterial cell under the experimental conditions.

Activity of selected mutants for methyl-parathion hydrolysis

clone	Activity* $mOD_{405}/nin/OD_{550}$	Activity Ratio (mutant/WT)
WT	28±6	
524	76±8	2.7
529	72±10	2.6
896	63±12	2.2
1040	93±4	3.3
1085	113±12	4.0

* Average of three independent assays

Figure 6. Summary of screening *opd* mutants library. A library of 1100 clones, generated following a first cycle of random mutagenesis of *opd*, was screened for mutants in activity towards hydrolysis of methyl-parathion. The data is presented for five selected mutants.

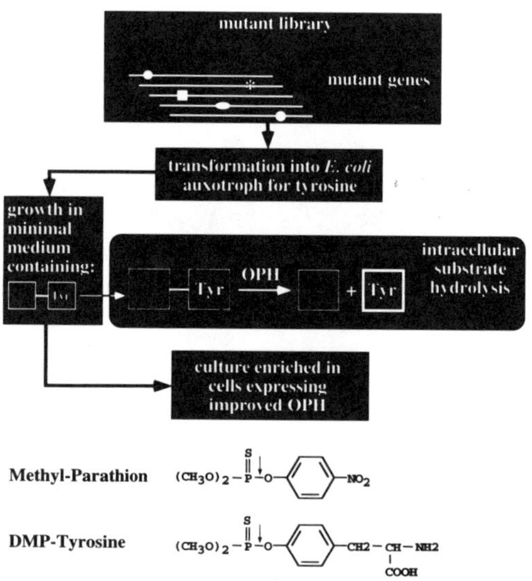

Figure 7. Scheme of a designed selection system for *opd* mutants with improved activity towards hydrolysis of methyl-parathion. The system is based on generating the OPH mutant library using a strain auxotroph for tyrosine, and growing it in a defined medium, containing a synthetic OPH substrate, which links tyrosine to methyl-parathion "core". See also text.

Growing cells, which are auxotroph for the essential factor and which harbor the mutant gene library, in a defined medium containing the synthetic substrate would yield the selective pressure to obtain a culture enriched in cells producing "improved" enzymes. This enriched culture could then be screened for the desirable mutants. Fig. 7 presents a scheme for a specific selection system designed for gaining mutant OPH enzymes, improved in their ability to hydrolyze methyl-parathion. The system is based on using *E. coli* strain, auxotroph to tyrosine, and a synthetic substrate, in which tyrosine is linked to the "core" of methyl-parathion. Evaluation of this system is in progress.

ACKNOWLEDGMENT

We would like to thank Dr. S. Cohen for fruitful discussions. The excellent technical assistance of G. Friedman, T. Sery, and D. Stein is highly appreciated.

REFERENCES

Arnold F.H. and Moore J.C. (1997) Optimizing industrial enzymes by directed evolution. In T. Scheper (ed.), *Advances in Biochemical Engineering/Biotechnology* Vol. 58, pp. 1–14.

Cadwell R.C. and Joyce G.F. (1992) Randomization of genes by PCR mutagenesis. *PCR Methods and Applications* 2: 28–33.

Dave K.I., Phillips L., Luckow V.A., and Wild J.R. (1994) Expression and post-translational processing of a broad-spectrum organophosphorus-neurotoxin-degrading enzyme in insect tissue culture. *Biotechnol. Appl. Biochem.* 19: 271–284

Dumas D.P., Caldwell S.R., Wild J.R., and Raushel, F.M. (1989) Purification and properties of the phosphotriesterase from *Pseudomonas diminuta*. *J.Biol. Chem.* 264: 19659–19665.

Dumas P.D., Dupont Durst H., Landis W.G., Raushel F.M., and Wild (1990) Inactivation of organophosphorus nerve agents by the phosphotriesterase from *Pseudomonas diminuta*. *Archives of Biochem. and Biophys.* 277: 155–159.

Hancock R.E.W. and Wong P.G.W. (1984) Compounds which increase the permeability of the Pseudomonas aeruginosa outer membrane. *Antimicrobial Agents and Chemotherapy* 26: 48–52.

Harper L.H., McDaniel C.S., Miller C.E., and Wild J.R. (1988) Dissimilar plasmids isolated from *Pseudomonas diminuta* MG and a *Flavobacterium* sp (ATCC 27551) contain identical *opd* genes. *Appl. and Environ. Microbiol.* 54: 2586–2589.

Kuchner O. And Arnold F.H. (1997). Directed evolution of enzyme catalysts. *Trends in Biotechnol.* 15: 523–530.

Moore J.C., and Arnold F.H. (1997) Directed evolution of a *para*-nitrobenzyl esterase for aqueous-organic solvents. *Nature Biotechnol.* 14: 458–467.

Mulbry W.W., and Karns J.S. (1989) Parathion hydrolase specified by the Flavobacterium opd gene: relationship between the gene and protein. *J. Bacteriol.* 171: 6740–6746.

Steiert J.G., Pogell B.M., Speedie M.K., and Laredo J. (1989) A gene encoding for a membrane bound hydrolase is expressed as a secreted, soluble enzyme in *Streptomyces lividans*. *Biotechnology* 7: 65–68.

Stemmer W.P.C. (1994) DNA shuffling by random fragmentation and assembly: *In vitro* recombination for molecular evolution. *Proc. Natl. Acad. Sci. USA* 91: 10747–10751.

Timmis K.N., Steffan. J., and Unterman R. (1994). Designing microorganisms for the treatment of toxic wastes. *Annu. Rev. Microbiol.* 48: 525–557.

PROKARYOTIC WHOLE-CELL LIVING BIOREPORTERS EXPRESSING BIOLUMINESCENCE UPON THE PRESENCE OF BIOAVAILABLE CONCENTRATIONS OF SPECIFIC POLLUTANTS

An Overview

Marco C. M. Jaspers,[1,*] Hauke Harms,[1,2] and Jan Roelof van der Meer[1]

[1]Swiss Federal Institute for Environmental Science and Technology (EAWAG) and Swiss Federal Institute for Technology (ETH)
CH-8600 Dübendorf, Switzerland
[2]Ecole Polytechnique Fédérale de Lausanne (EPFL)
CH-1015 Lausanne, Switzerland

1. INTRODUCTION

1.1. The Need to Measure Bioavailability in Bioremediation Processes

Bioremediation has gained a lot of support for the clean up of contaminated soils and sediments because of its elegance and cost-effectiveness compared to chemical and physical methods (for reviews, see (7, 30)). Bioremediation is based on the capacity of many indigenous microorganisms to degrade polluting compounds (13, 39, 43), although many different factors determine the actual microbial activity in the environment (43). For example, (macro) physicochemical parameters (e.g., redox potential, pH, temperature, salinity, moisture content, availability of nutrients, the presence and nature of surfaces) modulate microbial activity. A key factor influencing the rates of biodegradation in the field is the extent to which a compound is accessible (bioavailable) to the microorganisms.

* Corresponding author. Mailing Address: EAWAG, Dept. of Microbiology, Überlandstrasse 133, CH-8600 Dübendorf. Phone: ++41-1-823 52 02. Fax: ++41-1-823 55 47. E-mail: jaspers@eawag. ch.

Novel Approaches for Bioremediation of Organic Pollution, edited by Fass *et al.* Kluwer Academic / Plenum Publishers, New York, 1999.

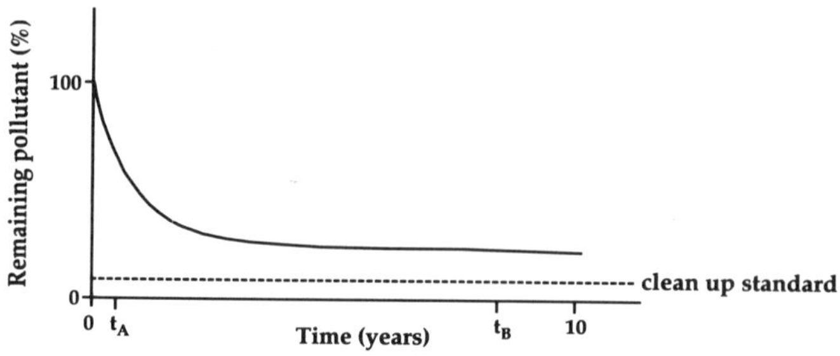

Figure 1. Time dependency of the pollutant concentration in a contaminated soil during a bioremediation process. The time-scale is arbitrarily set to 10 years.

Different studies pointed out that microorganisms can only degrade substrates which are dissolved in the water phase (26, 52, 67, 69). Therefore, to predict biodegradation rates there exists a clear need to measure the bioavailable pollutant concentrations in environmental sites as precisely as possible.

Mainly mass transfer processes determine the water-dissolved concentrations of environmental pollutants (25). Depending on the physical states (solid, non-aqueous phase liquid, dissolved) in which the pollutants were introduced into the environment, and the characteristics of the environment itself (soil, sediment, lakes), different mass transfer processes (or combinations thereof) prevail. Sorption equilibria and desorption kinetics, for example, describe the interactions of pollutants with the solid matrix. Dissolution is important to take into account when the pollutant is present as non-aqueous phase liquid or in crystalline form. These mass transfer processes connect the relatively big, but per se unavailable, reservoirs of the pollutant with its relatively small water-dissolved pool. The diffusion of the compound in the water phase to the bacterial cell, its uptake and metabolism are then the processes leading to removal of the compound. The overall biodegradation rate will therefore depend on the ratio of the supply rate to the water phase to the removal kinetics from the water phase.

We can illustrate the effect of different mass transfer processes on bioavailability with some examples. A first one describes sites with a long history of pollution that results in 'aging' of the pollution (1, 27). Sites with aged pollution often are assumed to have low bioavailable pollutant concentrations. Aging is probably the result of diffusion of pollutants into the pores of microporous soil particles (5, 75) and/or partitioning into the soil organic matter (8). This renders the compounds less accessible for microorganisms as microorganisms are too large to enter the micropores. To be transported to the microbial cells the compounds will need to diffuse through the pores (56). A second example of limiting bioavailability typically results from bioremediation itself. In the beginning the water-dissolved concentrations may be relatively high and in equilibrium with the sorbed compound. Degradation however diminishes the concentration in the water phase, and, when the kinetics of dissolution or desorption are too slow to follow the biodegradation, the water-dissolved concentration will constantly drop. Thirdly, in quite a few cases the characteristics of the pollutants themselves (e.g., solubility) are the factors limiting bioavailability. For example, dissolution of individual compounds from a non-aqueous phase oil droplet into water will be very fast for benzene, but rather slow for naphthalene, and the formation of semi-rigid films after prolonged time may extensively reduce mass transfer of remaining oil components to the water phase (47).

Different field studies pointed out that the mass transfer of the pollutant becomes limiting for the cells, leading to suboptimal biodegradation rates, despite sufficient capacity or metabolic activity of the microorganisms at that moment (Fig. 1, 2). For example, an

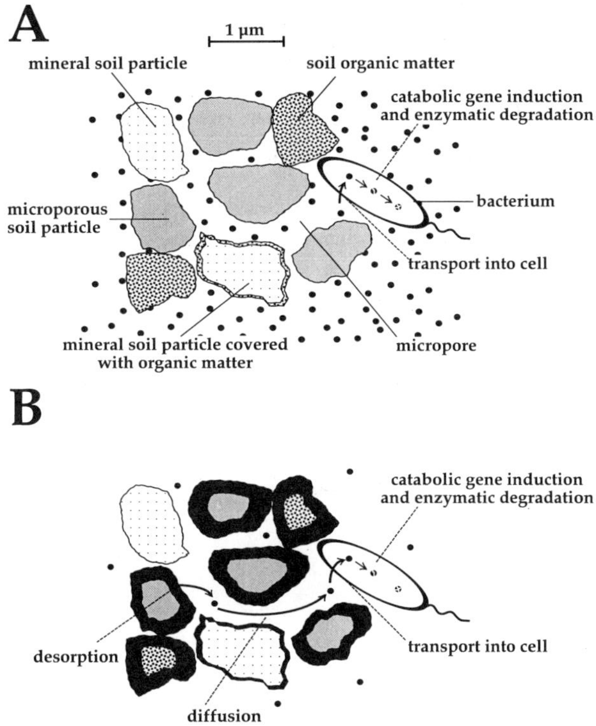

Figure 2. Schematic representation of pollutant bioavailability to microorganisms in a soil system after a short (A) and a prolonged (B) time period, corresponding to t_A, respectively t_B in Figure 1. The water soluble and absorbed/adsorbed fractions of the pollutant are depicted as dark gray spheres, respectively dark gray shadings. Dashed lines correspond to processes involved in the biodegradation of the pollutant. See text for more details.

agricultural topsoil with a history of 1,2-dibromoethane application contained an extractable 1,2-dibromoethane concentration of 130 ng/g soil which no longer diminished during a period of several years. Fresh application of an amount of 80 ng/g soil dissolved 1,2-dibromoethane to the soil resulted in immediate degradation of the compound. However, the previous 'residual' 1,2-dibromoethane concentration did not change (65). Similar observations were made for soils polluted with simazine (60) or with polycyclic aromatic hydrocarbons (22, 73).

1.2. Whole-Cell Living Bioreporters to Measure Bioavailable Pollutant Concentrations

1.2.1. General Principle. A good insight in the bioavailability of pollutants in the environment could be obtained by using microorganisms themselves. Considerable efforts are being undertaken to construct bacteria for measuring concentrations of specific compounds. To do so, specific bacterial strains are genetically modified to express an easy detectable signal upon sensing the presence of the compound, they were designed for (whole-cell living bioreporters). The sensing function is performed by one or more regulatory proteins which upon recognition of the compound can activate gene expression.

The scope of this review is to give an overview of bacterial whole-cell bioreporters for measuring the bioavailability of *specific* pollutants by emission of *bioluminescence*. Instead of bioluminescence, fluorescence could also serve as a suitable reporter signal. However, reports describing the use of the green fluorescent protein (GFP) (12, 54) from the jellyfish *Aequorea victoria* in whole-cell living bioreporters for the detection of spe-

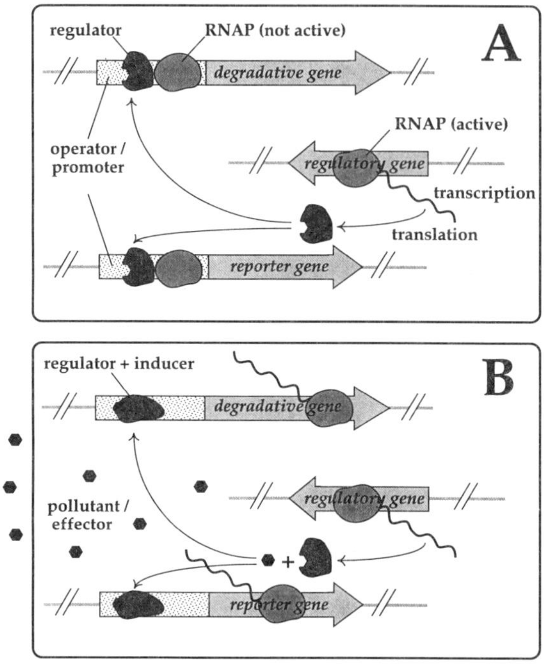

Figure 3. Schematic representation of a whole-cell living bioreporter in the absence (A) or presence (B) of a pollutant/effector. RNAP: RNA polymerase.

cific compounds are (still) scarce and will not be discussed here. The principles of reporter strain development and the requirements for a good bioreporter will be discussed and an overview will be presented of the different bioluminescence based reporter bacteria which have been constructed so far.

1.2.2. Using Degradative Pathways to Construct Whole-Cell Living Bioreporters. Bacteria which degrade environmental pollutants usually efficiently regulate the expression of genes coding for enzymes responsible for pollutant transformation. Representative examples are (i) the *xyl* genes on plasmid pWW0 for the degradation of toluene and xylenes (57), (ii) the *nah* genes on plasmid NAH7 for the degradation of naphthalene (32, 59), and (iii) the *tfd* genes on plasmid pJP4 for the degradation of 2,4-D (45). In the absence of the pollutant the degradative genes are not transcribed (Fig. 3A, upper half). Only when the pollutant is available for the cell, the genes for its degradation will be induced (Fig. 3B, upper half). This response is mediated by regulatory proteins, which upon sensing an effector molecule (the pollutant or one of its metabolites), interact with the operator/promoter region of the degradative genes in a way that promotes RNA polymerase (RNAP) to start transcription. The sensing is based on a molecular recognition between the regulatory protein and the effector.

A whole-cell living bioreporter contains the specific sensor element (e.g. the regulatory protein of a degradative pathway) in order to activate an easy measurable reporter element. This reporter element can be activated alone, or in addition to its native target degradative genes. Whole-cell living bioreporters are constructed by coupling a promoterless reporter gene, coding for an easy measurable gene product, to a copy of the operator/promoter region of the regulated degradative genes. The reporter fusion construct can be reintroduced into the original pollutant degrading bacterium on a plasmid or in the chromosome. Another approach which is often taken, is to introduce the reporter fusion

Table 1. Characteristics of bioluminescent reporter bacteria for organic pollutants

Pollutant/ Effector[a]	Luciferase gene(s)[b]	Operator/ promoter[c]	Regulatory gene	Host strain	Detected[d]	Reference
o-cresol[e]	*luxAB* (V.h.)	P_{dmp} (*dmpK*)	*dmpR*	*Pseudomonas putida*	3.2 μM	(63)
2-hydroxy-biphenyl[f]	*luxAB* (V.h.)	P_{hbpC} (*hbpC*)	*hbpR*	*Pseudomonas azelaica*	9 μM	(34)
Isopropyl-benzene[g]	*luxCDABE* (V.f.)	(*ibpA*)	*ibpR*	*Escherichia coli*	1 μM	(62)
Naphthalene[h]	*luxCDABE* (V.f.)	P_{sal} (*nahG*)	*nahR*	*Pseudomonas fluorescens*	0.35 μM	(29, 37)
n-octane[i]	*luxAB* (V.h.)	P_{alkB} (*alkB*)	*alkS*	*Escherichia coli*	24 nM	(66)
m-toluate	*luxAB* (V.h.)	P_m (*xylX*)	*xylS*	*Pseudomonas putida*	10 μM	(18)
Toluene[j]	*luxCDABE* (V.f.)	P_{tod} (*todX*)	*todST*[k]	*Pseudomonas putida*	0.33 μM	(3)
m-xylene[l]	*luc* (P.p.)	P_s (*xylS*)	*xylR*	*Escherichia coli*	5 μM	(38)

[a] The compounds listed here are the most effective inducers for the reporter systems. Other potential inducers are described in separate footnotes.
[b] Strain abbreviations in parentheses refer to the organism where the luciferase genes originate from. V.f.: *Vibrio fischeri*; V.h.: *Vibrio harveyi*; P.p.: *Photinus pyralis*.
[c] Symbols for the different operator/promoter regions are taken from the references. Gene symbols in parentheses refer to the first gene which is transcribed from the promoter.
[d] Lowest concentration tested that still gave a significant response compared with no inducer added.
[e] Other effectors: phenol, *m*-cresol, *p*-cresol, 3,4-dimethylphenol, 2,3-dimethylphenol, *o*-chlorophenol, *m*-chlorophenol, *p*-chlorophenol, *o*-ethylphenol, *m*-ethylphenol, catechol, 3-methylcatechol.
[f] Other effectors: 2,2'-dihydroxybiphenyl, 2-aminobiphenyl, 2-hydroxydiphenylmethane (Fig. 5).
[g] Other effectors: benzene, toluene, ethylbenzene, naphthalene, trichloroethylene.
[h] The actual effector is salicylate (2.9 μM detected), a degradation product of naphthalene (59).
[i] Other effectors: n-hexane, n-heptane, n-nonane, n-decane, 3-methylheptane.
[j] Other effectors: benzene, ethylbenzene, *m*-xylene, *p*-xylene, phenol.
[k] See reference (42).
[l] Other effectors: benzene, toluene, ethylbenzene, *o*-xylene, *p*-xylene, *m*-ethyltoluene, *o*-chlorotoluene, *m*-chlorotoluene, *p*-chlorotoluene.

into a heterologous host. This requires the provision of the gene coding for the regulatory protein. The constructed whole-cell living bioreporters will now activate the introduced reporter gene upon sensing their specific effector (Fig. 3B, bottom half). Examples of whole-cell living bioreporters for sensing and/or sensing plus degrading organic pollutants following the described approach are listed in Table 1.

2. WHOLE-CELL LIVING BIOREPORTERS APPLYING BIOLUMINESCENCE

2.1. General Characteristics of Bioluminescence as a Reporter Signal

Light can be measured noninvasively and sensitively without disrupting the bacterial cells and therefore enables *in situ* and on-line monitoring. Therefore many whole-cell living bioreporters are equipped with genes for bioluminescence. Bioluminescence is mediated by enzymes (luciferases) which can have a prokaryotic or eukaryotic origin (48). Because bioluminescence is a seldom trait, it can be used as a universal reporter signal in most bacterial species without any problem concerning native background activity. Bioluminescence can be quantified very sensitively by using luminometers, scintillation count-

Table 2. Characteristics of bioluminescent reporter bacteria for heavy metals

Metal/ effector[a]	Luciferase gene(s)[b]	Operator/ promoter[c]	Regulatory gene	Host strain	Detected[d]	Reference
Cd^{2+}	*luxAB* (V.h.)	P_{cad} (*cadC*)	ND[e]	*Staphylococcus aureus*	0.5 μM	(15)
CrO_4^{2-} [f]	*luxCDABE* (V.f.)	(*chrA*)	ND	*Alcaligenes eutrophus*	1 nM	(53)
Cu^{2+}	*luxCDABE* (V.f.)	ND	ND	*Escherichia coli*	1 μM	(31)
Cu^{2+}	*luxCDABE* (V.f.)	(*cupC*)	ND	*Alcaligenes eutrophus*	2 μM	(16, 17[g])
Hg^{2+}	*luxCDABE* (V.f.)	*mer* o/p (*merT*)	*merR*	*Escherichia coli*	0.5 nM[h]	(61)
Hg^{2+}	*luc* (P.p.)	*mer* o/p (*merT*)	*merR*	*Escherichia coli*	0.1 fM	(71)
SbO_2^- [i]	*luc* (P.p.)	P_{ars} (*arsR*)	*arsR*	*Staphylococcus aureus*	33 nM	(68)

[a,b,c,d] See corresponding notes in Table 1.
[e] ND, not described in the original reference.
[f] Other effectors: $Cr_2O_4^{2-}$, Cr^{3+}.
[g] Includes a list with bioreporters for Co, Cr, Ni, Tl, Zn not included in Table 2.
[h] The detection limit could be lowered to 10 pM by reducing biomass in the luciferase assay from 10^7 to 10^5 cells per ml (58).
[i] Other effectors: AsO_2^- (100 nM detected); Cd^{2+} (330 nM detected).

ers, or photographic films. The sensitivity even allows the detection of single bioluminescent bacterial cells by using a charge-coupled device (CCD) video camera connected to a microscope (64). Whole-cell reporter bacteria have also been immobilized on the tip of fiber-optic cables connected to a photomultiplier (28, 33). In this manner fluctuating salicylate and naphthalene concentrations were monitored on-line (28).

A listed overview of different bioluminescent whole-cell bioreporter bacteria for detecting specific organic pollutants or heavy metals is presented in Table 1 (organic pollutants) and Table 2 (heavy metals). Not included in these tables are bioreporters which react on environmental stress signals causing heat shock response (70), oxidative stress (6) or DNA damage (55, 72).

2.2. Prokaryotic Luciferases

2.2.1. The lux Genes. Two different strategies have been used for the construction of bacterial luciferase based reporter systems. One involves only the *luxAB* genes, the other one the complete *luxCDABE* genes (Tables 1,2). The former system requires external addition of the aldehyde to obtain bioluminescence, while the latter system generates the substrate for the luciferase itself. The *luxCDABE* genes were cloned from luminescent marine bacteria like *Vibrio fischeri* (4), *Vibrio harveyi* (14, 35, 36, 50, 51), and *Photobacterium leiognathi* (44). Bacterial luciferase is a heterodimeric (αβ) flavin monooxygenase with a molecular mass of 77 kDa. The α- (40 kDa) and β-subunits (37 kDa) are encoded by *luxA*, respectively *luxB* (49). In the light reaction, reduced riboflavin phosphate ($FMNH_2$) and a long chain fatty aldehyde containing 7 to 12 carbon atoms (RCHO) are oxidized in the presence of molecular oxygen, with the emission of blue-green light, according to:

$$FMNH_2 + RCHO + O_2 \rightarrow FMN + H_2O + RCOOH + h\nu \text{ (490 nm)}$$

The long chain fatty acids (RCOOH) generated in the light reaction are recycled to their corresponding aldehydes by a multienzyme fatty acid reductase complex. This complex contains a reductase (54 kDa), a transferase (33 kDa) and a synthetase (42 kDa) encoded by *luxC*, *luxD* and *luxE* , respectively (49).

2.2.2. Considerations when Using the lux Genes. Although the *lux* genes have often been applied in reporter bacteria they possess certain properties which under certain circumstances might be disadvantageous.

i. Bacterial luciferase is a very energy consuming enzyme with a quantum yield of only 10%, meaning that only 10% of the chemical energy will be converted into light energy. This corresponds to an energy cost of 60 ATP molecules per emitted photon (40). Metabolically inactive cells may have too little energy to drive an efficient light response. In the case of a bioreporter this would lead to underestimation of the bioavailable pollutant concentration.
ii. Overexpression of the *luxAB* genes may cause cell growth retardation (observed in *Bacillus subtilis* up to 70% of normal growth). The suggested reason for this was that bacterial luciferase binds to $FMNH_2$ even when no other substrates for the light reaction are present, thus reducing the pools of free coenzymes (41).
iii. The *luxAB* genes can influence the transcriptional activity from target promoters which are sensitive to DNA topology (24). An intrinsically curved DNA segment in the 5' coding sequence of the *luxA* gene was proposed to be responsible for this behavior.
iv. A higher bioluminescence response may result from whole-cell living reporters containing a complete *luxCDABE* operon fusion, when incubated in mixtures of effector and other noninducing hydrophobic compounds (e.g. solvents) as compared to only effector (2). This was probably due to perturbation of the membrane by the solvent, causing an increase in the cellular pools of long chain fatty acids. Since long chain fatty acids are the precursors for aldehydes in the light generating reaction, this could cause the higher observed 'unspecific' bioluminescence.
v. The luciferase from *V. harveyi* is a stable enzyme and therefore not suitable to follow changes in transcriptional activity from target promoters. However, the luciferase from *V. fischeri* decays rapidly at 38°C. Hence a mild heatshock (e.g. 30°C → 42°C for 2 min.) is expected to give a better responsiveness (28, 46).

2.3. Eukaryotic Luciferases

2.3.1. The luc Genes. Complementary DNA of the *luc* genes from the North American firefly *Photinus pyralis* (11, 20), the East European firefly *Luciola mingrelica* (21) and the luminous click beetle *Pyrophorus plagiothalamus* (11, 74) were successfully expressed in *E. coli* . Luciferase from *P. pyralis* has a molecular mass of 62 kDa and catalyses the oxidative decarboxylation of D-luciferin, a 6-hydroxybenzothiazole, to oxyluciferin in the presence of molecular oxygen, ATP, and Mg^{2+} (40) with the emission of yellow-green light, according to:

$$\text{D-luciferin} + \text{ATP} + O_2 \rightarrow \text{oxyluciferin} + \text{AMP} + CO_2 + \text{PPi} + h\nu \text{ (560 nm)}$$

2.3.2. Considerations when Using the luc Genes. The biggest advantage of using *luc* in stead of *lux* genes is the very high quantum yield (88%) of the firefly luciferase, corresponding to a cost of only one ATP per photon emitted (40). The difference in quantum yield between bacterial and firefly luciferase might have been the reason for a much lower detection limit (1 fM) in a *luc* based mercury bioreporter (71) than in an equivalent *lux* based bioreporter (58, 61) (lowest detectable concentration 10 pM). Firefly luciferases also have some drawbacks:

i. The substrate D-luciferin enters the cell membrane much less efficient than long chain aldehydes do. Therefore, sensitive measurements use cell extracts rather than whole cells, which makes the method more elaborate and limits on-line monitoring (40).
ii. Certain pollutants like 4-chlorophenol, pentachlorophenol, and Cu^{2+} exert a negative effect on the *in vitro* luciferase activity. In these cases, changes in the pollutant induced transcription from the target promoter are obscured by the direct effect of the pollutants on the luciferase activity itself (23).

3. REQUIREMENTS FOR WHOLE-CELL LIVING BIOREPORTERS

To serve as an effective tool for measuring pollutant bioavailability whole-cell living bioreporters should be sensitive, specific and respond in a quantitative manner. In addition, when on-line monitoring of (fluctuating) pollutant bioavailability is intended, the bioreporter needs to be responsive to changes in promoter activity.

3.1. Quantitative Responsiveness

Bioreporters should give a quantitative reproducible response with respect to the available pollutant concentration. In case the regulator is a cytoplasmic protein (which is the case for regulatory proteins enlisted in Table 1 and 2), it is expected that what the bacteria 'see', equals the intracellular pollutant (effector) concentration. For those reporter bacteria which cannot degrade the pollutant, the intra- and extracellular concentration will be the same. Reporter bacteria capable of simultaneously degrading a compound sense a lower intracellular concentration. However, the response of the reporter is still dependent on the extracellular concentration. Therefore, the reporter signal correlates in a specific manner with the extracellular concentration of the pollutant in the water phase. This also holds when a degradation intermediate of the pollutant is the actual effector (e.g. salicylate and not naphthalene is the effector for the *nah* system (59)). The concentration dependent reporter signal output is usually determined in calibration curves under standardized conditions. A typical calibration curve for an alkane bioreporter shows the dependency between bioluminescence response and octane concentration (Fig. 4). Light emission showed a saturation-type dependency on n-octane concentrations which could be fitted with a hyperbolic equation. At low octane concentrations (24.5 to 790 nM) a linear correlation was found to fit the data as well (66). The octane bioreporter could not degrade octane, and therefore the sensed intracellular concentration equals the extracellular concentration in the water phase. Indeed the maximum bioluminescence coincided with the maximum solubility of octane in water (6.3 μM).

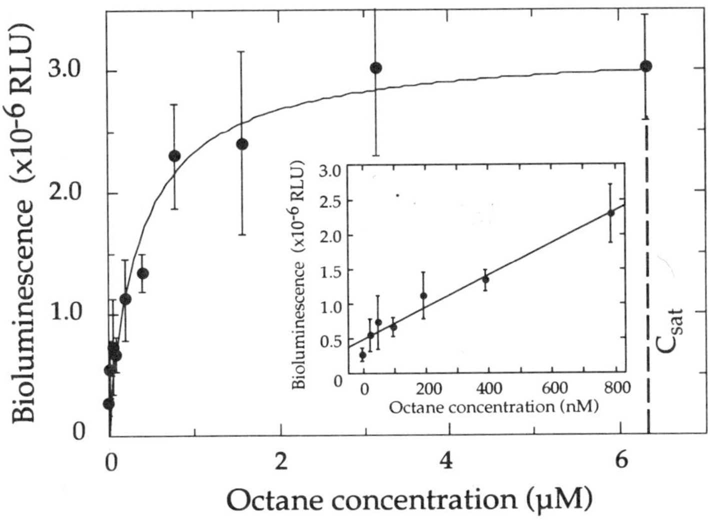

Figure 4. Bioluminescence from an *Escherichia coli* based bioreporter for the detection of linear middle-chain alkanes after a 60 minute induction period in the presence of different n-octane concentrations. C_{sat}: water-saturated n-octane concentration (6.3 μM; 720 μg/l). RLU: relative light units. (redrawn after (66)).

The performance of a whole-cell living bioreporter in environmental samples and the quantification of the bioluminescence signal is also dependent on the presence of other compounds which might either exert a stimulating (false positives) or an inhibitory (false negatives) effect on the reporter gene expression (10). False positives can be avoided by using a reporter system with a very narrow inducer spectrum range. Possible inhibitory effects can be addressed by spiking a known amount of the inducer/pollutant to an unknown sample and compare the expected reporter signal with the obtained one (10, 66). An alternative strategy, similar to the well known Microtox™ assay for monitoring the toxicity of aquatic samples (9), would be the use of isogenic reporter strains that have the luciferase genes under the control of a constitutive promoter. The degree of quenching of the bioluminescence would then be indicative for the negative effect the sample matrix exerts on the expression of the luciferase genes.

3.2. Sensitivity

Whole-cell living bioreporters should give a significant response to very low pollutant concentrations (nM-μM range). High indigenous levels of the reporter gene product in the absence of an inducer decrease the signal to noise ratio and should therefore be minimalised. To achieve this, promoter-reporter gene fusions should preferentially be present in mono-copy (e.g. integrated on the chromosome by using mini transposon vectors (19)) and be shielded from transcriptional read-through by transcription terminators (15, 66). Multi-copy plasmid-based fusions may reach a higher absolute activity in some strains but most often display a higher basal expression due to gene dosage effects. Moreover, the transcriptional regulator might be 'titrated out' in multicopy systems by the comparable high number of available operator/promoter sites. Upon induction only a limited number of promoters (those occupied by the regulatory protein) would be activated, whereas the

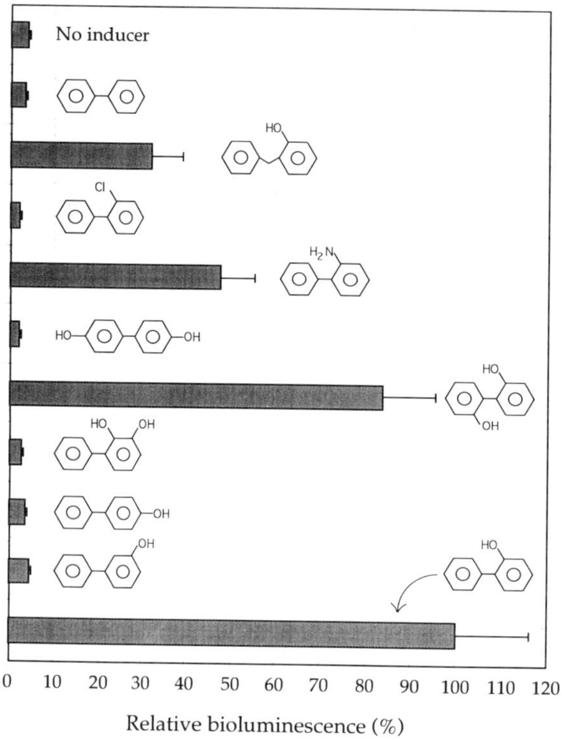

Figure 5. Bioluminescence from a *Pseudomonas azelaica* based bioreporter for 2-hydroxybiphenyl (2-HBP) in the presence of different compounds (0.2 mM). Bioluminescence was related to 2-HBP induced light emission (set 100%). Data taken from (34).

rest of the promoters would keep their basal expression level. An overview of the lowest pollutant concentrations which still gave a significant response in different whole-cell living bioreporters is presented in Tables 1 and 2. Interestingly, the lowest (tested) detection limits for organic pollutants are in the μM range (with exception for octane), but in the lower than nM-range for heavy metals. This may be due to stronger interactions between metal ions and the binding pocket in the regulatory protein, in terms of binding constants.

3.3. Specificity

The bioreporter should only respond to a narrow spectrum of pollutant (inducer) molecules in order to overcome false positive results (10). The specificity is determined by the range of molecules which are recognized by the transcriptional regulator. An example of an inducer spectrum of a bioreporter which can degrade and detect 2-hydroxybiphenyl is given in Figure 5. In this case, the bioreporter reacts mainly with a relatively small group of compounds, exemplified by a benzene ring with an hydroxyl (-OH) or amino (-NH_2) group at the first ring position and a fenylgroup (-C_6H_5) or benzylgroup (-CH_2-C_6H_5) at the *ortho* position. Other bioreporters, however, may respond to a wider range of compounds (Tables 1,2).

3.4. Adaptation to Changing Bioavailable Concentrations

For monitoring of fluctuating bioavailable concentrations whole-cell living bioreporters need to be responsive to changes in transcriptional activity from the target promoter. In that case a stable reporter gene product would be disadvantageous because it would accumulate in the cell and be indicative for the total transcriptional activity from the promoter instead of the current transcriptional activity (46). Therefore the reporter gene product should have a limited half-life. Alternatively, the reporter system should be manipulatable to conditions transiently increasing turnover rates of the reporter protein without negatively influencing transcription form the target promoter (e.g. by a mild heat shock). However, responsiveness competes with sensitivity, and a more labile reporter gene product may result in lower observable concentrations (46).

4. CONCLUDING REMARKS

Whole-cell living bioreporters are bacteria, which upon sensing the presence of a specific pollutant in the environment, express an easy detectable signal, such as bioluminescence. However, a good concept as to whether such bioreporters are actually measuring bioavailable concentrations, is still lacking. Important in this respect is to distinguish between reporter strains capable of degrading the compound and reporting simultaneously, and those which 'only' sense a compound and report. Whole-cell living bioreporters can be used for assessing the success of new bioremediation operations, and for fast and cost-effective monitoring of on-going bioremediation processes, complementing data obtained by traditional chemical analyses (GC, HPLC, AES, ICP etc.), thereby improving external process management.

Now that essentially the most significant environmental pollutants can be principally monitored by whole-cell living bioreporters, we should start using them more in field situations to improve our understanding of bioavailability. In fact, here lies a strong potential for bioreporters, which hopefully will be exploited in the near future.

ACKNOWLEDGMENTS

The work of M.C.M. J. was supported by grant 5001–044754 from the Swiss Priority Program Environment.

REFERENCES

1. Alexander, M. 1995. How toxic are toxic chemicals in soil? Environ. Sci. Technol. **29:**2713–2717.
2. Applegate, B., C. Kelly, L. Lackey, J. McPherson, S. Kehrmeyer, F.-M. Menn, P. Bienkowski, and G. Sayler. 1997. *Pseudomonas putida* B2: a *tod-lux* bioluminescent reporter for toluene and trichloroethylene co-metabolism. J. Ind. Microbiol. **18:**4–9.
3. Applegate, B. M., S. R. Kehrmeyer, and G. S. Sayler. 1998. A chromosomally based *tod-luxCDABE* whole-cell reporter for benzene, toluene, ethylbenzene, and xylene (BTEX) sensing. Appl. Environ. Microbiol. **64:**2730–2735.
4. Baldwin, T. O., J. H. Devine, R. C. Heckel, J. W. Lin, and G. S. Shadel. 1989. The complete nucleotide sequence of the *lux* regulon of *Vibrio fischeri* and the *luxABN* region of *Photobacterium leiognathi* and the mechanism of control of bacterial bioluminescence. J. Biolumin. Chemilumin. **4:**326–341.
5. Ball, W. P., and P. V. Roberts. 1991. Long-term sorption of halogenated organic chemicals by aquifer material. 2. Intraparticle diffusion. Environ. Sci. Technol. **25:**1237–1249.
6. Belkin, S., D. R. Smulski, A. C. Vollmer, T. K. Van Dyk, and R. A. LaRossa. 1996. Oxidative stress detection with *Escherichia coli* harboring a *katG'::lux* fusion. Appl. Environ. Microbiol. **62:**2252–2256.
7. Bouwer, E. J., and A. J. B. Zehnder. 1993. Bioremediation of organic compounds: putting microbial metabolism to work. Trends Biotechnol. **11:**360–367.

8. Brusseau, M. L., R. E. Jessup, and P. S. C. Rao. 1991. Nonequilibrium sorption of organic chemicals: Elucidation of rate-limiting processes. Environ. Sci. Technol. **25:**134–142.
9. Bulich, A. A. 1982. A practical and reliable method for monitoring the toxicity of aquatic samples. Process Biochem. **17:**45–47.
10. Burlage, R. S., A. V. Palumbo, A. Heitzer, and G. Sayler. 1994. Bioluminescent reporter bacteria detect contaminants in soil samples. Appl. Biochem. Biotechnol. **46:**731–740.
11. Cebolla, A., M. E. Vázquez, and A. J. Palomares. 1995. Expression vectors for the use of eukaryotic luciferases as bacterial markers with different colors of luminescence. Appl. Environ. Microbiol. **61:**660–668.
12. Chalfie, M., Y. Tu, G. Euskirchen, W. W. Ward, and D. C. Prasher. 1994. Green fluorescent protein as a marker for gene expression. Science. **263:**802–805.
13. Chaudhry, G. R., and S. Chapalamadugu. 1991. Biodegradation of halogenated organic compounds. Microbiol. Rev. **55:**59–79.
14. Cohn, D. H., A. J. Mileham, M. I. Simon, K. H. Nealson, S. K. Rausch, D. Bonam, and T. O. Baldwin. 1985. Nucleotide sequence of the *luxA* gene of *Vibrio harveyi* and the complete amino acid sequence of the α subunit of bacterial luciferase. J. Biol. Chem. **260:**6139–6146.
15. Corbisier, P., G. Ji, G. Nuyts, M. Mergeay, and S. Silver. 1993. *luxAB* gene fusions with the arsenic and cadmium resistance operons of *Staphylococcus aureus* plasmid pI258. FEMS Microbiol. Lett. **110:**231–238.
16. Corbisier, P., E. Thiry, and L. Diels. 1996. Bacterial biosensors for the toxicity assessment of solid wastes. Environ. Tox. Water Qual. **11:**171–177.
17. Corbisier, P., E. Thiry, A. Masolijn, and L. Diels. 1994. Construction and development of metal ion biosensors, p. 151–155. *In* A. K. Campbell, L. J. Kricka, and P. E. Stanley (ed.), Bioluminescence and Chemiluminescence: Fundamentals and Applied Aspects. John Wiley & Sons, Chichester, New York, Brisbane, Toronto, Singapore.
18. de Lorenzo, V., S. Fernández, M. Herrero, U. Jakubzik, and K. N. Timmis. 1993. Engineering of alkyl- and haloaromatic-responsive gene expression with mini-transposons containing regulated promoters of biodegradative pathways of *Pseudomonas*. Gene. **130:**41–46.
19. de Lorenzo, V., M. Herrero, U. Jakubzik, and K. N. Timmis. 1990. Mini-Tn*5* transposon derivatives for insertion mutagenesis, promoter probing, and chromosomal insertion of cloned DNA in Gram-negative eubacteria. J. Bacteriol. **172:**6568–6572.
20. de Wet, J. R., K. V. Wood, D. R. Helinski, and M. DeLuca. 1985. Cloning of firefly luciferase cDNA and the expression of active luciferase in *Escherichia coli*. Proc. Natl. Acad. Sci. USA. **82:**7870–7873.
21. Devine, J. H., G. D. Kutuzova, V. A. Green, N. N. Ugarova, and T. O. Baldwin. 1993. Luciferase from the east European firefly *Luciola mingrelica*: cloning and nucleotide sequence of the cDNA, overexpression in *Escherichia coli* and purification of the enzyme. Biochim. Biophys. Acta. **1173:**121–132.
22. Erickson, D. C., R. C. Loehr, and E. F. Neuhauser. 1993. PAH loss during bioremediation of manufactured gas plant site soils. Wat. Res. **27:**911–919.
23. Folkert, F. 1996. Ph.D. thesis. Rijksuniversiteit Groningen, Groningen, The Netherlands.
24. Forsberg, Å. J., G. D. Pavitt, and C. F. Higgins. 1994. Use of transcriptional fusions to monitor gene expression: A cautionary tale. J. Bacteriol. **176:**2128–2132.
25. Harms, H., and T. N. P. Bosma. 1997. Mass transfer limitation of microbial growth and pollutant degradation. J. Ind. Microbiol. **18:**97–105.
26. Harms, H., and A. J. B. Zehnder. 1995. Bioavailability of sorbed 3-chlorodibenzofuran. Appl. Environ. Microbiol. **61:**27–33.
27. Hatzinger, P. B., and M. Alexander. 1995. Effect of aging of chemicals in soil on their biodegradability and extractability. Environ. Sci. Technol. **29:**537–545.
28. Heitzer, A., K. Malachowsky, J. E. Thonnard, P. R. Bienkowski, D. C. White, and G. S. Sayler. 1994. Optical biosensor for environmental on-line monitoring of naphthalene and salicylate bioavailability with an immobilized bioluminescent catabolic reporter bacterium. Appl. Environ. Microbiol. **60:**1487–1494.
29. Heitzer, A., O. F. Webb, J. E. Thonnard, and G. S. Sayler. 1992. Specific and quantitative assessment of naphthalene and salicylate bioavailability by using a bioluminescent catabolic reporter bacterium. Appl. Environ. Microbiol. **58:**1839–1846.
30. Holliger, C., S. Gaspard, G. Glod, C. Heijman, W. Schumacher, R. P. Schwarzenbach, and F. Vazquez. 1997. Contaminated environments in the subsurface and bioremediation: organic contaminants. FEMS Microbiol. Rev. **20:**517–523.
31. Holmes, D. S., S. K. Dubey, and S. Gangolli. 1994. Development of biosensors for the detection of mercury and copper ions. Environ. Geochem. Health. **16:**229–233.

32. Huang, J. Z., and M. A. Schell. 1991. In vivo interactions of the NahR transcriptional activator with its target sequences. Inducer-mediated changes resulting in transcription activation. J. Biol. Chem. **266:**10830–10838.
33. Ikariyama, Y., S. Nishiguchi, T. Koyama, E. Kobatake, M. Aizawa, M. Tsuda, and T. Nakazawa. 1997. Fiber-optic-based biomonitoring of benzene derivatives by recombinant *E. coli* bearing luciferase gene-fused TOL-plasmid immobilized on the fiber-optic end. Anal. Chem. **69:**2600–2605.
34. Jaspers, M. C. M., D. H. M. Goslings, A. Schmid, W. A. Suske, H. E. Kohler, and J. R. van der Meer. HbpR, a novel member of the XylR/DmpR family of transcriptional activators, regulates the *hbpCAD* genes which encode 2-hydroxybiphenyl degradation in *Pseudomonas azelaica* HBP1. In preparation.
35. Johnston, T. C., K. S. Hruska, and L. F. Adams. 1989. The nucleotide sequence of the *luxE* gene of *Vibrio harveyi* and a comparison of the amino acid sequences of the acyl-protein synthetases from *V. harveyi* and *V. fischeri*. Biochem. Biophys. Res. Commun. **163:**93–101.
36. Johnston, T. C., R. B. Thompson, and T. O. Baldwin. 1986. Nucleotide sequence of the *luxB* gene of *Vibrio harveyi* and the complete amino acid sequence of the β subunit of bacterial luciferase. J. Biol. Chem. **261:**4805–4811.
37. King, J. M. H., P. M. DiGrazia, B. Applegate, R. Burlage, J. Sanseverino, P. Dunbar, F. Larimer, and G. S. Sayler. 1990. Rapid, sensitive bioluminescent reporter technology for naphthalene exposure and biodegradation. Science. **249:**778–781.
38. Kobatake, E., T. Niimi, T. Haruyama, Y. Ikariyama, and M. Aizawa. 1995. Biosensing of benzene derivatives in the environment by luminescent *Escherichia coli*. Biosensors Bioelectronics. **10:**601–605.
39. Kobayashi, H., and B. E. Rittmann. 1982. Microbial removal of hazardous organic compounds. Environ. Sci. Technol. **16:**170A-183A.
40. Koncz, C., W. H. R. Langridge, O. Olsson, J. Schell, and A. A. Szalay. 1990. Bacterial and firefly luciferase genes in transgenic plants: Advantages and disadvantages of a reporter gene. Dev. Genet. **11:**224–232.
41. Lampinen, J., L. Koivisto, M. Wahlsten, P. Mäntsälä, and M. Karp. 1992. Expression of luciferase genes from different origins in *Bacillus subtilis*. Mol. Gen. Genet. **232:**498–504.
42. Lau, P. C. K., Y. Wang, A. Patel, D. Labbé, H. Bergeron, R. Brousseau, Y. Konishi, and M. Rawlings. 1997. A bacterial basic region leucine zipper histidine kinase regulating toluene degradation. Proc. Natl. Acad. Sci. USA. **94:**1453–1458.
43. Leahy, J. G., and R. R. Colwell. 1990. Microbial degradation of hydrocarbons in the environment. Microbiol. Rev. **54:**305–315.
44. Lee, C. Y., R. B. Szittner, and E. A. Meighen. 1991. The *lux* genes of the luminous bacterial symbiont, *Photobacterium leiognathi*, of the ponyfish. Nucleotide sequence, difference in gene organization, and high expression in mutant *Escherichia coli*. Eur. J. Biochem. **201:**161–167.
45. Leveau, J. H. J., and J. R. van der Meer. 1996. The *tfdR* gene product can successfully take over the role of the insertion element-inactivated TfdT protein as a transcriptional activator of the *tfdCDEF* gene cluster, which encodes chlorocatechol degradation in *Ralstonia eutropha* JMP134(pJP4). J. Bacteriol. **178:**6824–6832.
46. Loper, J. E., and S. E. Lindow. 1997. Reporter gene systems useful in evaluating in situ gene expression by soil- and plant-associated bacteria, p. 482–492. *In* C. J. Hurst, G. R. Knudsen, M. J. McInerney, L. D. Stetzenbach, and M. V. Walter (ed.), Manual of environmental microbiology. ASM Press, Washington, D. C.
47. Luthy, R. G., A. Ramaswami, S. Ghoshal, and W. Merkel. 1993. Interfacial films in coal tar nonaqueous-phase liquid-water systems. Environ. Sci. Technol. **27:**2914–2918.
48. Meighen, E. A. 1993. Bacterial bioluminescence: organization, regulation, and application of the *lux* genes. FASEB. **7:**1016–1022.
49. Meighen, E. A. 1991. Molecular biology of bacterial bioluminescence. Microbiol. Rev. **55:**123–42.
50. Miyamoto, C. M., M. Boylan, A. F. Graham, and E. A. Meighen. 1988. Organization of the *lux* structural genes of *Vibrio harveyi*. Expression under the T7 bacteriophage promoter, mRNA analysis, and nucleotide sequence of the *luxD* gene. J. Biol. Chem. **263:**13393–13399.
51. Miyamoto, C. M., A. F. Graham, and E. A. Meighen. 1988. Nucleotide sequence of the *luxC* gene and the upstream DNA from the bioluminescent system of *Vibrio harveyi*. Nucleic Acids Res. **16:**1551–1562.
52. Ogram, A. V., R. E. Jessup, L. T. Ou, and P. S. C. Rao. 1985. Effects of sorption on biological degradation rates of (2,4-dichlorophenoxy)acetic acid in soils. Appl. Environ. Microbiol. **49:**582–587.
53. Peitzsch, N., G. Eberz, and D. H. Nies. 1998. *Alcaligenes eutrophus* as a bacterial chromate sensor. Appl. Environ. Microbiol. **64:**453–458.
54. Prasher, D. C., V. K. Eckenrode, W. W. Ward, F. G. Prendergast, and M. J. Cormier. 1992. Primary structure of the *Aequorea victoria* green-fluorescent protein. Gene. **111:**229–233.

55. Ptitsyn, L. R., G. Horneck, O. Komova, S. Kozubek, E. A. Krasavin, M. Bonev, and P. Rettberg. 1997. A biosensor for environmental genotoxin screening based on an SOS *lux* assay in recombinant *Escherichia coli* cells. Appl. Environ. Microbiol. **63:**4377–4384.
56. Ramaswami, A., and R. G. Luthy. 1997. Measuring and modeling physicochemical limitations to bioavailability and biodegradation, p. 721–729. *In* C. J. Hurst, G. R. Knudsen, M. J. McInerney, L. D. Stetzenbach, and M. V. Walter (ed.), Manual of environmental microbiology. ASM Press, Washington, D.C.
57. Ramos, J. L., S. Marqués, and K. N. Timmis. 1997. Transcriptional control of the *Pseudomonas* TOL plasmid catabolic operons is achieved through an interplay of host factors and plasmid-encoded regulators. Annu. Rev. Microbiol. **51:**341–373.
58. Rasmussen, L. D., R. R. Turner, and T. Barkay. 1997. Cell-density-dependent sensitivity of a *mer-lux* bioassay. Appl. Environ. Microbiol. **63:**3291–3293.
59. Schell, M. A. 1985. Transcriptional control of the *nah* and *sal* hydrocarbon-degradation operons by the *nahR* gene product. Gene. **36:**301–309.
60. Scribner, S. L., T. R. Benzing, S. Sun, and S. A. Boyd. 1992. Desorption and bioavailability of aged simazine residues in soil from a continuous corn field. J. Environ. Qual. **21:**115–120.
61. Selifonova, O., R. Burlage, and T. Barkay. 1993. Bioluminescent sensors for detection of bioavailable Hg(II) in the environment. Appl. Environ. Microbiol. **59:**3083–3090.
62. Selifonova, O. V., and R. W. Eaton. 1996. Use of an *ipb-lux* fusion to study regulation of the isopropylbenzene catabolism operon of *Pseudomonas putida* RE204 and to detect hydrophobic pollutants in the environment. Appl. Environ. Microbiol. **62:**778–783.
63. Shingler, V., and T. Moore. 1994. Sensing of aromatic compounds by the DmpR transcriptional activator of phenol-catabolizing *Pseudomonas* sp. strain CF600. J. Bacteriol. **176:**1555–1560.
64. Silcock, D. J., R. N. Waterhouse, L. A. Glover, J. I. Prosser, and K. Killham. 1992. Detection of a single genetically modified bacterial cell in soil by using charge coupled device-enhanced microscopy. Appl. Environ. Microbiol. **58:**2444–2448.
65. Steinberg, S. M., J. J. Pignatello, and B. L. Sawhney. 1987. Persistence of 1,2-dibromoethane in soils: Entrapment in intraparticle micropores. Environ. Sci. Technol. **21:**1201–1208.
66. Sticher, P., M. C. M. Jaspers, K. Stemmler, H. Harms, A. J. B. Zehnder, and J. R. van der Meer. 1997. Development and characterization of a whole-cell bioluminescent sensor for bioavailable middle-chain alkanes in contaminated groundwater samples. Appl. Environ. Microbiol. **63:**4053–4060.
67. Stucki, G., and M. Alexander. 1987. Role of dissolution rate and solubility in biodegradation of aromatic compounds. Appl. Environ. Microbiol. **53:**292–297.
68. Tauriainen, S., M. Karp, W. Chang, and M. Virta. 1997. Recombinant luminescent bacteria for measuring bioavailable arsenite and antimonite. Appl. Environ. Microbiol. **63:**4456–4461.
69. Thomas, J. M., J. R. Yordy, J. A. Amador, and M. Alexander. 1986. Rates of dissolution and biodegradation of water-insoluble organic compounds. Appl. Environ. Microbiol. **52:**290–296.
70. Van Dyk, T. K., W. R. Majarian, K. B. Konstantinov, R. M. Young, P. S. Dhurjati, and R. A. LaRossa. 1994. Rapid and sensitive pollutant detection by induction of heat shock gene-bioluminescence gene fusions. Appl. Environ. Microbiol. **60:**1414–1420.
71. Virta, M., J. Lampinen, and M. Karp. 1995. A luminescence-based mercury biosensor. Anal. Chem. **67:**667–669.
72. Vollmer, A. C., S. Belkin, D. R. Smulski, T. K. Van Dyk, and R. A. LaRossa. 1997. Detection of DNA damage by use of *Escherichia coli* carrying *recA'*::*lux*, *uvrA'*::*lux*, or *alkA'*::*lux* reporter plasmids. Appl. Environ. Microbiol. **63:**2566–2571.
73. Weissenfels, W. D., H. Klewer, and J. Langhoff. 1992. Adsorption of polycyclic aromatic hydrocarbons (PAHs) by soil particles: influence on biodegradability and biotoxicity. Appl. Microbiol. Biotechnol. **36:**689–696.
74. Wood, K. V., Y. Amy Lam, H. H. Seliger, and W. D. McElroy. 1989. Complementary DNA coding click beetle luciferases can elicit bioluminescence of different colors. Science. **244:**700–702.
75. Wu, S., and P. M. Gschwend. 1986. Sorption kinetics of hydrophobic compounds to natural sediments and soils. Environ. Sci. Technol. **20:**717–725.

15

ACTIVE BIOLOGICAL CONTAINMENT FOR BIOREMEDIATION IN THE RHIZOSPHERE

Cayo Ramos,[1,2] Lázaro Molina,[2] M. Carmen Ronchel,[2] Bjarke Christensen,[1] Juan L. Ramos,[2] and Søren Molin[1]

[1]Department of Microbiology, The Technical University of Denmark
Building 301, 2800 Lyngby, Denmark
[2]Department of Biochemistry and Molecular and Cellular Biology of Plants
Estación Experimental del Zaidín
Consejo Superior de Investigaciones Científicas
Apdo. 419, 18008 Granada, Spain

1. INTRODUCTION

Mineralization of organic molecules by microbes is essential for the carbon cycle to operate. The massive deposition of organic compounds in the environment, and the introduction of xenobiotics into the biosphere, lead to unidirectional fluxes, which result in the persistence of a number of chemicals in the biosphere, and thus constitute a source of pollution. Molecular biology offers tools to optimize the biodegradative capacity of microorganisms and accelerate the evolution of "new" activities (Ramos *et al.*, 1994). In several cases, genetically modified microorganims (GMMs) have been designed for the bioremediation of specific xenobiotic compounds and their deliberate release into the environment is expected (Ramos *et al.*, 1995).

The use of plant-microbial systems for bioremediation is a rapidly developing technology which, in some cases, may offer solutions for *in situ* treatment of contaminated soils. There are indications that plants may enhance the biodegradation of xenobiotics by increasing microbial growth and activity in the rhizosphere as a result of utilization of nutrients provided by root exudates (Reddy and Sethunathan, 1983; Walton and Anderson, 1990; Alvey and Crowley, 1996). Although the benefits for biological cleaning of a polluted site are obvious, the long-term behaviour of the released microbes in open environments is unknown. This issue raises a series of public and scientific concerns. The successful and "safe" application of plant-microbial systems for environmental clean up of contaminated sites requires that the beneficial effect/s of the rhizosphere on the survival of the introduced microbial strain and the persistence of the microbes after the biodegradative activity has been completed, are both taken into account.

Novel Approaches for Bioremediation of Organic Pollution, edited by
Fass *et al.* Kluwer Academic / Plenum Publishers, New York, 1999.

2. BIODEGRADATIVE POTENTIAL OF *Pseudomonas putida* STRAINS

Bacteria belonging to the genus *Pseudomonas* exibit a wide range of metabolic activities against natural and xenobiotic compound (Ramos *et al.*, 1994). *Pseudomonas putida* KT2440 is a DNA restriction negative strain derived from the soil bacterium *P. putida* mt-2, the natural host of the archetypal TOL plasmid pWWO (Worsey and Williams, 1974). This strain grows on toluene and *m*- and *p*-xylene among other hydrocarbons widely used as organic solvents, and has been shown to be highly versatile in its acquisition of recombinat DNA for expansion of the catabolic potential to remove pollutants (Ramos *et al.*, 1994). Furthermore, the ability of *P. putida* KT2440 and its derivatives to survive well in soil (Ramos *et al.*,1991; Duque *et al.*, 1993), and the availability of active biological containment systems for this group of bacteria (Jensen *et al.*, 1993; Ramos *et al.*, 1995; Ronchel *et al.*, 1995) make them of special interest for the development of microbial inoculants to be use in field assays for the bioremediation of polluted sites.

We have used *P. putida* strains able to degrade toluene as model strains to study bacterial survival in the rhizosphere under different conditions. The studies performed, from laboratory tests to field assays, have mainly concentrated on evaluating the colonization and survival of the strains in nonvegetated soil and in the rhizosphere of different plants, and on the construction and evaluation of a model system of chemical induction of suicide as a biological containment principle for the biodegradation of xenobiotics. The results obtained are summarized below.

3. RHIZOSPHERE COLONIZATION BY *P. putida*

In order to better understand the response to chemical pollution of microbial communities in the rhizosphere, information about survival, activity and spatial distribution of the different subpopulations colonizing this ecosystem is necessary. A number of very sensitive and specific molecular methods, complementary to the isolation of bacteria on selective media, have been developed during the last years for monitoring bacterial cells at the site they are colonizing. *In situ* non-destructive luminescence-based techniques detect labeled cells at the site of colonization (for reviews, see Stewart and Williams, 1992; Prosser, 1994). During the last years, several authors have reported the application of this technique to localize bacterial strains in the rhizosphere of different plants (de Weger, 1991; Rattray, 1995; Chabot, 1996). More recently, the green fluorescent protein (GFP) from the jellyfish *Aequorea victoria* (Chalfie *et al.*, 1994), and fluorescence-labeled probes such as antibodies and rRNA-targeted oligonucleotides, have been introduced as tools in microbial ecology. Scanning confocal laser microscopy (SCLM) have been used for the identification and localization of microbes in the rhizosphere combined with the use of fluorescent antibodies (Schloter *et al.*, 1993; Assmus *et al.*, 1997; Hansen *et al.*, 1997), oligonucleotides probes (Assmus *et al.*, 1995; Assmus *et al.*, 1997), or GFP-tagged cells (Bloemberg *et al.*, 1997).

3.1. Survival and Localization of a Bioluminescent *P. putida* Strain in Soil and Rhizosphere

Pseudomonas putida strain P1 carries a mini-Tn*5*-*lux*AB transposon inserted in the chromosome making the cells luminescent after exposure to *n*-decyl aldehyde. The strain

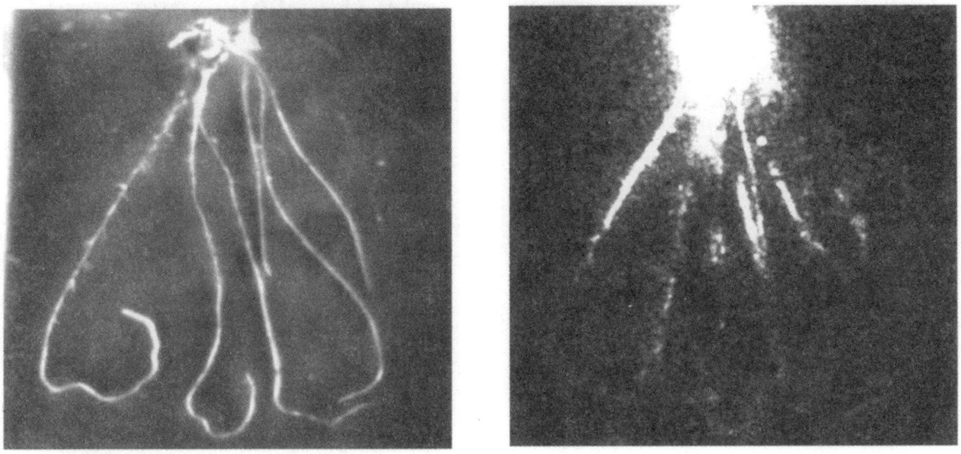

Figure 1. *In situ* visualization of *P. putida* P1 colonizing a barely root, by photon count imaging with a charged-couple device camera. Bright-field exposure (3 s) (A) and dark-field exposure (60 s) (B) of a 12-day-old root after the addition of *n*-decyl aldehyde. Bright- and dark-field images were processed with Adobe Photoshop software.

otherwise shows wild-type growth characteristics under laboratory conditions (Sternberg *et al.*, 1997). Under controlled temperature and humidity conditions, the behaviour of *P. putida* P1 and its parental strain KT2442 in nonsterile soil was similar. *P. putida* strains P1 or KT2442 were introduced at a cell load of 10^5 CFU/g of soil in pots containing a fluvisol soil relatively rich in organic matter. Under these conditions, both strains maintained constant cell numbers for at least 3 months. However, under open environmental conditions, the number of *P. putida* P1 cells decreased from an initial density of about 10^6 CFU/g of soil to about 10^2–10^3 CFU/g of soil after 50 days. At later times, the number of cells was below our detection limits by plating, although bioluminescence could be detected after selective enrichment of soil samples in LB medium containing the appropriate antibiotics.

Root colonization by *P. putida* P1 was assayed by allowing seeds coated with the strain to germinate in vermiculite. Detection of luminescence from *P. putida* P1 cells in the rhizosphere of 7- to 14-days-old barley plants was assessed by photon count imaging. Light emission was most intense in the upper root system, although other zones of intense light emission were observed throughout the entire root system (Fig. 1). *P. putida* P1 also colonized the roots of tomato, spinach, bean (*Vicia faba*) and corn.

Survival of *P. putida* in the rhizosphere under open environmental conditions in a field release assay was performed during the Winter of 1996–1997. Seeds of *Vicia faba* coated with approximately 10^6 CFU/seed of *P. putida* KT2442 were planted on a controlled site located in Granada, Spain. After the appearance of the first true leaves, bacteria were counted in the plant rhizosphere and in the surrounding soil (bulk soil). The number of *P. putida* cells in the rhizosphere soil was in the order of 2–5 x 10^6 CFU throughout the investigation. The cell density of the introduced strain in bulk soil was about two orders of magnitude below that found in the rhizosphere soil. These results indicate that *P. putida* strain KT2442 and its bioluminescent derivative *P. putida* P1 colonize effectively the rhizosphere of different plants, and under open environmental conditions the parental strain become established better in the rhizosphere than in the surrounding soil.

3.2. Localization of Single Cells and Spatial Distribution of *P. putida* in the Rhizosphere

Studies of the spatial location of bacteria in the rizosphere were performed using the GFP reporter system and scanning confocal laser microscopy. The delivery plasmid, pJBA29 (Andersen *et al.*, 1998), carrying a mini-Tn*5*-*gfp* transposon (Km^R), was transferred from *E. coli* MW1190-λpir to *P. putida* SMØ116 (Nal^R), the main toluene degrader isolated from a biofilm established for the removal of this aromatic compound (Møller *et al.*, 1996). Nal^R Km^R transconjugants of *P. putida* SMØ116 appeared with a frequency of about 10^{-5} per recipient. Among the transconjugants showing a Gfp^+ phenotype, a clone containing a highly stable insertion of the mini-Tn*5*-*gfp* on the chromosome, called *P. putida* CRR300, was selected for further studies. Green fluorescence from cells of *P. putida* strain CRR300 was easily visualized on the root surface following barley seedlings inoculation. After 2–7 days of seedling growth on an agar system, green fluorescence was detected throughout the entire root system. Single cells, as well as cell assemblages expressing GFP were detected at the border of adjacent plant cells (Fig. 2) and at the root hair zone. The red autofluorescence exhibited by the root material was used to visualize the plant material.

4. BIOLOGICALLY CONTAINED *P. putida* FOR ENVIRONMENTAL APPLICATIONS IN BIOREMEDIATION

The full exploitation of the recombinant derivatives of *P. putida* strains for bioremediation will inevitably involve their release into the environment, where non-controllable conditions prevail. One way to decrease the persistence of microbes in the environment is to provide them with active biological containment (ABC) systems. ABC systems are based on the control of the expression of a lethal function (i.e., a pore forming protein or a nuclease) via sensory systems that recognize physical or chemical signals from the surrounding environment (Ramos *et al.*, 1995). Contreras *et al.* (1991), designed the first circuit for the biological containment of bacteria that degrade pollutants. The system was based on the well-characterized regulatory circuit for the expression of the *meta*-cleavage pathway of *P. putida* TOL

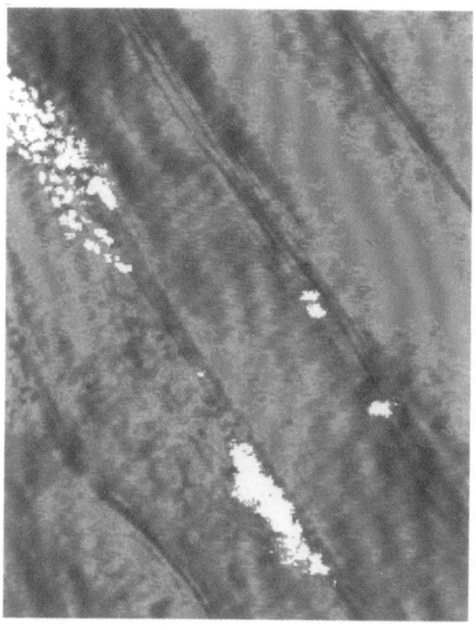

Figure 2. *In situ* visualization of GFP expression (white color) from cells of *P. putida* CRR300 at the border of adjacent root cells of a 3-day-old barley seedling, magnification X630. Microscopic observation and image adquisition were performed on a Leica TCS4D confocal microscope equipped with a argon-neon laser and 2 detectors for simultaneously monitoring of green and red fluorescence. Multichannel SFP (*s*imulated *f*luorescence *p*rojection) images were generated using IMARIS software running on a SiliconGraphics Indigo 2 workstation. Images were processed for display using Adobe Photoshop software. The red autofluorescence exhibited by the root material was used to visualize the root surface (grey tones).

plasmid pWWO, the *xylS* gene and its cognate promoter Pm (Ramos *et al.*, 1997), and a fusion of the lactose operon promoter to the *gef* gene of *Escherichia coli*, which encodes a membrane protein that collapses the cell membrane potential by generating pores (Molin *et al.*, 1993). In the presence of XylS effectors, i.e. alkyl-, chloro- and other halosubtituted benzoates (Ramos *et al.*, 1986), expression of the Gef protein is prevented, and the strains survive and remove the pollutant. Upon exhaustion of the target compound, or spread of the microbe to a nonpolluted site, the lack of induction of Pm results in the absence of the LacI protein in the host cell, which leads to the expression of the lethal protein and cell death. This model system has been shown to be functional in *E. coli* (Contreras *et al.*, 1991) and *P. putida* (Jensen *et al.*, 1993; Ronchel *et al.*, 1995) under laboratory conditions. The survival of a *P. putida* strain carrying this biological containment system have also been tested in soil microcosms, both in the presence and in the absence of the XylS effector *m*-methylbenzoate. The results obtained indicated that the system was also functional in soil under controlled conditions (Jensen *et al.*, 1993; Ronchel *et al.*, 1995).

The behaviour of the biologically contained *P. putida* strain CMC4, and its control strain *P. putida* EEZ32 was also analyzed under open conditions during the Autumn-Winter period of 1995–1996 in non-planted soil and soil planted with broad bean (*Vicia faba*), both in the presence and in the absence of *m*-methylbenzoate. In soil without plants the numbers of CFU/g of soil for both strains decreased with time. The decrease in the number of CFU/g of soil of the control strain was independent of the presence of *m*-methylbenzoate, and eventually was below the detection limit by plating (i.e., 100 CFU/g of soil) about 40 days after the start of the assay. The contained strain, *P. putida* CMC4, went below our detection limit faster in soil without *m*-methylbenzoate (15 days after the initiation of the assay) than in soil containing the aromatic compound (35 days after the start of the assay).

The control strain *P. putida* EEZ32 established at a level of about 10^3–10^6 CFU/g of rhizosphere soil, both in the absence and presence of *m*-methylbenzoate throughout the time of the assay. However, the number of CFU/g of rhizosphere soil of the contained strain decreased with time, as this was more pronounced in the absence of *m*-methylbenzoate, so that after 100 days the number of CFU fell below the detection limit. In the presence of the aromatic, the number of CFU/g rhizosphere soil was 1 to 2 orders of magnitude higher in any given sample, and at the end of the assay we found about 10^3 CFU/g of rhizosphere soil. These results indicate that the system tested also function as expected under open environmental conditions during the Autum-Winter period in the rhizophere of broad bean.

ACKNOWLEDGMENTS

This work was partly supported by GX-Biosystems funds, and by a grant from the Spanish CICYT (BIO97-0641). The work by C.R. in Denmark during 1995 was supported by EMBO fellowship ASTF8056, and during 1997 by CEC grant BIO4CT975038.

REFERENCES

Alvey, S. and Crowley, D.E. (1996). Survival and activity of an antrazine-mineralizing bacterial consortium in rhizosphere soil. Environ. Sci. Technol. **30**, 1596–1603.

Andersen, J.B., Sternberg, C., Poulsen, L.K., Bjørn, S.P., Givskov, M., and Molin, S. (1998). New unstable variants of green fluorescent protein for studies of transient gene expression in bacteria. Appl. Environ. Microbiol., in the press.

Assmus, B., Hutzler, P., Kirchof, G., Amann, R., Lawrence, J.R., and Hartmann, A.(1995). *In situ* localization of *Azospirilum brasilensis* in the rhizosphere of wheat with fluorescently labeled rRNA-targeted oligonucleotide probes and scanning confocal laser microscopy. Appl. Environ. Microbiol. **61**, 1013–1019.

Assmus, B., Scholter, M., Kirchhof, G., Hutzler, P., and Hartmann, A. (1997). Improved *in situ* tracking of rhizosphere bacteria using dual staining with fluorescence-labeled antibodies and rRNA-targeted oligonucleotides. Microb. Ecol. **33**, 32–40.

Bloemberg, G.V., O´Toole G.A., Lugtenberg B.J.J., and Kolter, R. (1997). Green fluorescent protein as a marker for *Pseudomonas* spp. Appl. Environ. Microbiol. 63, 4543–4551.

Chabot, R., Antoun, H., Kloepper, J.W., and Beauchamp, C.J. (1996). Root colonization of maize and lettuce by bioluminescent *Rhizobium leguminosarum* biovar phaseoli. Appl. Environ. Microbiol. **62**, 2767–2772.

Chalfie, M., Tu, Y., Euskirchen, G., Ward, W.W., and Prasher, D.C. (1994). Green fluorescent protein as a marker for gene expression. Science **263**, 802–805.

Contreras, A., Molin, S., and Ramos, J.L. (1991). Conditional-suicide containment system for bacteria which mineralize aromatics. Appl. Environ. Microbiol. **57**, 1504–1508.

Duque, E., Marqués, S., and Ramos, J.L. (1993). Mineralization of *p*-methyl-^{14}C-benzoate in soils by *Pseudomonas putida* (pWW0). Microb. Releases **2**, 175–177.

Hansen, M., Kragelund, L., Nybroe, O., and Sørensen, J. (1997). Early colonization of barley roots by *Psedomonas fluorescens* studied by immunofluorescence technique and scanning confocal laser microscopy. FEMS Microbiol. Ecol. **23**, 353–360.

Jensen, L.B., Ramos, J.L., Kaneva, Z., and Molin, S. (1993). A substrate dependant biological containment system for *Pseudomonas putida* based on the *Escherichia coli gef* gene. Appl. Environ. Microbiol. **57**, 85–92.

Molin, S., Boe, L., Jensen, L.B., Kristensen, C.S., Givskov, M., Ramos, J.L. and Bej, A.K. (1993). Suicidal genetic elements and their use in biological containment of bacteria. Ann. Rev. Microbiol. **47**, 139–166.

Møller, S., Pedersen, A.R., Poulsen, L.K., Arvin, E., and Molin, S. (1996). Activity and three-dimensional distribution of toluene-degrading *Pseudomonas putida* in a multispecies biofilm assessed by quantitative *in situ* hybridization and scanning confocal laser microscopy. Appl. Environ. Microbiol. **62**, 4632–4640.

Prosser, J.I. (1994). Molecular marker systems for detection of geneticlly engineered micro-organims in the environment. Microbiology **140**, 5–17.

Ramos, J.L., Stolz, A., Reineke, W., and Timmis, K.N. (1986). Altered effector specificities in regulators of gene expression: TOL plasmid *xyl*S mutants and their use to engineer expansion of the range of aromatics degraded by bacteria. Proc. Natl. Acad. Sci. USA **83**, 8467–8471.

Ramos, J.L., Duque, E., and Ramos-González, M.I. (1991). Survival in soils of an herbicide-resistant *Pseudomonas putida* strain bearing a recombinant TOL plasmid. Appl. Environ. Microbiol. **57**, 260–266.

Ramos, J.L., Díaz, E., Dowling, D., de Lorenzo, V., Molin, S., O'Gara, F., Ramos, C., and Timmis, K.N. (1994). The behavior of bacteria designed for biodegradation. Bio/Technology **12**, 1349–1356.

Ramos, J.L., Andersson, P., Jensen, L.B., Ramos, C., Ronchel, M.C., Díaz, E., Timmis, K.N., and Molin, S. (1995). Suicide microbes on the loose. Bio/Technology **13**, 35–37.

Ramos, J.L., Marqués, S., and Timmis, K.N. (1997). Transcriptional control of the *Pseudomonas* TOL plasmid catabolic operons is achieved through an interplay of host factors and plasmid encoded regulators. Ann. Rev. Microbiol. **51**, 341–373.

Rattray, E.A.S., Prosser, J. I. L., Glover, A., and Killham, K. (1995). Characterization of rhizosphere colonization by luminescent *Enterobacter cloacae* at the population and single-cell levels. Appl. Environ. Microbiol. **61**, 2950–2957.

Reddy, B.R., and Sethunathan, N. (1983). Mineralization of parathion in the rice rhizosphere. Appl. Environ. Microbiol. **45**: 826–829.

Ronchel, M.C., Ramos, C., Jensen, L.B., Molin, S., and Ramos, J.L. (1995). Construction and behaviour of biologically contained bacteria for environmental applications in bioremediation. Appl. Environ. Microbiol. **61**, 2990–2994.

Schloter, M., Borlinghaus, R., Bode, W., and Hartmann, A. (1993). Direct identification, and localization of *Azospirillum* in the rhizosphere of wheat using fluorescence-labelled monoclonal antibodies and confocal scanning laser microscopy. J. Microscopy **171**, 173–177.

Sternberg, C., Eberl, L., Poulsen, L.K., and Molin, S. (1997). Detection of bioluminescence from individual bacterial cells: a comparison of two different low-light imaging systems. J. Biolumin. Chemilumin. **12**, 7–13.

Stewart, G.S.A.B. and Williams, P. (1992). *lux* genes and the applications of bacterial bioluminescence. J. Gen. Microbiol. **138**, 1289–1300.

Walton, B.T. and Anderson, T.A. (1990). Microbial degradayion of trichloroethylene in the rhizopshere: potential application to biological remediation of waste sites. Appl. Environ. Microbiol. **56**: 1012–1016.

de Weger, L.A., Dunbar, P., Mahafee, W.F., Lugtenberg, B.J.J., and Sayler, G.S. (1991). Use of bioluminescence markers to detect *Pseudomonas* spp. in the rhizosphere. Appl. Environ. Microbiol. **57**, 3641–3644.

Worsey, M.J. and Williams, P.A. (1974). Metabolism of toluene and xylenes by *Pseudomonas putida* (*arvilla*) *mt*-2: evidence for a new function of the TOL plasmid. J. Bacteriol. **124**, 7–13.

16

BIOREMEDIATION OF HIGH MOLECULAR WEIGHT PAHs

Application of Techniques in Bioaugmentation and Bioavailability Enhancement

P. H. Pritchard,[1] J. Jones-Meehan,[1] J.G. Mueller,[2] and W. Straube[1]

[1]Naval Research Laboratory
Environmental Quality Science Section
Washington D.C. 20375
[2]Dames and Moore
Rolling Meadows, Illinois 60008

ABSTRACT

Land farming is a common treatment option for PAH-contaminated soils. The process is based on the degradation of PAHs by natural microbial communities in the presence of aeration (tilling), fertilizer addition, and physical mixing. In general, considerable degradation of total PAHs occurs but the removal of the high molecular weight (HMW) PAHs (four or more aromatic rings), those that are of most concern from a regulatory and remediation standpoint, is very problematic. This is due many times to the absence of a key metabolic activity, namely the ability to cometabolize HMW PAHs, and to the lack of availability of the PAHs to the degrading microorganisms. The problem can be corrected, we believe, by bioaugmentation with two key microorganisms; one that adds significant cometabolic capability and another that produces a biosurfactant. The characteristics of the organisms and the practicality of introducing them into soil are discussed. Although bioaugmentation will add costs to the treatment process, it is likely that the improved reliability for removing HMW PAHs to below regulatory limits will eventually justify the added cost.

INTRODUCTION

Bioremediation offers an effective and relatively inexpensive technology for the clean up of polyaromatic hydrocarbon (PAH)-contaminated soils. The technology has been

Novel Approaches for Bioremediation of Organic Pollution, edited by Fass *et al.* Kluwer Academic / Plenum Publishers, New York, 1999.

applied to a variety of soils and contaminant types (creosote, coal tar, petroleum) and is generally preferred over nonbiological approaches such as stabilization, chemical oxidation and incineration (Mueller et al., 1995). Despite considerable success, bioremediation of PAH-contaminated soil has not performed consistently because of difficulties in degrading PAHs composed of four or more fused aromatic rings (referred to hereafter as high molecular weight ,HMW, PAHs), or those that are potentially carcinogenic. This is due, in part, to the problems of unpredictable metabolic capabilities of natural microbial communities and availability of the HMW PAHs to the degrading microorganisms. We have pursued, therefore, the use of bioaugmentation as a potential solution to these problems since we believe that supplementations to the activities of natural microbial communities can build reliability and effectiveness into the biological treatment process for soil. This paper describes our concepts, justifications and current supporting data for the use of bioaugmentation in land farming treatment of PAH-contaminated soils.

To this end, we have been studying microorganisms that grow with HMW PAHs and other hydrocarbons with the idea that selecting and introducing the right combination of organisms into the treatment system will provide the needed consistency and eventually justify the added cost. Given this potential, we must thoroughly understand the mechanisms by which microbial communities degrade HMW PAHs. We believe these mechanisms depend heavily on fortuitous or cometabolic process, although this has not been firmly established (Mueller et al., 1995). If we are to optimize the safe and effective performance of treatability systems based on cometabolic processes, particularly at field-scale levels, then extensive information is needed on how to control, enhance and maintain these cometabolic processes with microbial communities.

In addition, it is also clear that the bioavailability of HMW PAHs plays a key role in their degradation and in bioremediation strategies. As a contaminated soil ages, PAHs tend to move into the deeper recesses of soil particles, soil aggregates, and the organic matter sorbed to soil particle surfaces (Pignatello and Xing, 1996; Zhang et al., 1998). Consequently desorption is usually described as a rapid initial release of PAHs that are close to the surface and a very slow release of PAHs that are more deeply sorbed. Although it appears that considerable amounts of sorbed PAHs will eventually leach out over long periods (years), this time frame is usually too long for shorter term remediation techniques, such as land treatment (months), to be cost effective. However, if strategic modifications of bioremediation techniques can be made to increase desorption rates over the shorter treatment term, then the added amount of degradation from bioaugmentation and bioavailability enhancement may very well mean the difference between meeting cleanup goals or not in a reasonable time period.

NATURAL ATTENUATION AND LAND FARMING

Treatment of PAH contaminated soils using natural attenuation generally employs either land farming (solid phase treatment) and reactors (slurry phase treatment) (Mueller et al., 1991 a,b, 1993; Harmsen, 1991; Yare, 1991). In both cases, the main emphasis is to stimulate the degradation capabilities of natural microbial communities by providing oxygen and nitrogen, and to encourage physical mixing of the soil to distribute the contaminants over a greater surface area. The resulting degradation usually takes place in two phases; an initial phase of rapid degradation (weeks to months) involving the most soluble or bioavailable hydrocarbons, usually resulting in a significant reduction in total PAHs, and a second phase of much slower degradation (years) that can eventually remove some

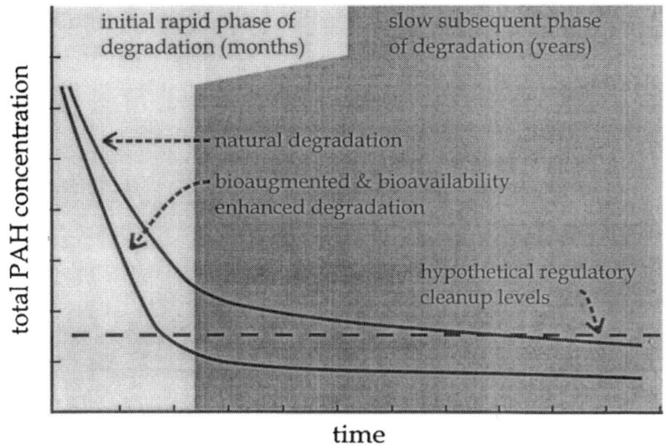

Figure 1. Conceptual illustration of PAH degradation during a land farming treatment showing a comparison between the potential effect of bioaugmentation and enhancement of bioavailability and natural degradation.

of the HMW PAHs (Brown et al., 1995, Mueller et al., 1991a,b; Pollard et al., 1994). This is conceptually illustrated in Figure 1. The two-phase degradation process is typical of both land farming and slurry reactor treatment strategies. Although the slurry reactor typically gives the fastest removal of PAHs during the initial phase, both treatment strategies give similar rates of degradation during the slow phase (Mueller et al., 1991a,b). Problems arise from a regulatory point of view when the residual concentrations of certain target HMW PAHs (i.e., those with toxicological and tumorgenic implications) remain above regulatory clean up criteria following the initial rapid degradation phase. Slurry reactors are more likely to push concentrations of these PAHs below regulatory limits but treating large amounts of soil in this manner is considerably more expensive than land farming. Nonetheless, both treatment approaches suffer from inconsistent and unpredictable extents of HMW PAH removal and this can often convert a straight forward and inexpensive treatment strategy into a complicated, long term approach with correspondingly higher costs.

If the first phase of degradation does not achieve regulatory closure, then implementing additional short term treatment options (e.g., chemical oxidation, surfactants, physical restructuring, etc.) or impounding the treated soils and allowing natural degradation processes to eventually degrade the HMW PAHs, can be considered. With the impoundment option, leachable hydrocarbons, mainly the low molecular weight PAHs, are not likely to cause environmental problems since they have been largely removed during the initial treatment phase. In essence, these soils can be considered as "biostablized" (Mueller et al, 1998), the concept being that the slow leaching of the residual HMW PAHs is counterbalanced by degradation capabilities of the indigenous microbial communities. This requires, consequently, that the impounded soil be carefully managed and monitored over periods of years, a task that can add considerable long term costs.

However, to avoid the possibility of long term passive treatment or the use of additional active treatments technologies, efforts are being made to enhance the initial degradation phase and provide a greater, more predictable degradation of HMW PAHs. Our knowledge of HMW PAH degradation mechanisms suggests that improvements in metabolic capabilities of natural microbial communities and in bioavailability are reasonable and feasible during this initial phase of degradation. In fact, cometabolic processes, which are likely responsible for the degradation of HMW PAHs, are better managed when other low molecular weight PAHs are being metabolized by natural microbial communities in the early phase of bioremediation. Our current field efforts will attempt to verify this approach and examine, ultimately, the cost benefit relationships.

APPROACH

Our approach to the use of bioaugmentation in the land farming treatment of PAH-contaminated soils includes the following; (a.) the use of an inoculation processes that is designed to supplement and accelerate the activities of natural microbial communities rather than replace or ignore them, (b.) the introduction of metabolic capability that is designed to promote cometabolism of high molecular weight PAHs, (c.) bioavailability enhancement procedures using the generation of biosurfactants *in situ* from introduced microorganisms and (d.) the use of an oil phase to feed the biosurfactant producer and sequester PAHs in a more available form, and (e.) the use of vermiculite as an inexpensive and "self-sustaining" carrier for the microorganisms as they introduced into the soil. A discussion of each of these aspects is important to justify the added cost and operational details that will accompany a bioaugmentation strategy.

Inoculation Process

The past success of bioaugmentation at field scale in conjunction with bioremediation has not been great because, in many cases, too much has been expected. For example, if a mixture of contaminants such as hydrocarbons is to be treated, it is unreasonable to expect that the addition of one or several organisms will have any more overall effect than the natural microbial communities, assuming that the latter is stimulated with additions of oxygen and fertilizer. In general, bioaugmentation under these circumstances requires too much biomass of the introduced organisms to have a significant impact. In addition, the rapid die off of introduced organisms invariably occurs in the field and therefore one should not expect to impact degradation processes weeks after treatment initiation. Reinoculation is possible, but the cost benefit ratio is likely to be unreasonable. Finally, bioaugmentation works best if nature is supplemented as an event, rather that trying to replace nature with a long term management approach. That is, it is best to use bioaugmentation as a boost to something that nature is slow to generate or unable to produce consistently.

In the case of land farming treatment, a lot happens in the first few weeks of treatment, the so called active treatment phase described above, when tilling is designed to maximally aerate and mix the soil and microbial activities are optimally stimulated due to high nitrogen abundance and utilization of the most available and degradable hydrocarbons. It is, therefore, important to capture as much degradation capability as possible during this initial phase. This is particularly true of cometabolism of high molecular weight PAHs, which, to best of our knowledge, depends on microorganisms that are rapidly degrading the low molecular weight PAHs. It is important to stimulate the cometabolism as much as possible. The stimulation does not have to be great because inconsistencies in land treatment are due, many times, to a cessation in the removal of high molecular weight PAHs just prior to reaching acceptable limits. In Figure 1, the hypothetical decay curve for total PAHs following bioaugmentation and bioavailability enhancement is designed to show that rates and extents of degradation need be only slightly greater than the nonsupplemented system since the objective is to reduce the high molecular weight PAHs to just below the regulatory acceptable limits, not to completely remove them, the latter being virtually impossible anyway. Thus reduction of HMW PAHs, in many cases, may require only a few percentage points greater degradation extent or 10's of mg/kg of soil lower concentration. The important point is to go below some predetermined level as quickly and as consistently as possible. Often, although the activities of the natural microbial com-

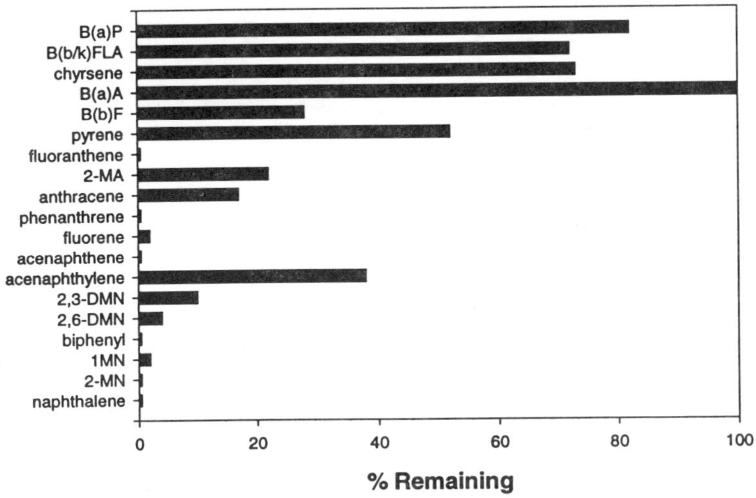

Figure 2. Degradation of PAHs in a creosote-PAH fraction in soil slurries supplemented with surfactant (0.1% Triton X-100) and EPA 505 and incubated for 28 days.

munities are sufficient to reach the target PAH concentrations, they can't be relied upon from site to site or from laboratory test to field scale, just because of natural variability. Management of the activities during the initial phases of land farming treatment, we believe, offers a greater consistency in treatment and reducing the chances of more expensive long term treatment. In this regard, using bioaugmentation as a follow up procedure depending on how close you get to a clean up goal after the initial treatment phase (natural degradation, tilling and fertilization) is unlikely to be effective because interactions between substrates and microorganisms will be considerably reduced.

Absence of Appropriate Metabolic Capability

We now know that bacteria can be isolated which degrade HMW PAHs such as fluoranthene and pyrene as source of carbon and energy (Mueller et al., 1990; Weissenfels et al., 1990; Heitkamp et al., 1988) but to date no isolates have been obtained that will grow on 5- and 6-ringed PAHs. This may simply be a limitation of our current enrichment techniques or it may be that the very low water solubility of these PAHs prevents microorganisms from ever adapting to utilizing them for growth. Since the 5-ring PAHs often disappear to some extent in certain treatment scenarios, we assume cometabolism by the enzymes present in microorganisms that are actively growing on low molecular weight PAHs is occurring. In fact, this is probably true. An isolate that we have been studying, *Sphingomonas paucimobilis,* strain EPA 505, can significantly reduce concentrations of HMW PAHs that are not growth substrates when the isolate is exposed to a mixture of PAHs (the PAH fraction of creosote). This is shown in Figure 2. After incubation of the PAH mixture with EPA 505 and Triton X-100 (0.01%) for 28 days, reductions in pyrene (48%), benzo(b)fluorene (73%), chrysene (26%), benzo(b/k)fluoranthene (28%), and benzo(a)pyrene (17%) were considerable.

In addition, Ye, et al. (1996), using the same strain, showed similar degradation of HMW PAHs, albeit to an even greater extent, when studied as individual compounds. In preliminary mechanistic studies we have performed with pyrene, a nongrowth PAH for

Table 1. Amount of PAHs remaining (% Degradation) following biodegradation of PAH fraction by selected strains of PAH degraders[1]

Isolate/Strain	2-Ring PAHs	3-Ring PAHs	4,5,6-Ring PAHs	Total PAHs
Initial	149.5 (0)	168.1 (0)	60.0 (0)	377.6 (NA)
Uninoculated	71.3 (NA)	107.0 (NA)	55.9 (NA)	234.1 (NA)
CRE 7[2]	1.7 (98)	14.8 (86)	47.2 (16)	63.7 (73)
N3P2[3]	0.5 (99)	4.9 (95)	46.1 (18)	51.5 (78)
UN1P1[3]	0.5 (99)	3.4 (97)	43.0 (23)	46.9 (80)
EPA 505[3]	0.3 (100)	2.2 (98)	18.6 (67)	21.1 (91)
PRY-1[4]	4.8 (93)	70.6 (34)	40.3 (28)	115.7 (51)

[1]Modified from Mueller, et al., *Antonie van Leewenhoek J.* **71**:329-338 (1997).
[2] Phenanthrene degrader
[3] Fluoranthene degraders
[4] Pyrene cometabolizer

EPA 505, it appears that at least two different partial degradation products of pyrene are produced, both representing the opening of an aromatic ring and both indicating an inability to cleave carbon moieties to use for growth. Thus this strain appears to have a remarkable breadth of cometabolic capability, quite possibly due to very loose specificity of the dioxygenase enzymes that are responsible for metabolizing fluoranthene.

However, this cometabolic capability is not typical of all PAH degrading strains studied. Table 1 (modified from Mueller et al., 1997) shows the effectiveness of several isolates in their ability to degrade different PAH fractions of creosote. The CRE strains were isolated for their ability to grow on phenanthrene and the UNP/NP strains were isolated for their ability to grow on fluoranthene. It is clear that EPA 505 was the most effective in terms of removing the 4,5,6-ring PAHs (67% removal compared to 28 % removal for next best strain, PRY-1, a *Mycobacterium* strain able to cometabolize pyrene; Heitkamp et al., 1988) and in producing the greatest removal of PAHs overall (91% removal). Thus, just because a strain grows well on particular PAH, does not mean that it will have broad cometabolic capabilities.

These results question how frequently an organism like EPA 505 occurred in PAH-contaminated soils; in other words, is this metabolic capability to be expected in most soils. Isolates that were able to grow on fluoranthene were therefore obtained by incubating samples of PAH-contaminated soil in dilute complex medium for 24 hours and then plating dilutions on agar containing low concentrations of complex medium. Once colonies formed, the plates were sprayed with fluoranthene to create an opaque film of the PAH over the surface of the agar (Kiyohara et al., 1982). Colonies that showed clearing around the colony were picked and further characterized for their ability to degrade PAHs, as determined by clearing around colonies when sprayed with the appropriate PAH. Where clearing was observed, individual isolates were tested for growth on the PAH. None of the isolates grew on the high molecular weight PAHs (pyrene and above) and so clearing around the colonies suggested that cometabolism was occurring. Most of these isolates were also able to degrade low molecular weight PAHs. Table 2 compares the cometabolic capability of 30 different fluoranthene degraders randomly isolated from 15 different contaminated soils. Very few strains compared favorably with the metabolic capability of EPA 505. That is, most strains were unable to cometabolize pyrene, chrysene, benzo(b)anthracene, and benzo(a)pyrene in the absence of phenanthrene (a growth substrate for all isolates) and a somewhat greater number were able to do so in the presence of phenanthrene. Perhaps 2–3 isolates showed capabilities equal to EPA 505. These results serve to demonstrate that good cometabolic capability

Table 2. Number of Fluoranthene-degrading isolates, out of 30, causing clearing around colony in PAH spray plate method

PAH Substrate	Not induced	Phenanthrene induced
Fluorene	18	19
Phenanthrene	30	30
Fluoranthene	30	30
Pyrene	16*	21*
Anthracene	17	21**
Benzo(b)fluorene	0	15**
Benzo(a)anthracene	11	20**
Chrysene	0	1 **
Benzo(a)pyrene	0	4**

*Weak clearing but strong color formation.
** Strain EPA 505 included

is unlikely in most PAH-contaminated soils therefore, in a land farming context, it is problematic that HMW PAHs will be effectively reduced to regulatory significant levels. The value of bioaugmentation, in this case, is apparent.

Bioavailability Enhancement

The ability of microorganisms to produce surface-active compounds, referred to as biosurfactants, when they utilize insoluble or immiscible compounds such as hydrocarbons is well known (Desai and Banat, 1997; Willumsen et al., 1998). Biosurfactants exhibit low interfacial tension, low critical micelle concentration values and are produced by a variety of bacteria, yeast and fungi. Contradictory results on the effects of biologically produced surfactants on PAH biodegradation are found in the literature (Miller, 1995), but many reports indicate that they can enhance hydrocarbon biodegradation by increasing microbial accessibility to insoluble substrates (Willumsen, 1998; Zhang and Miller, 1994; Zhang et al., 1997). This would make them potentially useful for bioremediation. However, producing large quantities of biosurfactant to apply during a bioremediation process is currently cost prohibitive and logistically complex. Alternatively, it is quite reasonable to add a microorganism that produces a biosurfactant, particularly if the organism will grow on one of the contaminants in the soil to be treated. We have been examining different hydrocarbon degrading strains of bacteria for their biosurfactant producing capability and have focused our attention on a *P. aeruginosa* strain #64, which produces three different rhamnolipid biosurfactants. Biosurfactant production in this strain is dependent on the growth medium, with good production when grown on light oil, hexadecane or glycerol. Large-scale production of *P. aeruginosa* #64 rhamnolipid biosurfactant shows very high activity such that a 1:4 dilution of the concentrate gives a surface tension measurement of 24.0 dynes/cm. Solubilization of phenanthrene by the rhamnolipid was about the same as that obtained with Triton X-100, a commercially available chemical surfactant.

Since many PAH-contaminated soils also contain other hydrocarbons, it is reasonable to expect that adding strain #64 to soil will allow, in essence, in situ production of the biosurfactant. Inexpensive petroleum products could also be added to the soil to further enhance growth of the organisms. We are currently testing conditions for this possibility. Adding oil to the soil would act as a solvent, possibly helping to make HMW PAHs more bioavailable. It is hoped that the enhanced bioavailability and increased metabolic capability resulting from bioaugmentation will ultimately create lower concentrations of PAHs in

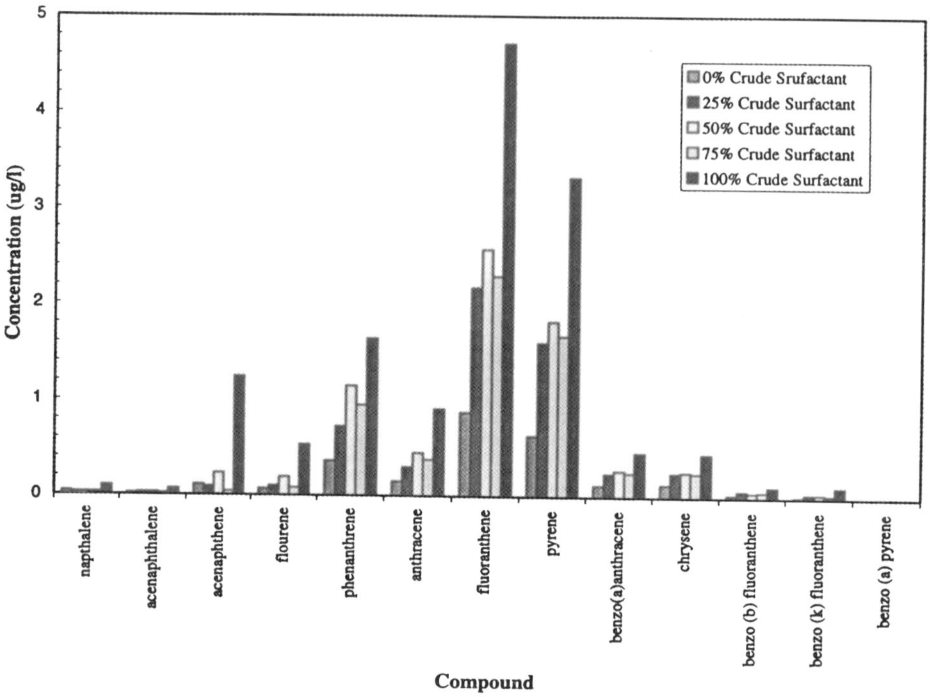

Figure 3. Effect of culture supernatant (biosurfactant) from strain #64 on the aqueous extraction of PAHs from soil.

the aqueous phase and thereby develop steep concentration gradients that will lead to greater amounts of PAH desorbed. Again, it may not take much enhancement of bioavailability to eventually drive concentrations HMW PAHs below clean up criteria.

To test the potential effectiveness of the biosurfactant a preliminary test was performed with contaminated soil. Strain #64 was grown in broth containing glycerol (a carbon source that gives good surfactant production) and after removing cells, the supernatant at different concentrations was added to creosote-contaminated soil and the amount of additional PAHs that could be extracted with water was determined (Figure 3). The biosurfactant caused considerable increases in the amount of PAHs that could be extracted, especially some of the HMW PAHs, indicating that they would be consequently more available to the degrading microorganisms. Studies are currently underway to inoculate strain #64 into soil and determine how the production of biosurfactant effects degradation of the PAHs.

Bioaugmentation with Inert Carriers

An approach for improving the effectiveness of bioaugmentation in bioremediation systems is to introduce the inoculum on an inert carrier. This involves immobilizing bacterial cells in a material that will enhance survivability of the inoculum in soil and allow it to express it's metabolic activity. We have had considerable success with vermiculite, an inexpensive inorganic material that is safe to add to the environment. Since dry vermiculite adsorbs several times it's weight in liquid, it is possible to add microbial growth medium to sterilized vermiculite. Inoculation with organisms then allows the cells to, in essence, grow "into" the vermiculite where they are shielded from direct competition with

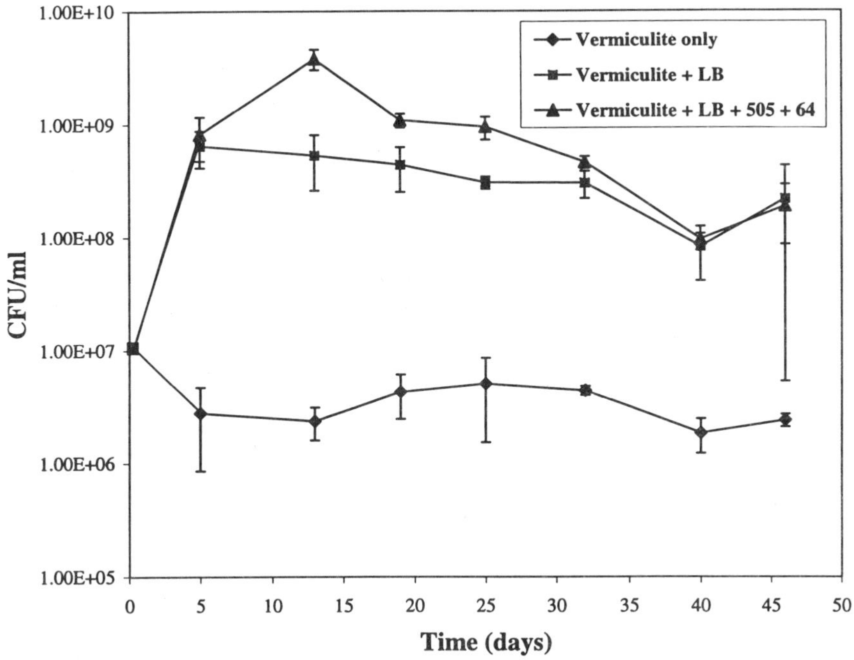

Figure 4. Changes in total colony forming units in PAH contaminated soil following the introduction of vermiculite alone and vermiculite spike with LB broth and LB broth with strain #64 and EPA 505.

indigenous organisms in the soil and where they can slowly grow in the vermiculite, potentially "shedding" cells continuously into the surrounding soil. Studies in the laboratory have demonstrated that vermiculite–carried cells remain viable in soils months after inoculation (Pritchard et al., 1995). In addition, studies with EPA 505 have shown that it's ability to mineralize fluoranthene in soil is not effected by immobilization on vermiculite (Pritchard et al., 1995). Thus the vermiculite appears to be an excellent biocarrier, assuring some degree of staying power for the inoculum.

Furthermore, vermiculite also serves as a bulking agent for soil, giving better soil mixing characteristics and consistency in land treatment units. To determine effectiveness of bioaugmentation with the vermiculite, PAH-contaminated soil was spiked with 2% by weight vermiculite treated in two different ways. In one case the vermiculite was moistened with Luria-Delbruck (LB) broth, a common growth medium for bacteria, and in the second case vermiculite was treated with both LB broth and a mixture of strain #64 and strain EPA 505 (approximately $1x10^6$ cells/ml of each). These were added to contaminated soil and the total numbers of bacteria present in the soil over time compared to vermiculite only-amended soil (no supplements added to the vermiculite). The results are shown in Figure 4. Total counts of bacteria increased by three orders of magnitude in the soil supplemented with the amended vermiculite, suggesting that addition of LB broth had a stimulating effect on the background microbial population. But most importantly, the highest number of cells was observed with EPA 505 and strain #64, suggesting that these organisms were a major part of the population. There dominance lasted for at least 30 days. Strain #64, which is a fluorescent pseudomonad, can be identified by the green pigment elaborated into the agar medium from colonies. Pigment producing colonies were observed up to about day 25, further suggesting that strain #64 was present for considerable periods of time as a result of the bioaugmentation on vermiculite. Strain EPA 505 was not tracked during this experiment. Thus it appears that bacterial strains can be effectively introduced into soil when carried on vermiculite. Inoculated vermiculite can therefore be prepared and stored in large quantities prior to us in a land farming remediation, a far simpler task than producing "naked cells" just prior to inoculation into the soil. In addition, the physical bulking effect of the vermiculite on the soil, makes the soil considerably easier to work with during land farming operations.

TECHNICAL FEASIBILITY

We are planning to conduct land farming microcosm experiments and eventually field experiments using the bioaugmentation (with vermiculite carried cells) and bioavailability enhancing approach out lined above. The objective will be to determine if out approach will lead to consistently greater reductions of the high molecular weight PAHs in soil. The microcosms have been designed and tested. Soil will be tilled by hand, moisture will be held to approximately 60% soil holding capacity, fertilizer will periodically added, and any leachate produced will be recycled back into the microcosms, as we envision it will done in the field. The leachate recycling could potentially act as a reinoculation step and it would allow any PAHs that are released as a result of the bioavailability enhancement to be degraded in the microcosms.

To verify that the bioaugmentation process will work in principle, we report here on some similar studies that were performed on sediments from Boston Harbor. In these sediment slurry flask experiments, EPA 505 was added to the sediments (with no carrier) along with 0.1% DMSO as basically a surrogate surfactant. All flasks were supplemented with the same amount of fertilizer (basal salts medium). The results are shown in Table 3 and Figure 5. Table 3 shows that the most degradation of the 4–5 ring PAHs occurred when all factors were optimized; that is, bioaugmentation, surfactant (DMSO) addition, and nutrient amendment. No treatment alone was as effective as the combined treatment. These results are encouraging for the application to land farming treatment. But most importantly, the combination treatment, also had pronounced effect on the degradation kinetics of certain HMW PAHs. The effects on benzo(a)anthracene (B(a)A), a PAH that is

Table 3. Final concentrations of selected PAHs in Boston Harbor sediment slurries (mg compound/kg dry wt. sediment) following 80 days incubation in shake flasks

	Final concentration in slurries (ppm)			
Selected PAHs	No additions	Nutrients[1] + surfactant[2]	Nutrients + microbes	Nutrients + microbes[3] + surfactant
Phenanthrene	37	41	22	16
Fluoranthene	1175	1258	104	24
Pyrene	712	968	784	338
Total 2-Ring PAHs	0	10	0	0
Total 3-Ring PAHs	539	394	365	146
Total 4,5-Ring PAHs	2931	3946	2102	814
Total PAHs	3470	4350	2467	960

[1] Minimal salts medium
[2] DMSO (0.1%)
[3] Mixture of phenanthrene degraders and EPA 505

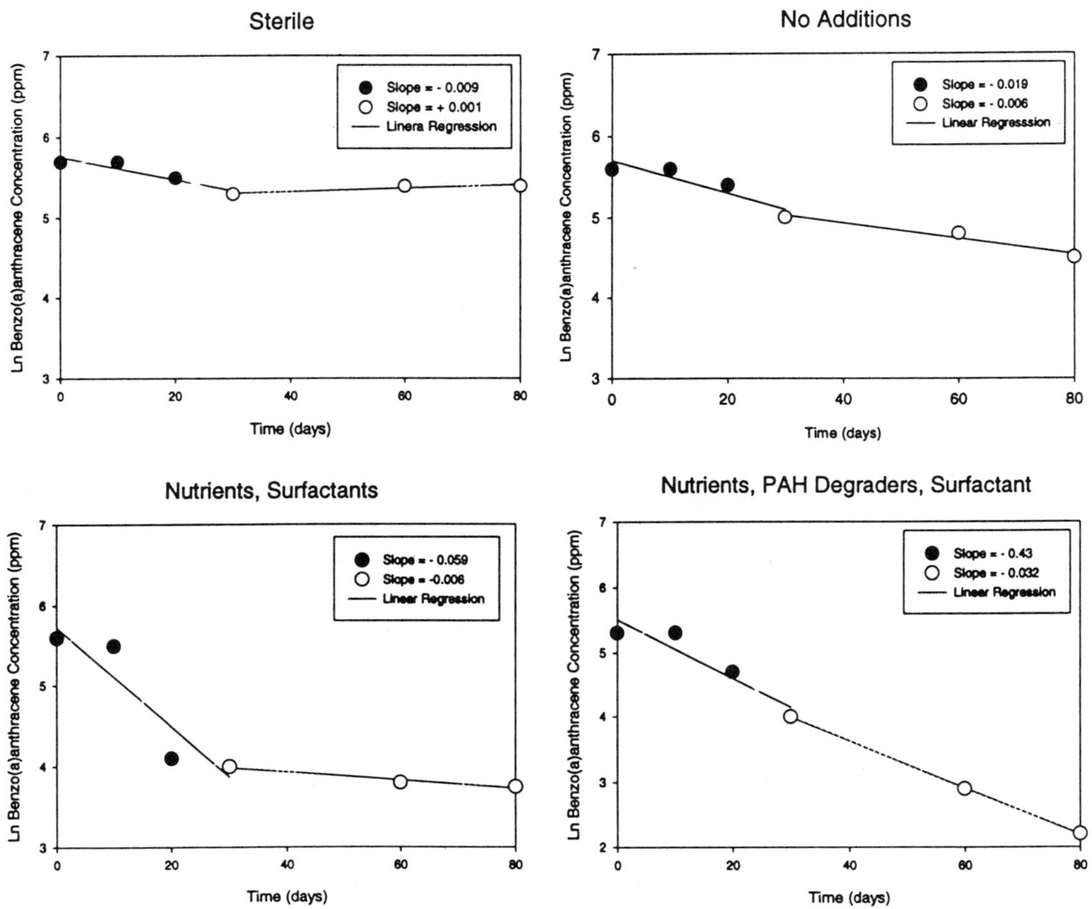

Figure 5. Degradation rates of benzo(a)anthracene in slurry flasks with Boston Harbor sediments supplemented with inorganic nutrients and different combinations of microbes and DMSO (surrogate surfactant).

likely cometabolized, is shown in Figure 5. By considering the decay curves as biphasic, a fast initial phase, followed by a slower second phase (there are not enough data points to make this division statistically significant), a situation that is synonymous with the removal of PAHs from soil during land farming operations was obtained. The onset of the slower secondary phase is the troublesome aspect because it suggests that B(a)A degradation may, under most conditions, moved into the second phase too early, leaving residuals that could be unacceptable from a clean up standpoint. However, the addition of the inoculum and surfactant together, in essence, eliminated the biphasic response, presumably by sustaining enhanced degradation over a longer period. In fact, at the termination of the experiment it appeared that the B(a)A was still continuing to decline as part of the more rapid initial phase.

Thus, it does seem feasible to effect HMW PAH degradation by providing the appropriate metabolic activity and improving bioavailability. Future studies will verify that this approach can work at field scale. If success is obtained, technical guidance documents will be developed for use with land farming treatment on a broad scale.

ACKNOWLEDGMENTS

We wish to acknowledge the competent and dedicated technical support of Charlotte Martin, Dava Dalton, Richard Cripe. Parts of this study were supported by the Flask to Field Program of the Strategic Environmental Research and Development Program (SERDP).

REFERENCES

Brown, R.A., Hicks, P.M., Hicks, R.J., and Leahy M.C. (1995). Postremediation Bioremediation, p. 77–84. In R.E. Hinchee, J.T. Wilson, and D.C. Downey (eds.), *Intrinsic Bioremediation*, Battelle Press, Columbus, OH.

Desai, J.D., and Banat, I.M. (1997). Microbial production of surfactants and their commercial potential. *Microbiol. Mol. Biol. Rev.* **61**:47–64.

Harmsen, J. (1991). Possibilities and limitations of landfarming for cleaning contaminanted soils, p. 255–272. In R.E. Hinchee and R.F. Olfenbuttel (eds.), *In Situ Bioreclamation; Applications and Investigations of Hydrocarbon Contaminated Site remediation*, Butterworth-Heinemann, London.

Heitkamp, M.A., Freeman, J.P., Miller, D.W., and Cerniglia, C.E. (1988). Pyrene degradation by a Mycobacterium sp.: identification of ring oxidation and ring fission products. *Appl Environ Microbiol.* **54**:2556–2565.

Kiyohara, H, Nagao, K., Yana, K. (1982). Rapid screen for bacteria degrading water insoluble, solid hydrocarbons on agar plates. *Appl Environ Microbiol.* **43**:454–457.

Miller, R.M. (1995). Surfactant-enhanced bioavailability of slighty soluble organic compounds. P.33–54. In H.D. Skipper, and R.F. Turco (eds.) *Bioremediation Science and Application*, Soil Sci. Soc. Amer., Madison WI.

Mueller, J.G., Chapman, P.J., Blattmann,B.O. and Pritchard, P.H. (1990). Isolation and Characterization of a Fluoranthene-Utilizing Strain of Pseudomonas paucimobilis. *Appl. Environ. Microbiol.*, **56**:1079–1086.

Mueller, J.G., Lantz, S.E., Blattmann, B.O. and Chapman, P.J. 1991a. Bench-Scale Evaluation of Alternative Biological Treatment Processes for the Remediation of Pentachlorophenol- and Creosote-contaminanted materials: slurry-phase bioremediation. *Environ. Sci. Technol.* **25**:1055–1061.

Mueller, J.G., Lantz, S.E., Blattmann, B.O. and Chapman, P.J. 1991b. Bench-Scale Evaluation of Alternative Biological Treatment Processes for the Remediation of Pentachlorophenol- and Creosote-contaminated materials: solid-phase bioremediation. *Environ. Sci. Technol.* **25**:1045–1055.

Mueller, J.G., Cerniglia, C.E. and Pritchard, P.H., "Bioremediation of Environments Contaminated by Polycyclic Aromatic Hydrocarbons," In: "Bioremediation, Principles and Applications. R. Crawford and D. Crawford (eds). Cambridge Press 1995.

Mueller, J.G., Devereux, R., Santavy, D.S., Lanrz, S.E., Willis, S.G. and Pritchard, P.H., (1997) Phylogenetic and Physiological Comparsions of PAH-Degrading Microorganisms from Geographically Diverse Soils, *Antonie van Leewenhoek J.* **71**:329–338.

Mueller, J.G., Pritchard, P.H., Tischuk, M., Brourman, M., Swallow, P., Tabe, M., and Smith, J. (1998). Biostablization technology for treating PAH-and PCP-impacted soil to environmentally acceptable endpoints: ten-year field assessment. Proceeding of 42nd OHOLO Conference, May 3–6, 1998, Eilat, Israel.

Pignatello, J.J., and Xing B. (1996). Mechanisms of slow sorption of organic chemicals to natural particles. *Environ. Sci. Technol.* **30**:1–11.

Pollard, S.J., Hrudey, S.E., and Fedorak, P.M. (1994). Bioremediation of petroleum- and creosote-contaminated soils: A review of constraints. *Waste Mange. and Res.* **12**:173–194.

Pritchard, P.H., Lin, J-E, Lantz, S. and Mueller, J.G. (1995) The Use of Microbial Encapsulation/Immobilization for Biodegradation of PAHs, p. 211–220. In R.E. Hinchee and S.K.Ong (eds.) *Bioremediation of Polycyclic Aromatic Hydrocarbons*, Lewis Publishers, Boca Raton.

Weissenfels, W.D., Beyer, M., Klein, J., and Rehm, H.J. (1991). Microbial metabolism of fluoranthene: isolation and identification of ring fission products. *Appl. Microbiol Biotechnol* **34**:528–535.

Willumsen, P.A., Karlson U., and Pritchard, P.H. (1998). Response of Fluoranthene-Degrading Bacteria to Surfactants. Accepted *Appl. Microbiol. Biotechnol.*

Yare, B. (1991). A comparison of soil-phase and slurry-phase bioremediation of PNA-containing soils, p. 173–1187. In R.E. Hinchee and R.F. Olfenbuttel (eds.), *In Situ Bioreclamation; Applications and Investigations of Hydrocarbon Contaminated Site remediation*, Butterworth-Heinemann, London.

Ye, D., Siddiqi, M.D., Maccubbin, A.E., Kumar, S., and Sikka, H.C. 1996. Degradation of polynuclear aromatic hydrocarbons by *Sphingomonas paucimobilis*. *Environ. Sci. Technol.* **30:**136–142.

Zhang W., Bouwer, E.J., and Ball, W.P. (1998). Bioavailability of hydrophobic organic contaminants: Effects and Implications of sorption-relatd mass transfer on bioremediation. *Ground Water Monit. Remediat.* **18**: 126–138.

Zhang, Y., and Miller, R.M. (1994). Effect of *Pseudomonas* rhamnolipid biosurfactant on cell hydrophobicity and biodegradation of octadecane. *Appl. Environ. Microbiol.* **60**:2101–2106.

Zhang, Y., Maier, W.J., and Miller, R.M. (1997). Effect of rhamnolipids on the dissolution, bioavailability, and biodegradation of phenanthrene. *Environ. Sci. Technol.* **31**:2211–2217.

17

ROLE OF *Acinetobacter* BIOEMULSANS IN PETROLEUM DEGRADATION

E. Rosenberg,[1] T. Barkay,[1,2] S. Navon-Venezia,[1] and E. Z. Ron[1]

[1]Department of Molecular Microbiology and Biotechnology
Tel Aviv University, Ramat Aviv, Israel
[2]Center for Environmental Diagnostics and Bioremediation
University of West Florida, Pensacola, Florida

1. ABSTRACT

Bacterial degradation of hydrocarbons is a multiphase reaction, involving oxygen gas, water-insoluble hydrocarbons, water, dissolved salts and microorganisms. The fact that the first step in hydrocarbon catabolism involves a membrane-bound oxygenase makes it essential for microorganisms to come into direct contact with the hydrocarbon substrate. Growth then proceeds on the hydrocarbon/water interface. Bacteria have developed two general strategies for enhancing contact with water-insoluble hydrocarbons: adhesion mechanisms and production of extracellular surfactants. Since petroleum is a complex mixture of many different classes of hydrocarbons, of which any particular microorganism has the potential to degrade only part, it follows that the microorganisms must also have a mechanism for desorbing from 'used' oil droplets. *Acinetobacter* species produce a wide variety of extracellular emulsifiers that are high molecular weight complexes of anionic polysaccharides and proteins. These bioemulsans can enhance the biodegradation of hydrocarbons by increasing the surface area of the hydrocarbon substrate, desorbing the bacteria from crude oil after depleting the oil of utilizable components and pseudosolubilizing insoluble hydrocarbons. One of these bioemulsans, alasan, increased the apparent solubility of polycyclic aromatics by 5–26 fold and increased the rate of bacterial degradation of fluoranthene more than two-fold.

2. *Acinetobacter* BIOEMULSANS

Production of extracellular high molecular weight bioemulsifiers is a widespread phenomenon in *Acinetobacter* species (Rosenberg and Kaplan, 1986). Of 16 different *Acinetobacter* strains examined, eight produced high levels of bioemulsans when grown on minimal salts medium containing ethanol as the carbon source (Sar and Rosenberg, 1983).

Novel Approaches for Bioremediation of Organic Pollution, edited by
Fass *et al.* Kluwer Academic / Plenum Publishers, New York, 1999.

Table 1. Properties of *Acinetobacter* bioemulsans

	Emulsan RAG-1	Emulsan BD4	Alasan
Producing bacterium	*A. calcoaceticus* RAG-1	*A. calcoaceticus* BD4	*A. radioresistens* KA53
Chemical properties	Polysaccharide containing fatty acid esters	Polysaccharide: protein complex	Polysaccharide with N-linked alanine: protein complex
Substrate specificity	Requires both aliphatic and aromatic hydrocarbons	Similar to RAG-1	Long chain aliphatics and aromatics containing at least 10 carbon atoms
Stability to heat & alkali	Unstable	Unstable	Activated by heat; stable to alkali

All of the emusifiers contained polysaccharide as their major component. Three of these *Acinetobacter* bioemulsans have been studied extensively (Table 1): RAG-1 emuslan, BD4 emulsan and alasan.

2.1. RAG-1 Emulsan

A. calcoaceticus RAG-1 (ATCC 31012) was the first *Acinetobacter* strain that was shown to produce a bioemulsan (Reisfeld et al., 1973) and has been extensively studied (for Reviews see Rosenberg and Kaplan, 1986; Rosenberg et al., 1998 and Rosenberg and Ron, 1997). The chemical and physical properties of RAG-1 emulsan are summarized in Table 2. The unique feature of RAG-1 emulsan is the presence of long chain fatty acids covalently attached to the anionic polysaccharide backbone via O-ester and N-acyl linkages (Belsky et al., 1979). The hydrophobic groups are distributed across the entire molecule (Shoham and Rosenberg, 1983), similar to hydrophobically modified, comb-type polymers. Mild alkaline conditions cause hydrolysis of the ester linkages in emulsan and loss of its ability to stabilize oil-in-water emulsions. Although not essential for its activity, proteins associated with the emulsan polysaccharide also contribute to its emulsifying properties (Zosim et al., 1982)

2.2. BD4 Emulsan

A. calcoaceticus BD4, initially isolated and characterized by Taylor and Juni (1961), produces a large polysaccharide capsule. *A. calcoaceticus* BD413, a mini capsulated mu-

Table 2. Chemical and physical properties of RAG-1 emulsan

A. Chemical composition
1. N-acetyl hexosamines (70%)
D-galactosamine
L-galactosamine uronic acid
Dideoxy-diaminohexose
2. Fatty acids (15%)
3. Water and ash (15%)
B. Physical properties
1. Molecular weight =
9.8×10^5 (weight avg.)
9.4×10^5 (viscosity avg.)
2. Axial ratio ~ 50

→3)-a-L-Rha-(1 → 3)-α-D-Man-(1 → 3)-α-L-Rha-(1 → 3)-α-L-Rha-(1 → 3)-β-D-Glc-(1→
2
↑
1
α-L-Rha-(1 →4)-β-D-GlcUA

Figure 1. The repeating heptasaccharide structure of *A. calcoaceticus* BD4 bioemulsan polysaccharide (Kaplan et al., 1985).

tant of BD4 (Juni and Janick, 1969), produces a much reduced capsule. When grown on glucose-minimal medium, BD413 excretes a polysaccharide-containing emuslan, whereas BD4 exhibits relatively low levels of emulsifying activity (Kaplan and Rosenberg, 1982). BD413 excretes 45% of the polysaccharide it produces, whereas BD4 excretes only 11%. In contrast to RAG-1, strain BD413 produces its extracellular polysaccharide mainly during cell growth; only 20% of the emulsan is produced after the onset of the stationary phase (Kaplan and Rosenberg, 1982). Under certain growth conditions (e.g., 2% ethanol as the carbon and energy source), an enhanced release of emulsan by strain BD4 was obtained (Sar and Rosenberg, 1983). However, in contrast to RAG-1, no decrease in capsular polysaccharide was observed; rather, capsular polysaccharide levels remained constant when enhanced levels of extracellular polysaccharide were obtained. A BD4 capsule-specific bacteriophage, termed SL-1, was used to select for BD4 phage-resistant mutants. These capsule-negative mutants produced no extracellular emulsan. The main evidence of the origin of BD4 and BD413 emulsans from capsular polysaccharide came from elucidation of the primary structure of these polysaccharides (Fig. 1). When the BD4 crude emulsan was applied to a Sepharose 4B column, it eluted in a single peak containing a polysaccharide-protein complex. Separation of the complex and reconstitution of the emulsifying activity is described in Table III. The polysaccharide component was obtained by deproteinization and purification of the BD4 capsule or BD413 emuslan. The protein component was obtained from the extracellular supernatant fluid of strain BD4-R7, a capsule-negative mutant of BD4 that produces no extracellular polysaccharide. The polysaccharide and protein components by themselves had no emulsifying activity. Mixing the protein fraction with the polysaccharide fraction restored the original emulsifying activity.

2.3. Alasan

The chemical and physical properties of alasan, produced by a strain of *Acinetobacter radioresistens*, is summarized in Table IV. The two unusual features of this bioemulsan are (1) the anionic heteropolysaccharide contains 8% alanine, covalently bound via an amide linkage, and (2) this polysaccharide/protein complex undergoes a dramatic change in confirmation when heated and cooled. As seen in Fig. 2, heating alasan to 40 C or 50 C caused a large increase in specific viscosity without any effect on emulsifying activity. When alasan was heated from 60 C to 90 C, there was a decrease in viscosity and a large increase in emulsifying activity. These changes involve interactions between the proteins and polysaccharides, since the viscosity of the polysaccharide by itself (apoemulsan) was not affected by these heat treatments. As was the case with the BD4 emulsan, the proteins, which comprise 15–20% of the weight of alasan, are essential for its emulsifying activity. Deproteinization with hot phenol or treatment with proteases resulted in a loss of emulsifying activity (Navon-Venezia et al., 1997). One of the three major proteins (45 kD) showed homology to outer membrane proteins (Navon-Venezia, unpublished).

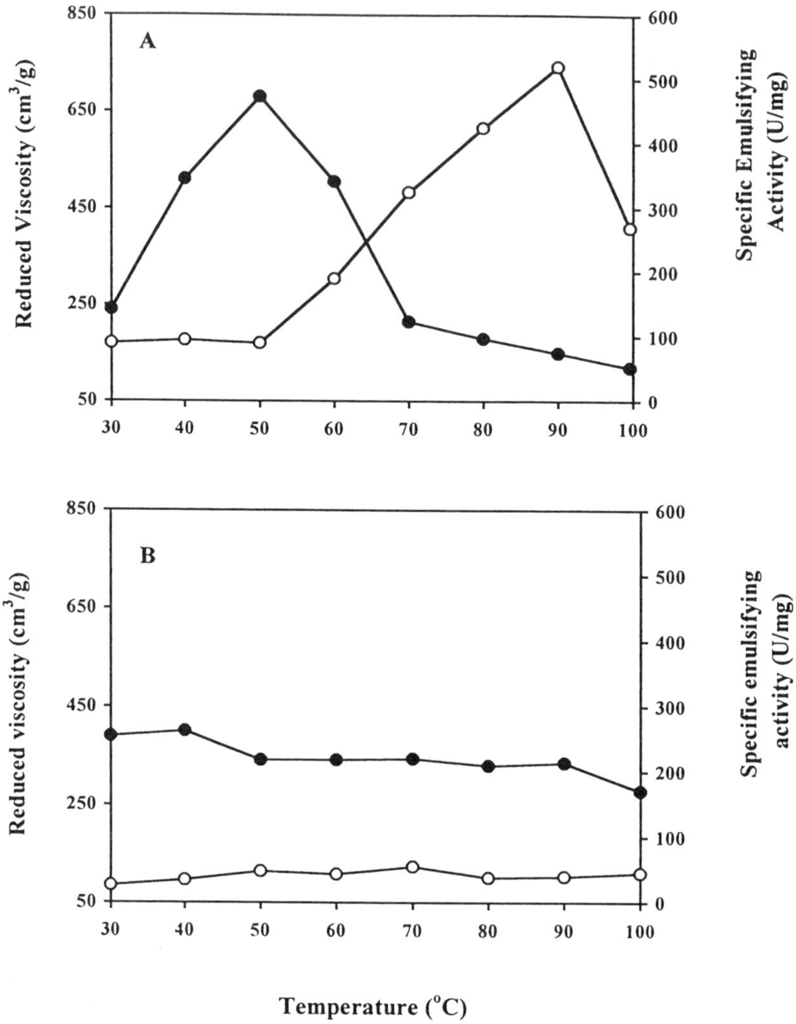

Figure 2. Effect of temperature treatment on viscosity and emulsifying activity of alasan (panel A) and apo-alasan (panel B). Solutions of alasan and apo-alasan (2 mg/ml in 20 mM Tris-HCl buffer, pH 7.0) were placed in a thermoblock and incubated for 10 min at 30° C. After treatment, samples were taken for measuring emulsiyfing actitivy (○). and reduced viscosity (●) The solutions were then heated at the next highest temperature for 10 min. The procedure was repeated up to 100° C. All the measurements of viscosity and emusifying activity were performed at 30° C.

3. EMULSIFYING ACTIVITIES

One of the interesting properties of bioemulsans is that they exhibit considerable substrate specificity with regard to which hydrocarbons they emulsify (Table V). RAG-1 emulsan does not emulsify pure aliphatic, aromatic, or cyclic hydrocarbons. However, all mixtures that contain an aliphatic and aromatic hydrocarbon (e.g., hexadecane and 2-methylnaphthalene) are emulsified. Long chain alkylbenzenes behave like mixtures of alkanes and aromatics. Alasan showed a different hydrocarbon substrate specificity than

RAG-1 emuslan. Alasan emulsified efficiently all pure hydrocarbons tested that contained ten or more carbons. Bioemulsans are not very efficient in reducing interfacial or surface tensions and require an energy input, such as mechanical agitation or sonic oscillation, to initiate the formation of small droplets. However, bioemulsans are highly efficient in stabilizing these droplets once they are produced. Studies of the mechanism of emulsion stabilization by RAG-1 emulsan (Zosim et al., 1982) have shown that oil-in-water emulsions formed by sonic oscillation with an emulsan to hydrocarbon ratio of 1:155 consist of stabilized hexadecane droplets with an average diameter of 4.0 μm. These droplets rise slowly over a period of several days to form a cream layer, containing approximately 65% hexadecane, 34% water and 1% emulsan (over 95% of the input emuslan). The emulsified droplets never coalesce, because they contain a strong surface film of emulsan. It was calculated that this film was approximately 20 thick.

4. NATURAL ROLE OF BIOEMULSANS

The biosynthesis and excretion of bioemulsans by *Acinetobacter* strains requires a large input of carbon, nitrogen and energy, suggesting that they play an important role in the bacteria's physiology. In hypothesizing what this role might be, it is useful to consider two facts. First, the bioemulsans are produced primarily when the cells become starved (Rubinovitz et al., 1982; Navon-Venezia et al., 1995). Second, the bioemulsans are found both on the cell surface and in the extracellular fluid. Most of the bioemulsan is extracellular when the cells reach stationary phase (Goldman et al., 1982). It is possible that bioemulsans have evolved exclusively to serve as capsular materials and that their surface-active and emulsifying properties are purely coincidental. If the cell lyses or the material is excreted due to unbalanced growth, the amphipathic nature of bioemulsans gives them the properties of a surfactant. However, at least in the cases of RAG-1 emulsan and alasan, the data suggest that they play fundamental roles in growth on hydrocarbons.

4.1. Emulsification

Many hydrocarbon-degrading microorganisms produce extracellular emulsifying agents, implying that emulsification plays a role in growth on water insoluble substrates. What is the evidence that emulsification plays a role in hydrocarbon degradation? First, theoretical and experimental studies on the production of single-cell protein from hydrocarbons in a fermentor demonstrated that growth rate can be limited by the interfacial surface area between water and oil (Erickson and Nakahara, 1975). Second, in several cases, emulsifier production is induced by growth on water-insoluble substrates exclusively (Ito and Inouye, 1982; Hisatsuka et al., 1971, 1972; Zhang and Miller, 1992). Third, mutants that do not produce the emulsifier grew poorly on hydrocarbons (Itoh and Suzuki, 1972). All these data pertain to low molecular weight emulsifiers. There is no direct evidence that the natural role of bioemulsans involve emulsification of the substrate.

4.2. Desorption

In the case of the two *Acinetobacter* exopolysaccharides that have been studied, RAG-1 and BD4, the emulsans are antiadhesion factors, at least for hydrophobic surfaces. After microbial attachment to and growth on a surface for a period of time, the conditions on the surface often became unfavorable for further development. For example, toxins

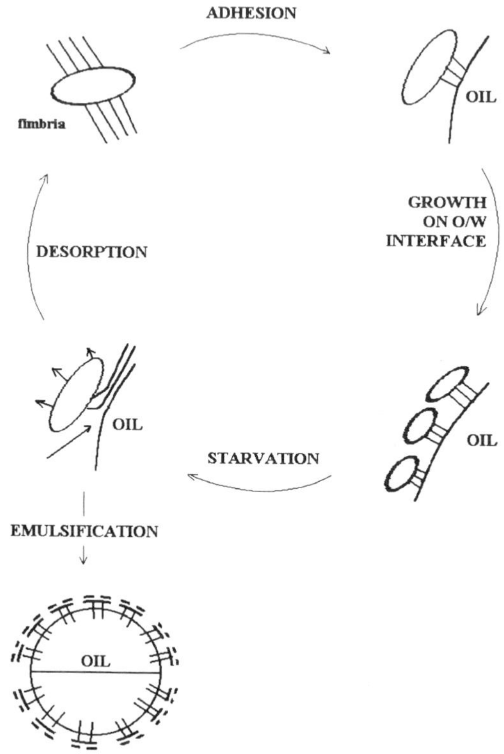

Figure 3. Model for the growth of *A. calcoaceticus* RAG-1 on crude oil.

Table 3. Reconstitution of BD4 emulsan activity[a]

Fraction	Polysaccharide (mg/ml)	Protein (mg/ml)	Emulsifying activity (units/ml)
BD4 emulsan[b]	3.1	0.55	302
Pure protein[c]	0.04	3.5	5
Pure polysaccharide[d]	2.9	0.07	0
Protein & polysaccharide	3.0	0.40	380

[a] Data adapted from Kaplan et al., 1987
[b] Purified on a Sepharose 4B column
[c] Obtained from a polysaccharide minus mutant
[d] Obtained by phenol extraction

may accumulate and crowded conditions may prevent transport of necessary nutrients. It would then be advantageous for the microorganisms to desorb and find a new habitat. In fact, mechanisms for detachment seem to be essential for all attached microorganisms in order to facilitate dispersion and colonization of new surfaces (Shilo, 1982).

One of the natural roles of emulsan may be in regulating desorption of the producing strain from hydrophobic surfaces. Consider *A. calacoaceticus* RAG-1 growing on the surface of crude oil in seawater (Fig. 3). During the growth phase, the bacteria accumulate emulsan in the form of a minicapsule on the cell surface. This bacterium uses only relatively long-chain n-alkanes for growth. After these components are used, RAG-1 becomes starved, although it is still attached (via fimbriae) to the oil droplet, enriched in aromatics and cyclic paraffins. Starvation of RAG-1 causes release of the minicapsule of emulsan. It is postulated that this released emulsan forms a polymeric film on the n-alkane-depleted oil droplet, thereby desorbing the starved cell. In effect, the "emulsifier" frees the cell to find fresh substrate. At the same time, the depleted oil droplet has been marked as used by now having a hydrophilic outer surface to which the bacterium cannot attach.

4.3. Solubilization

The following data support the hypothesis that alasan is able to solubilize polycyclic aromatic hydrocarbons (PAHs) by binding them in a reversible manner and thereby making the PAHs more readily degradable by bacteria. As seen in Table 6, 0.5 mg alasan per ml increased the solubility of phenanthrene, fluoranthene and pyrene by 5.2, 26 and 19 fold, respectively. The amount of PAH solubilized was proportional to alasan concentration. Although the solubility data presented in Table 6 were after shaking the PAH crystals with a solution of alasan overnight, independent experiments indicate that the rate of solubilization was rapid. For example, half maximum solubility of phenanthrene was reached in less than 15 min. The most likely mechanism for the increased solubility of PAHs in the presence of alasan is that the PAH binds to one or more of the proteins in the alasan complex. Deproteinized alasan (apoemulsan) did not solubilize the PAHs. The amount of phenanthrene solubilized (bound) was 13.4 μg per mg alasan complex. This corresponds to ca. 75 molecules phenanthrene per one unit of alasan (MW=10^6).

Table 4. Properties of Alasan[a]

A. Chemical composition
1. Anionic heteropolysaccharide containing:
N-acetylglucosamine
N-acetygalactosamine
N-acetyaminouronic acid
D-glucose
L-alanine
2. Proteins
24 kD, 31 kD and 45 kD
B. Physical properties
1. Molecular weight ~ 1 x 10^6
2. Viscosity
$\eta_R = 245\ cm^3/g$
$\eta_R = 655\ cm^3/g$ (after heating)

[a] Data adapted from Navon-Venezia et al., 1995.

Table 5. Hydrocarbon substrate specifities of bioemulsans

Hydrocarbon substrate	Emulsion turbidity (KU)[a]	
	RAG–1 emulsan	Alasan
Alkanes		
$C_5 - C_8$	<20	0–27
$C_{10} - C_{15}$	26–30	240–310
$C_{16} - C_{18}$	30–40	330–570
Aromatics		
Benzene	<20	<20
Toluene	44	120
Xylenes	44	100–170
Ethylbenzene	24	220
Propylbenzene	24	540
Hexylbenzene	196	420
Heptylbenzene	210	1180
2-methylnaphthalene	<20	590
Mixtures		
Hexadecane + 2-methylnaphthalene[b]	488	950
Hexadecane + cyclohexane	162	320
Crude oil	1300	1900

[a]Data from Rosenberg et al., 1979 and Navon–Venezia et al., 1995.
[b]Ratio of 1:1 (vol/vol).

The binding of phenanthrene appears to be slowly reversible because 60% of the alasan-solubilized phenanthrene passed the dialysis membrane after 4 h, whereas 70% of free phenanthrene (control) passed the membrane in 1 h. This further indicates an interaction between the PAH and the non-dialyzable alasan. Attempts to purify the phenanthrene-alasan complex by size exclusion chromatography on Sephadex G15 was not successful, probably because of this slow dissociation.

From a bioremediation point of view, it is significant that alasan stimulated the bacterial breakdown of PAHs. For example, in the presence of 300 μg alasan per ml, the initial mineralization rate of fluoranthene by *Sphyngomonas paucimobilis* (Mueller et al., 1990) increased from 7.9 to 16.7 μg per h (Fig. 4). After 32 h, 270 μg fluoranthene were

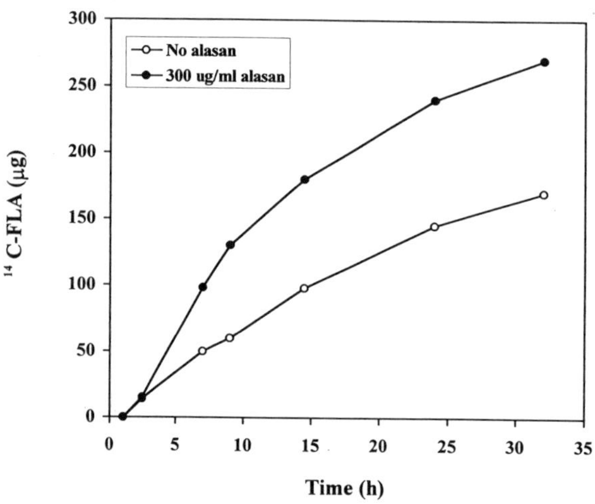

Figure 4. Effect of alasan on the mineralization of ^{14}C-fluoranthene by *Sphyngomonas paucimobilis* EPA505. Mineralization rate was determined at 30° C by measuring release of $^{14}CO_2$. A total of 1 mg of ^{14}C-fluoranthene was added to 50 ml mineral salts medium.

Table 6. Solubilization of polycyclic aromatic hydrocarbons by Alasan[a]

Alasan (mg/ml)	Aqueous solubility (μg/ml) of: Phenanthrene	Fluoranthene	Pyrene
0	1.6	0.27	0.18[b]
0.05	2.0	0.80	0.20
0.10	2.5	1.4	0.49
0.20	4.8	2.6	1.4
0.30	5.1	4.6	1.9
0.40	7.6	5.1	—
0.50	8.3	7.0	3.5

[a]Solubility was measured in 10 mM Tris buffer, pH 7 at 25 C, after shaking overnight and passing the mixture through a Whatman No. 1 filter paper.
[b]Taken from Klevens, 1954.

mineralized in the presence of alasan. By comparison, 165 μg fluoranthene per ml were mineralized in the absence of alasan, corresponding to 38% of the input PAH. Thus, alasan not only increases the solubility of PAHs, but it also makes them more available for biodegradation. Studies on the biodegradation of alkanes by *A. radioresistens* in the presence of alasan are now being performed.

ACKNOWLEDGMENTS

This work was supported in part by the Pasha Gol Chair for Applied Microbiology, the Manja and Morris Leigh Chair for Biophysics and Biotechnology, and the Israel Ministry of Science Grant No. 8867197, and by Grant No.RP8021-10 from the Electric Power Research Institute.

REFERENCES

Belsky, I., Gutnick, D.L., and Rosenberg, E. (1979). Emulsifier of *Arthrobacter* RAG-1: determination of emulsifier-bound fatty acids. *FEBS Lett.* **101**: 175–178.

Erickson, L.E., and Nakahara, T. (1975). Growth in cultures with two liquid phases: hydrocarbon uptake and transport. *Process. Biochem.*, **10**:9–13.

Goldman, S., Shabtai, Y., Rubinowitz, C., Rosenberg, E., and Gutnick, D.L. (1982). Emulsan in *Acinetobacter calcoaceticus* RAG-1: distribution of cell-free and cell-associated cross-reacting materials. *Appl. Environ. Microbiol.* **44**: 165–170.

Hisatsuka, K., Nakahara, T., Sano, N., and Yamada, K. (1971). Formation of rhamnolipid by *Pseudomonas aeruginosa* and its function in hydrocarbon fermentation. *Agr. Biol. Chem.*, **35**: 686–692.

Hisatsuka, K., Nakahara, T., and Yamada, K. (1972). Protein-like activator for n-alkane oxidation by *Pseudomonas aeruginosa* S_7B_1. *Agr. Biol. Chem.*, **36**: 1361–1369.

Ito, S., and Inoue, S. (1982). Sophorolipids from *Torulopsis bombicola* as microbial surfactants in alkane fermentations. *Appl. Environ. Microbiol.* **43**: 1278–1283.

Itoh, S., and Suzuki, T. (1972). Effect of rhamnolipids on growth of *Pseudomonas aeruginosa* mutant deficient in n-paraffin-utilizing ability. *Agr. Biol. Chem.*, **36**: 2233–2235.

Juni, E., and Janick, A. (1969). Transformation of *Acinetobacter calcoaceticus* (*Bacterium anitratum*). *J. Bacteriol.* **98**:281–288.

Kaplan, N., and Rosenberg, E. (1982). Exopolysaccharide distribution and bioemulsifier production in *Acinetobacter calcoaceticus* BD4 and BD413. *Appl. Environ. Microbiol.* **44**: 1335–1341.

Kaplan, N., Jann, B., and Jann, K. (1985). Structural studies on the capsular polysaccharide of *Acinetobacter calcoaceticus* BD4. *Eur. J. Biochem.* **152**: 453–458.

Kaplan, N., Zosim, Z., and Rosenberg, E. (1987). *Acinetobacter calcoaceticus* BD4 emulsan: reconstitution of emulsifying activity with pure polysaccharide and protein. *Appl. Environ. Microbiol.* **53**: 440–446.

Klevens, H.B. (1954). Solubilization of polycyclic hydrocarbons. *J. Phys. Colloid Chem.* **54**: 283–298.

Mueller, J.A., Chapman, P.J., Blattmann, B.O. and Pritchard, H.P. (1990). Isolation and characterization of a fluoranthene-utilizing strain of *Pseudomonas paucimobilis*. Appl. Environ. Microbiol. **56**: 1079–1086.

Navon-Venezia, S., Banin, E., Ron, E.Z., and Rosenberg, E. (1998). The bioemulsifier alasan: role of protein in maintaining structure and activity. *Appl. Microbiol. & Biotechnol.* (in press).

Navon-Venezia, S., Zosim, Z., Gottlieb, A., Legmann, R., Carmeli, S., Ron, E.Z., and Rosenberg, E. (1995). Alasan, a new bioemulsifier from *Acinetobacter radioresistens*. *Appl. Environ. Microbiol.* **61**: 3240–3244.

Reisfeld, A., Rosenberg, E., and Gutnick, D. (1972). Microbial degradation of crude oil: factors affecting the dispersion in sea water by mixed and pure cultures. *Appl. Microbiol.* **24**: 363–368.

Rosenberg, E., and Kaplan, N. (1986). Surface-active properties of *Acinetobacter* exopolysaccharides. p. 311–341. In M. Inouye (ed.), *Bacterial Outer Membranes and Model Systems*. John Wiley & Sons, Inc., New York.

Rosenberg, E., Navon-Venezia, S., Zilber-Rosenberg, I., and Ron, E.Z. 1998. Rate-limiting steps in the microbial degradation of petroleum hydrocarbons. p. 159–172. In H. Rubin, N. Narkis and J. Carberry (eds.), *Soil and Aquifier Pollution*, Springer-Verlag, Berlin, Heidelberg.

Rosenberg, E., Perry, A., Gibson, D.T., and Gutnick, D.L. (1979). Emulsifier of *Arthrobacter* RAG-1: specificity of hydrocarbon substrate. *Appl. Environ. Microbiol.* **37**: 409–413.

Rosenberg, E., and Ron, Eliora Z. (1997). Bioemulsans: microbial polymeric emulsifiers. *Curr. Opin. in Biotechnol.* **8**: 313–316.

Rubinovitz, C., Gutnick, D.L., and Rosenberg, E. (1982). Emulsan production by *Acinetobacter calcoaceticus* in the presence of chloramphenicol. *J. Bacteriol.* **152**: 126–132.

Sar, N., and Rosenberg, E. (1983). Emulsifier production by *Acinetobacter calcoaceticus* strains. *Curr. Microbiol.* **9**: 309–314.

Shilo, M. (1982). Photosynthetic microbial communities in aquatic ecosystems. *Philos. Trans. R. Soc. Lond. (Biol.)* **297**: 565–574.

Shoham, Y., and Rosenberg, E. (1983). Enzymatic depolymerization of emulsan. *J. Bacteriol.* **156**: 161–167.

Taylor, W.H., and Juni, E. (1961). Pathways for biosynthesis of a bacterial polysaccharide. I. Characterization of the organism and polysaccharide. *J. Bacteriol.* **81**: 688–693.

Zhang, Y., and Miller, R.M. (1992). Enhanced octadecane dispersion and biodegradation by a *Pseudomonas* rhamnolipid surfactant (biosurfactant). *Appl. Environ. Microbiol.* **58**: 3276–3282.

Zosim, Z., Gutnick, D., and Rosenberg, E. (1982). Properties of hydrocarbon-in-water emulsions stabilized by *Acinetobacter* RAG-1 emulsion. *Biotechnol. Bioeng.* **24**: 281–292.

18

BACTERIAL MONITORING OF A CRUDE OIL-CONTAMINATED SOIL UNDERGOING LABORATORY-SCALE BIOREMEDIATION

Pierre Wattiau,[1,*] Dirk Springael,[2] Johan Gemoets,[2] Ludo Diels,[2] and Guy Cornelis[1]

[1]Université Catholique de Louvain (UCL)
Microbial Pathogenesis Unit
av. Hippocrate 74 box 7449 B-1200 Brussels, Belgium
[2]Vlaamse Instelling voor Technologische Onderzoek (VITO)
Environmental Technology
Boeretang 200, B-2400 Mol, Belgium

1. INTRODUCTION

Environmental bacteria are able to face many chemical pollution's. For example, soil's bacteria challenged with petroleum-derived hydrocarbons become very often able to eliminate these pollutants after an adaptive period. Their aerobic metabolism can be stimulated by the addition of nutrients (nitrogen and phosphorous sources) at physiological pH and by supplying sufficient oxygen concentrations. This natural elimination of the pollutants is known as biological remediation or "bioremediation". Despite substantial improvements in the technologies, little attention has been paid to the micro-organisms themselves. In particular, no accurate data are available concerning the nature and the fate of bacterial populations inside a polluted biotope. The aim of our work was to identify, describe and monitor oil-degrading bacterial populations associated with oil-polluted biotopes before, during and after laboratory-scale bioremediation treatments. We believe that some knowledge of the bioremediating organisms of a particular soil based on such small scale treatments can help to set up and monitor larger on-site or in situ treatments. In addition, the isolation and study of indigenous oil-degrading organisms may provide material and data for future bioaugmentations. A soil that has been polluted by crude oil for many years and that will be referred hereafter as "F-soil" was used in our experiments.

* Corresponding author: Tel +32 2 7647449 Fax +32 2 7647498 e-mail wattiau@mipa.ucl.ac.be.

Novel Approaches for Bioremediation of Organic Pollution, edited by Fass *et al.* Kluwer Academic / Plenum Publishers, New York, 1999.

2. RESULTS

2.1. Initial Characterisation of the Soil

The mineral oil content of the F-soil was analysed after either freon or Acetone/petroleum ether extraction. Both Infra Red and Gas Chromatography analysis methods showed an oil contamination of about 50 gr/kg of dry soil and a main hydrocarbon range between C10 and C40. The mineral nitrogen content was low (C/N ratio of 68/1) whereas the phosphate content was below the detection limit (< 2.5 mg/kg). Polyaromatic compounds were present at concentrations that were under the European norms for industrial sites. The F-soil was a sandy soil (mean diameter of the particles : 142μm) and had a pH close to the physiological value (7.30).

Given its historical pollution and despite insufficient nitrogen and phosphorous sources, the F-soil was expected to contain hydrocarbon-degrading bacteria. However, a very small number of such bacteria could be isolated at start (<10 CFU/gram). To demonstrate a bioremediation potential and a diverse initial microflora, F-soil samples were used to inoculate a series of liquid cultures consisting of a mineral growth medium with no other carbon source than just crude oil or diesel fuel. After 2 weeks at 25°C, cultures were stopped and analysed. All cultures were positive, suggesting that a hydrocrabon degrading microflora can easily be enriched from the F-soil. 10 colonies recovered from each test and chosen according to their peculiar macroscopic aspect (*i.e.* as different as possible) were analysed by REP-PCR (Versalovic et al., 1991) and the individual patterns were compared. Table 1 shows that in most cases, different bacterial species were recovered from each culture, whereas the number of different bacterial species recovered from each test was never greater than 4.

Table 1. Comparison of hydrocarbon utilising bacteria isolated from enrichment cultures and by direct plating

	Inoculum				
	Concentrated aqueous soil extract	Soil sample	Soil sample	Soil sample	
	Method				
	Direct plating on solid medium	Liquid enrichment (1 culture)	Liquid enrichment (6 subcultures)	Liquid enrichment (1 culture)	
	C-source				
Strain	Crude oil	Crude oil	Crude oil	Diesel fuel	Genus
VM121	10	10	-	-	*Rhodococcus* sp.
VM122	80	40	-	-	*Stenotrophomonas sp.*
VM123	10	-	-	-	ND
VM168	-	50	-	-	ND
VM277	-	-	90	-	ND
VM278	-	-	10	-	ND
VM265	-	-	-	10	ND
VM266	-	-	-	30	ND
VM270	-	-	-	20	ND
VM271	-	-	-	40	ND

The values indicate the percentage of the isolates displaying identical genotypic patterns (as analysed by REP-PCR). Bacterial genus was determined by partial sequencing of the 16SrRNA gene. ND, not determined

The F-soil thus contains many different bacteria with interesting oil-degrading capacity, but the number of viable bacteria is very low at start. The reason for such a scarce initial colonisation by oil-degrading bacteria is unknown and, as will be demonstrated later, cannot be completely explained by the scarcity of nutrient concentrations or by the presence of toxic compounds. Fortunately enough, this situation provides excellent "blank" conditions for a dynamic study of bacterial colonisation.

2.2. Monitoring of Dry Solid Reactor Biotreatments

To simulate a bioremediating soil system, a few kilos of the F-soil were treated in sterile dry solid reactors (DSR), which we estimated would mimic on a reasonable time scale a soil undergoing an aerated biotreatment, though in quite more idealised conditions. This type of reactor allows continuous mixing of dry soils together with on-line monitoring of several physico-chemical parameters (Springael et al., 1995). Two DSRs were run with about 2.5 kg of F-soil and a water content of 25%. One of the reactors was supplemented with mineral nutrients (KH_2PO_4 and NH_4Cl) so that a C:N:P ratio of 100:10:1 was reached, whereas the other reactor was not amended. Soil samples were retrieved from both reactors every 7 days during 1 month and analysed for their bacterial and oil content. Chemical monitoring revealed that almost no oil degradation occurred during this treatment, as analysed by Infra Red spectrometry (data not shown). Bacteria were recovered from the soil samples by either aqueous or hydrophobic solvent extraction. The hydrophobic solvent we used was diesel fuel. This product contains middle-distillated hydrocarbons resembling those contaminating the F-soil. Diesel fuel was apparently harmless for the local bacteria since we noticed that they could survive in this solvent for at least several days. Once recovered in the form of diesel fuel suspensions, the so-called " hydrophobic " bacterial extracts were plated on mineral M284 solid medium (Mergeay et al., 1985) so that their hydrocarbon-degrading capacity was directly assayed. The hydrocarbon mixture used to feed bacteria recovered by aqueous extraction was also obtained by extracting the initial F-soil's hydrocarbons by diesel-fuel. This diesel-extracted soil mixture was cleared from both soil's particles and micro-organisms by filtration through appropriate filters and plated on M284 medium to feed bacteria. Specific growth on oil-hydrocarbons was controlled by plating aqueous bacterial suspensions on mineral medium M284 without carbon source. The bacterial counts deduced from each soil sample is summarised in Table 2.

After 14 days of growth at 25°C, hydrocarbon-utilising colonies were sorted according to macroscopic criteria. 5 isolates of each pre-sorted category were then reisolated and further analysed by comparing either REP-PCR or 16S-23S-ITS-PCR profiles (Barry et al., 1991). One strain of each well-defined category was then analysed by partially sequencing a PCR product corresponding to the 16SrRNA gene. The results of this bacterial

Table 2. Counts of hydrocarbon-degrading bacteria recovered from dry solid reactor-treated F-soil (CFU/gr of soil)

	Non-amended DSR		DSR amended with nutrients	
Day	Hydrophilic	Hydrophobic	Hydrophilic	Hydrophobic
0	<10	438	<10	438
6	5×10^4	ND	ND	ND
14	1.1×10^6	2.2×10^3	1.2×10^4	1.4×10^4
20	2×10^6	6.9×10^3	5.3×10^6	1.7×10^5
28	5×10^6	5×10^3	2.1×10^7	1.6×10^5

"Hydrophilic" bacteria were isolated from soil samples extracted with NaCl (9 gr/L), "hydrophobic" bacteria were isolated by further extracting the NaCl-washed residual soil samples with diesel fuel. ND: Not Determined.

Table 3. Species identification of hydrocarbon-degrading bacteria recovered from dry solid reactor-treatments of F-soil, as determined by partial 16SrDNA sequencing and comparison with the GenBank database

Day	Strain	Closest species	Family	Similarity	% in the HL fraction	% in the HB fraction
6	VM193	*Pseudomonas* A177	Gram—(γ subdivision)	1249/1258	>95%	>95%
14	VM193	*Pseudomonas* sp. A177	Gram—(γ subdivision)	1249/1258	86%	52%
	VM209	*Actinomyces birnadii*	Firmicutes (Gram +)	185/188	6%	19%
	VM225	*Brevibacterium* sp.	Firmicutes (Gram +)	148/148	8%	20%
	VM214	*Dietzia maris*	Firmicutes (Gram +)	1468/1469	0	9%
20	VM214	*Dietzia maris*	Firmicutes (Gram +)	1468/1469	54%	94%
	VM225	*Brevibacterium* sp.	Firmicutes (Gram +)	148/148	19%	2%
	VM209	*Actinomyces birnadii*	Firmicutes (Gram +)	185/188	10%	4%
	VM236	*Brachybacterium conglomeratum*	Firmicutes (Gram +)	299/301	17%	0%
28	VM214	*Dietzia maris*	Firmicutes (Gram +)	1468/1469	90%	96%
	VM225	*Brevibacterium* sp.	Firmicutes (Gram +)	148/148	8%	2%
	VM209	*Actinomyces birnadii*	Firmicutes (Gram +)	185/188	2%	2%

Similarities are given as the fraction of identical sequenced residues. The species distribution in each sample is given as the fraction of each species for both hydrophilic (HL) or hydrophobic (HB) extract.

analysis is summarised in table 3. In the DSR not supplemented with nutrients, only a single *Pseudomonas* species with apparent hydrocarbon-degrading capacity could be recovered during the entire treatment. Though tolerant to the F-soil's conditions and intensively dividing, this strain utilised soil's substrates that could later not be determined by metabolic studies (Table 5). By contrast, in the course of the nutrient-supplemented DSR treatment, at least 5 different hydrocarbon-degrading species could be identified that mainly belonged to the Gram + bacterial family. Moreover, water and diesel fuel extraction's did not yield the same proportion of isolates for all the samples recovered from that reactor. The same *Pseudomonas* species recovered from the non-amended reactor also developed during the first 2 weeks in the nutrient-amended one and disappeared after day 14. This species was mainly associated with the water-extractable fraction. At that moment (14 days), 3 previously undetected species started to grow from which 2 were found preferentially associated with the hydrophobic phase of the soil (*Actinomyces* sp., *Brevibacterium* sp.) whereas one was found almost exclusively associated to that latest phase (*Dietzia* sp.). In the next 2 samples, the number of *Dietzia* colonies substantially increased and could be recovered by both diesel fuel and water-extraction. At day 28, the time that was chosen to stop the experiment, it had almost completely colonised the reactor.

2.3. Monitoring of Slurry Reactor Biotreatments

Preliminary results suggested that similar if not identical colonisation profiles could be observed when slurry reactors rather than DSRs were used for the bacterial monitoring

Table 4. Oil content (in mg per kg of dry soil) and counts of total hydrocarbon-degrading bacteria in F-soil during treatment in slurry reactors (CFU/gr of soil)

	Day 0		Day 32		Day 67	
Slurry reactor conditions	Oil content	Bacterial counts	Oil content	Bacterial counts	Oil content	Bacterial counts
Untreated	46250	< 10	-	-	-	-
A (+ nutrients)	ND	< 10	ND	1.13×10^6	18667	4.5×10^5
B (- nutrients)	ND	< 10	36500	3.9×10^5	31000	3.4×10^5
C (+ nutrients + bacteria)	ND	< 10	27000	6.4×10^5	14333	7.0×10^5
D (+ bacteria)	ND	< 10	37667	1.1×10^5	23333	7.5×10^5
E (+ NaN_3 = dead control)	ND	< 10	47333	< 10	44500	100

studies. Therefore, a series of slurry reactors were loaded with F-soil and treated under different conditions : one reactor was amended with nutrients (reactor A), another one was not (reactor B), a third one was amended with both nutrients and a bacterial inoculum originating from the DSR-treated, nutrient-amended soil (reactor C), another one was amended with the same inoculum but no nutrients (reactor D) and a control reactor was supplemented with the bactericidal compound sodium azide. These slurry reactors were run for 2 months. By contrast with the previous experiment, a decrease in oil contamination could be observed, as monitored by GC-FID (Table 4). The best results were obtained in reactor C, from which 40% of the oil content was eliminated within one month. After two months, the oil removal observed in reactor C was no longer different from that of reactor A (70% and 60% of oil removal, respectively), suggesting that the effect of the bacterial addition operated in reactor C took place in the course of the first month of treatment. Addition of nutrients had a substantial effect on oil removal whereas no significant difference could be observed in the bacterial counts of the different reactors analysed after either 1 or 2 months of incubation. Surprisingly, bacterial analysis after 2 months revealed that all the reactors were mainly colonised by a single species that belongs to the *Mycobacterium* genus as determined by partial 16SrRNA sequencing.

2.4. Metabolic Analysis of the Recovered Bacteria

To determine the nature of the substrates that were metabolised by the F-soil's microflora, a series of test tubes containing mineral M284 medium supplemented with just one of the hydrocarbons listed in Table 5 were separately inoculated with 3 different bacteria recovered from the biotreatment tests or with a diluted sample of the initial soil. The results shown in Table 5 yields kind of a "hydrocarbon metabolic fingerprint" of the bacteria. It can be noticed that most of the hydrocarbons utilised by the individual species belong to the alkane family. Compared to the isolated species, the metabolic potential of the soil's population is wider, suggesting that more (yet unisolated) bacteria with additional catabolic potential are likely to contribute to hydrocarbons' removal. This experiment also demonstrates the bioremediation potential of the initial soil's microflora that was not highlighted by direct plating (see chapter 2.1).

Southern blot experiments confirmed the existence of alkane degradation genes in the genome of the F-soil's bacteria. Genomic DNA from strains VM209, VM214, VM121 and VM122 probed with DNA fragments corresponding to the *alkST* or *alkBAC* genes of *Pseudomonas oleovorans* (Eggink et al., 1988) yielded very clear hybridisation signals, suggesting that these bacteria contain genes similar to those involved in alkane metabolism by *Pseudomonas oleovorans* (data not shown).

Table 5. Hydrocarbon "metabolic fingerprint" of 3 bacteria isolated during DSR treatment of F-soil and of a sample of the initial soil

Compound/Inoculum	*Actinomyces* sp. VM209	*Dietzia* sp. VM214	*Pseudomonas* sp. VM193	Diluted soil sample
Saturated alkanes				
n-pentane	-	+	-	+
n-heptane	+	+	-	+
n-octane	+	+	-	+
n-nonane	+	+	-	+
n-decane	+	+	-	+
n-undecane	+	+	-	+
n-dodecane	+	+	-	+
n-hexadecane	-	-	-	+
n-pentadecane	-	-	-	+
n-heptadecane	-	-	-	+
n-octadecane	+	+	-	+
n-nonadecane	-	+	-	+
n-eicosane	-	-	-	+
Alkenes				
1-octene	-	-	-	+
1-octadecene	-	-	-	+
trans-4-octene	+	+	-	+
Branched chains				
2,6,10,14-tetramethyl-pentadecane	-	-	-	+
Cycloalkanes				
cyclopentane	-	-	-	-
cyclohexane	-	-	-	-
PAHs				
chinoline	-	-	-	-
naphthalene	-	-	-	+
phenanthrene	-	-	-	+
fluorene	-	-	-	-
pyrene	-	-	-	-
fluoranthene	-	-	-	-
anthracene	-	-	-	-
biphenyl	-	-	-	-
Composite oils				
diesel oil	+	+	-	+
crude oil	+	+	-	+
Aromatic metabolites				
m-toluate	-	-	-	+
p-toluate	-	-	-	+
Benzoate	+	+	-	+
Phenol	-	-	-	+
Salicylate	-	-	-	+
LB medium	+	+	+	+

3. DISCUSSION

Bioremediation of soils contaminated by crude oil or distillated fuels is known to rely on the presence of indigenous hydrocarbon-degrading organisms and on the availability of oxygen and nutrients (nitrogen and phosphorous sources). In the present study of a crude-oil polluted soil, we focused on the monitoring of bacteria developing in small scale reactors (solid or liquid) eventually amended with nutrients and bacterial cultures derived from the soil. The soil used in our experiments initially contained a few number of hydrocarbon-degrading bacteria. However, a bacterial population able to degrade hydrocarbons could be easily enriched from the soil. Most of the initially cultivable hydrocarbon-degrading bacteria were never recovered in the course of the biotreatments. Oil removal was rather conducted by bacterial species that were initially undetectable, either because their initial number was too low or because they were in a non cultivable state. A remarkable example of this peculiar behaviour is given by a bacterial species resembling *Dietzia maris* (Rainey et al., 1995) that almost completely colonised a nutrient-amended DSR within 1 month, whereas it was undetectable at start. This bacterium seems to be associated with the hydrophobic phase of the soil and could degrade a wide variety of hydrocarbons.

We clearly observed that different species developped depending on whether nutrients are supplied to the soil or not. In our experiments, Gram positive bacteria from the Firmicutes family were exclusively recovered from the nutrient-amended soil. In addition, the natural remediation of the soil seems to fit an apparent colonisation chronology though we could not link the sequential development of individual species to the sequential removal of specific pollutants.

Enrichment cultures from the initial soil assayed with different hydrocarbon mixtures led to the development of bacterial species that were always different relative to each other. Thus, we could demonstrate a "self-formatting" capacity of the microflora , *i.e.* the preferential development of peculiar species according to the nature of the pollutants and to the soil's physiological conditions. Finally, chemical analysis of the soil and metabolic studies of the microflora suggest that alkanes (and possibly alkenes) are the hydrocarbons that are the most easily catabolised by the micro-organisms, although this has not been quantified. Bioaugmentations assayed with native bacteria derived from reactors run during one month with the same soil increased the rate of oil removal during the first weeks of the laboratory treatments, but had no influence on the later removal rate.

ACKNOWLEDGMENTS

We are very grateful to M.-E. Renard, A. Ryngaert and L. Hooyberghs for their excellent technical assistance and to J.-M. Debaty, S. Demeure and J.-F. Couvreur for their enthusiastic collaboration and suggestions. This work was supported by a grant from the PetroFina company (Brussels, Belgium).

REFERENCES

Barry, T., Colleran, G., Glennon, M., Dunican, L.K., and Gannon, F. (1991). The 16S/23S ribosomal spacer region as a target for DNA probes to identify eubacteria. *PCR Methods Appl* 1:51–56.

Eggink, G., Engel, H., Meijer, W.G., Otten, J., Kingma, J., and Witholt, B. (1988) Alkane utilization in *Pseudomonas oleovorans*. Structure and function of the regulatory locus *alkR*. *J. Biol. Chem.* **263**:13400–13405.

Mergeay, M., Nies. D., Schlegel, H.G., Gerits, J., Charles, P., and Van Gijsegem, F. (1985). *Alcaligenes eutrophus* CH34 is a facultative chemolithotroph with plasmid-bound resistance to heavy metals. *J. Bacteriol.* **162**:328–334.

Rainey, F.A,, Klatte, S., Kroppenstedt, R.M., and Stackebrandt, E. (1995). *Dietzia*, a new genus including *Dietzia maris* comb. nov., formerly *Rhodococcus maris*. *Int. J. Syst. Bacteriol.* 45:32–36.

Springael, D., Carpels, M., Hooybrechts, L., Vanhoutven, D., Bastiaens, L., Kinnaer, L., and Diels, L. (1995). Monitoring of the mineral oil degrading *Pseudomonas aeruginosa* LH220 tagged with mini-Tn*5-luxAB* during bioremediation of oil contaminated soil in a dry solid reactor. *Med. Fac. Landbouw. Univ. Gent.* **60**:2647–2650.

Versalovic, J., Koeuth, T., and Lupski, J.R. (1991). Distribution of repetitive DNA sequences in eubacteria and application to fingerprinting of bacterial genomes. *Nucleic Acids Res* **25**:6823–6831.

19

BIOAVAILABILITY OF 1,2,3,4-TETRACHLORODIBENZO-*p*-DIOXIN (TCDD) FOR DECHLORINATION BY AN ANAEROBIC MICROBIAL CONSORTIUM, IN THE PRESENCE OF DISSOLVED ORGANIC CARBON

Preliminary Results

M. Toussaint,* H. B. Krop, M. N. Grooteman, M. van Breugel, P. de Vries, and J. R. Parsons

Dept. Environmental and Toxicological Chemistry, ARISE
University of Amsterdam
Nieuwe Achtergracht 166, 1018 WV Amsterdam, The Netherlands

1. INTRODUCTION

The availability of persistent and hydrophobic compounds such as chlorinated dioxins (log K_{ow} 3.4.-13.08 Mackay *et al.*, 1992) for bacteria is often thought to be the limiting factor for their biodegradation and therefore also the limiting factor for the bioremediation of these compounds (Leahy and Colwell, 1990; Mihelcic *et al.*, 1993; Harms and Bosma, 1997). The fraction of these compounds freely dissolved in the water phase is commonly thought to be the amount readily bioavailable (Belfroid *et al.*, 1996). The fraction of the compound sorbed to sediment or to dissolved organic carbon (DOC) is thought to be either reversibly or irreversibly bound and therefore slowly or not at all available for biodegradation (Leahy and Colwell, 1990; Mihelcic *et al.*, 1993; Harms and Bosma, 1997). In this situation the fraction dissolved in the water phase and the rate of desorption of the sorbed fraction determines the degradation rate of the compound. Sediment or DOC is often excluded from experiments to determine the rate of degradation of such com-

* Corresponding author: E-mail: mtouss@mtc.chem.uva.nl

Novel Approaches for Bioremediation of Organic Pollution, edited by Fass *et al.* Kluwer Academic / Plenum Publishers, New York, 1999.

pounds by micro-organisms. However, natural systems always include sediment or soil and DOC, and moreover bacteria tend to produce, to excrete or to be associated with DOC (Wetzel, 1975; Stumm and Morgan, 1981). This is an argument why DOC (or sediment) should be included in degradation experiments when bioavailability or bioremediation is studied.

DOC is a rather undefined substance and is composed of different components that pass through a 0.45μm filter. It is formed by, among others, degradation (biotic or abiotic) of macromolecules from the environment. These macromolecules originate from flora and fauna in the (aquatic)ecosystem (Schnitzer and Khan, 1972; Wetzel, 1975; Suffet and MacCarty, 1989). Its composition therefore depends on its source and is therefore variable and often difficult to determine, although several attempts were made to define its chemical structure (Suffet and MacCarty, 1989; Schnitzer and Khan, 1972). DOC diminishes the water surface tension and therefore acts similar to a surfactant (Guetzloff and Rice, 1994). This means that DOC is capable of forming micelles, with a hydrophobic interior, above the critical micelle concentration (CMC), in which hydrophobic compounds such as dioxins could partition. Below this CMC, DOC could act as a cosolvent and thus increase the apparent solubility. The interaction of DOC with compounds such as dioxins is however poorly studied. Some studies have been carried out on the bioavailability of compounds for micro-organisms for degradation, in the presence of DOC or humic acids. The amount available is normally decreased when DOC or sediment is present (Parsons, 1992; Katayama *et al.*, 1993; Robinson and Novak, 1994).

Other influences on dechlorination may also occur when DOC or comparable compounds are present. These compounds may contain quinone structures or they could be associated to a metal in such a way that they can serve as an electron shuttle and therefore may be involved in redox reactions such as reductive dechlorination (Curtis and Reinhard, 1994; Barkovskii *et al.*, 1995; Lovely *et al.*, 1996a and b; Lovely, 1996b; Adriaens *et al.*, 1996). Comparable compounds with similar functions are found in the membranes of micro-organisms, where they function in an electron transfer chain (respiratory chain) and as such involved in the production of ATP (Dolfing, 1990; Gantzer and Wackett, 1991; Assaf Anid *et al.*, 1992; Schumacher and Holliger, 1996). These membrane electron shuttles could be involved in dechlorination of compounds such as dioxins by donating an electron to them. The process in which an electron and a proton is donated by the reductant (donor) to the oxidant (acceptor), thereby replacing a halogen atom by a proton, is called reductive dehalogenation (Mohn and Tiedje, 1992; Commandeur and Parsons, 1994)

We have previously shown that 1,2,3,4-tetrachlorodibenzo-*p*-dioxin (TCDD) is dechlorinated by an anaerobic mixed culture enriched from Lake Ketelmeer sediment which is contaminated with dioxins, PCBs and chlorobenzenes (Beurskens *et al.*, 1995). This culture has been found capable of dechlorinating 1,2,3,4-TCDD to 1,2,4-trichloro-*p*-dioxin (TrCDD) and 1,2,3-TrCDD, 1,3-dichlorodibenzo-*p*-dioxin (DCDD) and 2,3-DCDD, and even minor amounts of 1-monochlorodibenzo-*p*-dioxin (MCDD) and 2-MCDD (Beurskens *et al.*, 1995; Toussaint *et al.*, 1998). The dechlorination of 1,2,3,4-TCDD was relatively fast, with a $t_{1/2}$ of 15.5 days.

In this study we examined the effect of different concentrations of natural extracted DOC on the bioavailability of 1,2,3,4-TCDD, for reductive dechlorination. The total amount of metabolites reductively formed from 1,2,3,4-TCDD is taken as a measure of bioavailability of this compound. This is justified because the purpose of this study was to determine which amount is available for dechlorination of 1,2,3,4-TCDD by this culture. This may be an underestimation of the bioavailability of the compounds for uptake by the bacteria. Therefore the amount of metabolites found is at least an underestimation of the bioavailability for this culture.

2. MATERIALS AND METHODS

Four large (1.8 liter) anaerobic mixed cultures, originally from Lake Ketelmeer sediment, were enriched with hexachlorobenzene as an electron acceptor as described earlier (Beurskens *et al.*, 1995). Five different DOC concentrations were used (0, 10, 25, 100, 400 mg/l) and samples were taken at five time intervals (0, 1, 2, 6, 10 days) after adding the bacteria. Triplicate sterile and anaerobic bottles were used for each time interval and each DOC concentration. These were filled with 10 ml of a certain concentration of extracted DOC from a natural uncontaminated sediment (Schoonrewoerdsewiel, The Netherlands) to obtain the concentrations of DOC, mentioned above. To these bottle 11.84 μg of 1,2,3,4-TCDD was added in 200 μl aceton/methanol, to give a final concentration (after adding the bacteria) of 236.8 μg/l in each incubation. This mixture of DOC and 1,2,3,4-TCDD was left to equilibrate while shaking for one day before adding the bacteria. To each bottle, 40 ml of the bacterial culture from the large incubations which had been homogenized before, was added.

Additional bottles served as controls and were first pasteurized (75°C, 20 min.) and then, after 24 hours incubation at 23°C, autoclaved (121°C, 3 h), according to Van Dort and Bedard (1991), before adding 1,2,3,4-TCDD. Additional incubations without bacteria served as controls for abiotic dechlorination. Bottles were left shaking at room temperature until they were sampled. Each bottle was sampled (50 ml) in triplicate at the given time intervals. The bacterial activity was stopped by adding 3 ml KOH (10 M) and, after addition of a mixture of internal (recovery) standards (a separate internal standard for each compound), the samples were extracted with pentane and a clean-up was done as described previous by Beurskens *et al.* (1995). A second internal (injection) standard was added, just before analysis using GC-ECD.

Analyses were done on a Carlo Erba (GC 8000, MFC 800) gas chromatograph, using a cold on-column injector, and a DB-5 capillary column with a 0.53 mm ID retention gap (J&W, 30m x 0.323 mm) and a ^{63}Ni electron capture detector. Helium was used as the carrier gas and argon/methane as the make-up gas. The following temperature program was used: Initial temperature: 120°C, rate: 30°C/min to 180°C and 2°C/min to 190°C hold 12 min; rate 20°C/min to 320°C, hold 1 min. The sample size injected was 1 μl. Quantification was done by using authentic pure standards at 9 different concentrations.

Metabolites were identified on a GC-MSD (Hewlett Packard 5890 gas chromatograph with a Hewlett Packard 5970 mass selective detector, column: DB-5 (J&W, 30 m x 0.32 mm) and cold on column injection. The carrier gas was He, inlet pressure: 3 psi; temperature program: initial 100°C, rate 30°C/min to 150, rate 4°C/min to final 250°C). Selective ion monitoring of the M, M+2 and M+4 masses was used. Metabolite identification was confirmed by comparison of mass spectra with those of authentic standards.

The standards 1,2,3-TrCDD, 1,2,4-TrCDD and 2,3-DCDD were obtained from Amchro (Accustandard, 99% pure, Sulzbach/Taunus, Germany). 2-CDD was synthesized following the method of Pohland and Yang (1972) and 1,3-DCDD was synthesized following the method of Gray *et al.* (1972). All synthesized chemicals were purified on silica-gel columns eluted with distilled hexane (Janssen Chimica, Tilburg, The Netherlands) and also checked for contamination by GC-MSD and GC-ECD. Sulphuric acid was obtained from Janssen Chimica (Tilburg, The Netherlands). Pentane and 2,2,4-trimethylpentane (glass distilled grade) were obtained from Rathburn (Walkerburn, Scotland). NaOH was obtained from J.T. Baker (Deventer, The Netherlands). Silica gel 60 (70–230 mesh) was obtained from Merck (Amsterdam, The Netherlands).

3. RESULTS AND DISCUSSION

In this paper we present some of the preliminary results of this study. The complete results will be published elsewhere. The results of the incubations in the absence of DOC is plotted against time in figure 1. The metabolite production was relatively low and the mass balance is poor. This can be due to the unexpected relatively low activity of the bacteria. The total amount of metabolites was 0.252 nM after 10 days (fig.1). In contrast to this, the activity was much higher in the incubations where DOC was added: for 25 mg/l DOC (fig. 2) this was 0.386 nM and for 100mg/l DOC (fig. 3) this was 1.038. Figure 4 shows the total amount of metabolites formed at day 10 plotted against the different DOC concentrations. A trend is apparent towards increasing metabolite production with increasing DOC incubation. However, above 100 mg/l DOC the activity appears to decrease. The reduced bioavailability at 400 mg/l DOC may be due to the formation of micelles at DOC concentrations above the critical micelles concentration. This indicates that below the CMC the DOC may act as a cosolvent and thereby increase the bioavailability of 1,2,3,4-TCDD for the bacteria. However, when micelles are formed, above the CMC, the desorption from this separate phase may limit the bioavailability of the dioxin. The findings that DOC, at certain concentrations, could increase the bioavailability of hydrophobic compounds for these bacteria is in contrast to many other studies with other aquatic organisms (Kukkonen *et al.*, 1990; Parsons, 1992; Katayama *et al.*, 1993; Robinson and Novak, 1994; Belfroid *et al.*, 1996; Landrum *et al.*, 1997).

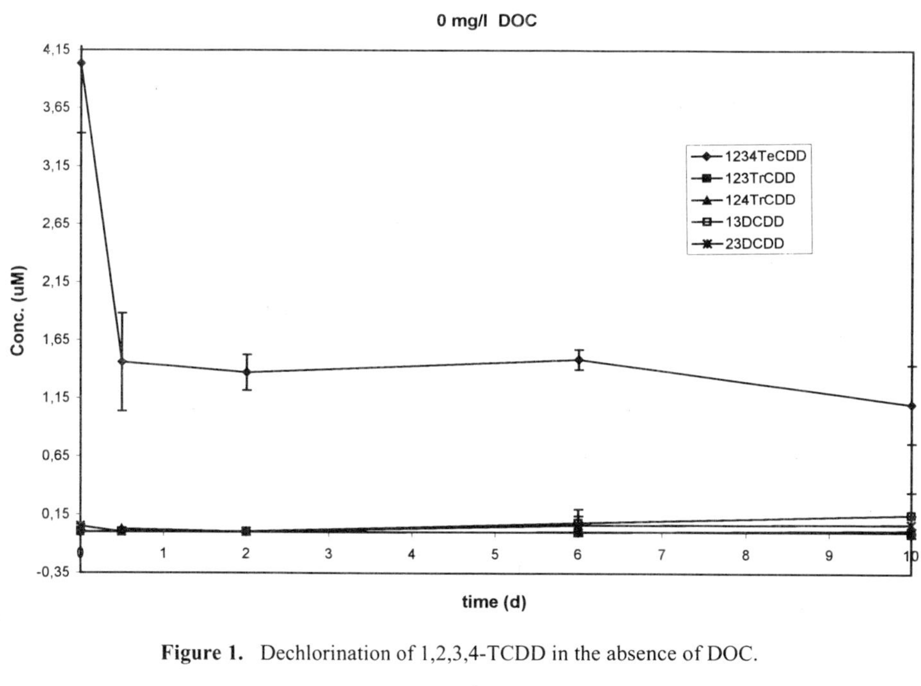

Figure 1. Dechlorination of 1,2,3,4-TCDD in the absence of DOC.

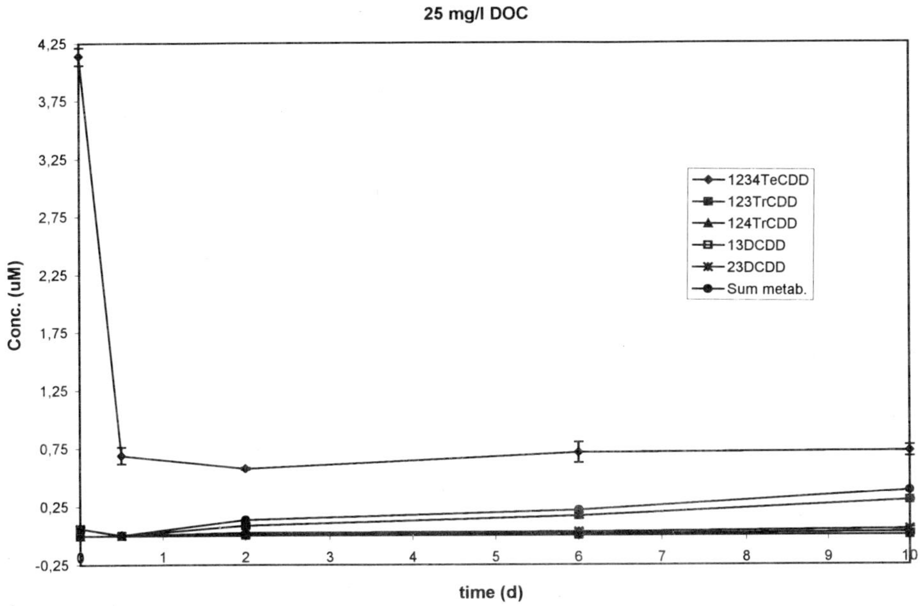

Figure 2. Dechlorination of 1,2,3,4-TCDD in the presence of 25 mg/l DOC.

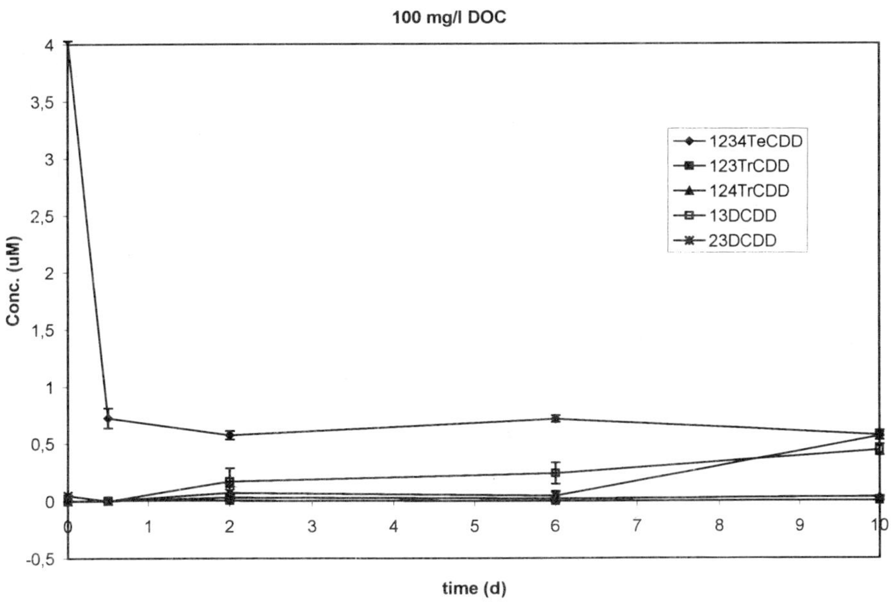

Figure 3. Dechlorination of 1,2,3,4-TCDD in the presence of 100 mg/l DOC.

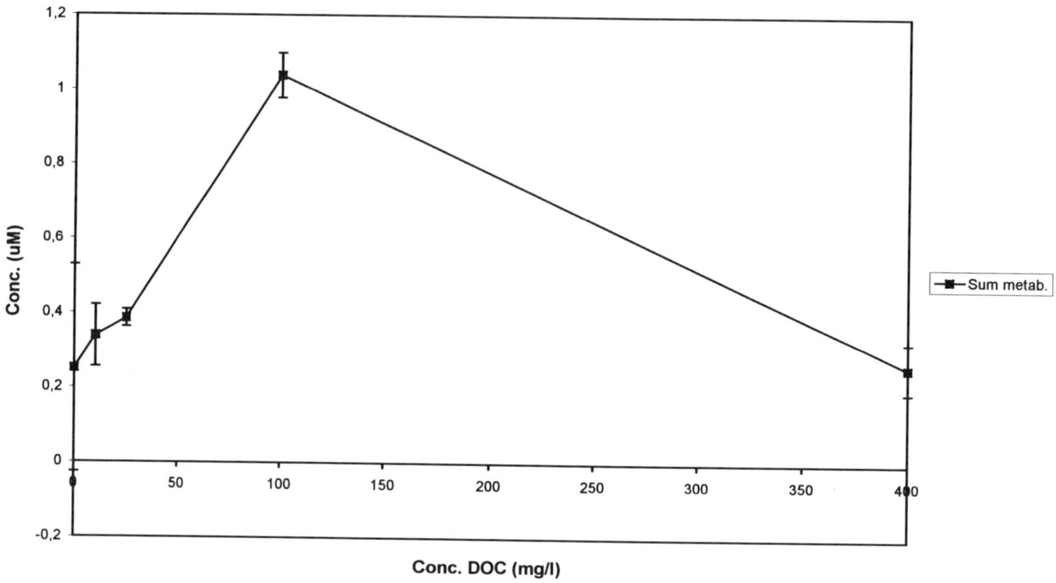

Figure 4. The effect of DOC on the dechlorination of 1,2,3,4-TCDD after 10 days.

These results clearly show that DOC does not always decrease the bioavailability of hydrophobic compounds for microorganisms and that DOC could play a role in bioremediation. Whether this role of DOC is as a cosolvent or involves the dechlorination process itself is unclear. Further research is needed to investigate this.

REFERENCES

Adriaens P., Chang P.R., and Barkovskii A.L. (1996). Dechlorination of PCDD/F by organic and inorganic electron transfer molecules in reduced environment. *Chemosphere* **32**: 433–441.

Assaf-Anid N. Nies L., and Vogel T.M. (1992). Reductive dechlorination of a polychlorinated biphenyl congener and hexachlorobenzene by vitamin B_{12}. *Appl. Environ. Microbiol.* **58**: 1057–1060.

Barkovskii A.L., Bouilant M.L., Monrozier, L.J., and Balandreau, J. (1995). *Azospirillum* strains use phenolic compounds as intermediates for electron transfer under oxygen-limiting conditions. *Microb. Ecol.* **29**: 99–114.

Belfroid, A.C., Sijm, D.T.H.M., and Van Gestel, C.A.M. (1996). Bioavailability and toxicokinetics of hydrophobic aromatic compounds in bentic and terrestrial invertebrates. *Environ. Rev.* **4**: 276–299.

Beurskens J.E.M., Toussaint M., De Wolf J., van der Steen, J.M.D., Slot, P.C., Commandeur, L.C.M., and Parsons, J.R. (1995). Dehalogenation of chlorinated dioxins by an anaerobic microbial consortium from sediment. *Environ. Toxicol. Chem.* **14**: 939–943.

Beurskens J.E.M., Mol G.A.J., and Barreveld H.L. (1993). Geochronology of priority pollutants in a sedimentation area of the Rhine river. *Environ. Toxicol. Chem.* **12**: 1549–1566.

Commandeur L.C.M., and Parsons JR. (1994). Biodegradation of halogenated aromatic compounds. In Ratledge C, ed. Biochemistry of Microbial Degradation. Dordrecht: Kluwer: 423–458.

Curtis G.P,, and Reinhard, M. (1994). Reductive dehalogenation of hexachloroethane, carbon tetrachloride and bromoform by anthrahydroquinone disulfonate and humic acid. *Environ Sci. Technol.* **28**: 2393–2401.

Dolfing J.(1990). Reductive dechlorination of 3-chlorobenzoate is coupled to ATP production and growth in an anaerobic bacterium, strain DCB1. *Arch. Microbiol.* **153**: 264–266.

Dort, van H.M., and Bedard, D.L. (1991). Reductive ortho and meta Dechlorination of a Polychlorinated biphenyl congener by anaerobic bacteria. *Appl. Environ. Microbiol.* **57,** 1576.

Gantzer C.J., and Wackett L.P. (1991). Reductive dechlorination catalyzed by bacterial transition-metal coenzymes. *Environ. Sci. Technol.* **25**: 715–722.

Gray, A.P., Cepa, S.P., Solomon I.J., and Aniline, O. (1976). Synthesis of specific polychlorinated dibenzo-p-dioxins. *J. Org. Chem.*, Vol.**41,** 2435.

Guetzloff, T.F., and Rice, J.A. (1994). Does humic form a micelle? *Sci. Tot. Environ.* **152**: 31–35.

Harms, H., and Bosma, T.N.P. (1997). Mass transfer limitations of microbial growth and pollutant degradation. *J. Ind. Microbiol. Biotechnol.* **18**: 97–105.

Katayama, A, Fujiama, Y., and Kuwatsuka, S. (1993). Microbial degradation of DDT at extremely low concentrations. *J. Pesticide Sci.* **18**: 289–298.

Kukkonen, J., McCarty, J.F., and Oikari, A. (1990). Effects of XAD-8 fractions of dissolved organic carbon on the sorption and bioavailability of organic micropollutants. *Arch. Environ. Contam. Toxicol.* **19**: 551–557.

Leahy, J.G., and Colwell, R.R. (1990). Microbial degradation of hydrocarbons in the environment. *Microbiol. Rev.* **54**: 305–315.

Landrum, P.F., Gossiaux, D.C., and Kukkonen, J. (1997). Sediment characteristics influencing the bioavailability of nonpolar organic contaminants to *Diporeia* spp. *Chem. Spec. Bioavail.* **9**(2): 43–55.

Lovley D.R, Coates J.D., Blunt-Harris E.L., Phillips, E.J.P., and Woodward, J.C., (1996a). Humic substances as electron acceptors for microbial respiration. *Nature*. **382**: 445–448.

Lovley D.R., Woodward J.C., and Chapelle F.H.. (1996b). Rapid anaerobic benzen oxidation with a variety of chelated Fe(III) forms. *Appl. Environ. Microbiol.* **62**: 288–291.

Mackay D., Shiu W.Y., and Ma K.C. (1992). Illustrated handbook of physical-chemical properties and environmental fate for organic chemicals. Chelsea (Michigan): Lewis publishers. **11**: 368–483.

Mihelcic, J.R., Lueking, D.R., Mitzell, R.J., and Stapleton, J.M. (1993). Bioavailability of sorbed- and seperate-phase chemicals. *Biodegradation*. **4**: 141- 153.

Mohn W.W., and Tiedje J.M. (1992). Microbial reductive dehalogenation. *Microbiol. Rev.* , **56**: 482–507.

Parsons, J.R. (1992). Influence of suspended sediment on the biodegradation of chlorinated dibenzo-*p*-dioxins. *Chemospere*. **25**: 1973–1980.

Pohland, A.E., and Yang, G.C. (1972). Preparation and characterization of chlorinated dibenzo-*p*-dioxins. *J. Agric. Food Chem.* **20**: 1093–1099.

Robinson, K.G., and Novak, J.T. (1994). Fate of 2,4,6-trichloro(C^{14})-phenol bound to dissolved humic acid. *Wat. Res.* **28**: 445–452.

Schnitzer, M., and Khan, S.U. (172). Humic substances in the environment. Marcel Dekker, INC. New York.

Schumacher W., and Holliger C. (1996). The proton/electron ratio of the menaquinone-dependent electron transport from dihydrogen to tetrachloroethene in "Dehalobacter restrictus". *J. Bacteriol.* **178**: 2328–2333.

Stumm, W., and Morgan, J.J. (1981). Aquatic Chemistry, 2nd edition, Wiley, New York.

Suffet, H.I., and MacCarty, P. (editors) (1989). AquaticHumic substances. Influence on fate and treadment of pollutants. American Chemical Society, Washington, DC.

Toussaint M., Beurskens J.E.M., De Wolf J., Grooteman, M.N., van der Steen, J.M.D., Slot, P.C., and Parsons, J.R. (1998). Elucidation of the dechlorination pathway of 1,2,3,4-TCDD by an anaerobic microbial mixed culture isolated from sediment. in prep.

Wetzel, R.G. (1975). Limnology. W.B. Saunders Company, Philadelphia.

20

BIOTIC AND ABIOTIC FACTORS INVOLVED IN BIOAVAILABILITY OF POLLUTANTS

Y. Henis

The Hebrew University of Jerusalem
Faculty of Agricultural, Food and Environmental Sciences
P.O.B. 12, Rehovot 76100, Israel

The principles which determine bioavailability of organic pollutants are similar to those employed with regard to biodegradable natural organic substances prevailing in natural environments. The factors involved in pollutant's biodegradation can be devided arbitrarily into four main categories: the degrading microflora; the pollutant; the indigenous, nondegrading microflora and microfauna; and the abiotic factors which comprise the environment (Fig.1).

These categories interact with each other: The degrading microorganisms are affected by the nature of the pollutant, by the physical and chemical conditions prevailing in the environment and by the indigenous microflora and microfauna which are not directly involved in the degradation.

Some properties of the contaminant which have a potential effect on its bioavailability are presented in Table 1.

Recalcitrant molecules having slow rates of degradation are also common among natural products, which originate from living systems. Thus, whereas glucose added to soil is degraded within minutes and macromolecules such as albumin and starch are degraded within hours, some proteins other than albumin- such as collagen, and some natural polysaccharides other than starch- such as cellulose, are degraded at a much slower rate. Among natural products, the ones most resistant to degradation are lignin and soil humus. These differences in bioavailability evolve from the molecular structure. Synthetic pollutants may share with natural recalcitrant substances the slow activity rates of enzymes capable of degrading them. In addition, microorganisms capable of degrading synthetic pollutants may be scarce or absent in the contaminated environment. Other factors affecting pollutants' biodegradability are high toxicity, low solubility, molecular size and shape. The presence of particular substituents, such as halogens and nitro-groups , and the molecular site of the substitution, may greatly affect bioavailability (Alexander,1994).

In addition to these comprehended factors, there are some interesting, yet not fully understood phenomena, such as aging and complexing, which frequently affect bioavail-

Novel Approaches for Bioremediation of Organic Pollution, edited by
Fass *et al.* Kluwer Academic / Plenum Publishers, New York, 1999.

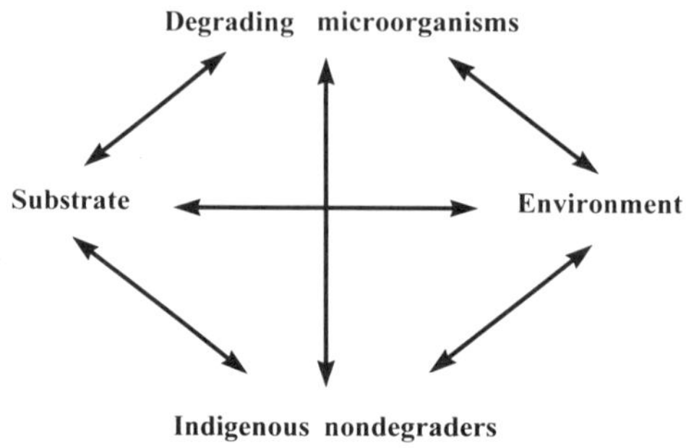

Figure 1. Bioremediation: Interactions between biotic and abiotic factors.

ability. Aging is defined as the decline in time of bioavailability of a particular pollutant in soil with no apparent change in its chemical or physical status in soil (Alexander, 1994). Some of the mechanisms which may be involved in aging are adsorption (retention by surfaces of solid materials) and absorption (retention within the mass of solid materials). Another phenomenon involved in the decline of a pollutant's bioavailability in soil is complexing with soil humus (Bollag, 1992).

Finally, the fate of a particular contaminant is frequently determined by the presence of other accompanying pollutants. A typical example of such a case is the contamination of terrestrial and aquatic environments by oil spill (Prince, 1992). Crude oil consists of many components, such as alkanes and polyaromatics, which differ in their bioavailability, affect each other's degradation, and are degraded by the microorganisms in sequence rather than in a simultaneous manner (Rosenberg and Ron, 1995). In the meantime, throughout the process of degradation, the presence of multiple substrates- many of which are quite toxic to each others' degraders- undoubtedly affects bioavailability .

A second category which affects biodegradation is the environment. Soil, as a terrestrial environment, (Table 2), is highly complicated and versatile, consisting of solid, liquid and gaseous phases.

Under natural conditions soil can be regarded as a noncontinuous system consisting of capillaries and spaces, some filled with water and some with gases. Both the contaminant and its potential degraders are randomly distributed in soil, forming patches and microbial microcolonies, respectively. The chance of their meeting each other depends on the contaminant's diffusibility and on the active and passive movement of the microbial cells, which in turn depends on the amount of free water present in the soil. According to Griffin

Table 1. Contaminant properties and bioavailability

1.	Natural products: Albumin *vs* collagen; starch *vs* cellulose *vs* lignin
2.	Synthetics *vs* natural: absence of enzymes
3.	Toxicity
4.	Solubility
5.	Molecular size
6.	Molecular shape (alkane-aromatic linkage point)
7.	Sequestering (aging): decline in bioavailability with time (micropores, sediments)
8.	Adsorption, absorption, complexing
9.	Substituents (halogens, NO_2)
10.	Single *vs* multiple components (fuel)

Table 2. The soil environment

1. Solid phase (sand, clay, humus, roots, residues)
2. Liquid phase (water)
3. Gaseous phase (H_2O vapors, O_2, CO_2, N_2)
4. Dissolved components: Inorganic and organic nutrients (N,P,S, Fe, microelements)
5. Physical parameters affecting microbial growth and bioavailability: Humidity, pH, temperature
6. Abiotic processes affected by soil type and conditions: adsorption, absorption, diffusion, complexing
7. Microbial movement and soil conditions: Griffin & Quail's law (Fig. 2)

and Quail (1968), movement of bacteria in a particular soil is possible only if the amount of the free water in that soil is at least 0.11 ml/g above the basic level of a matric potential of -bars=1.0. The heavier the soil, the greater is the amount of water that should be added to allow bacterial movement (Fig. 2; Griffin, 1972).

Because of the complexity of the soil structure, both diffusion and bacterial movement often take an indirect path (Fig. 3), resulting in a longer time period required for meeting than that precalculated for a homogenous environment (Alexander, 1994).

Biodegradation is affected by soil type, humidity, pH, temperature, and organic matter content. Degradation of organic contaminants is often enhanced by the addition of a

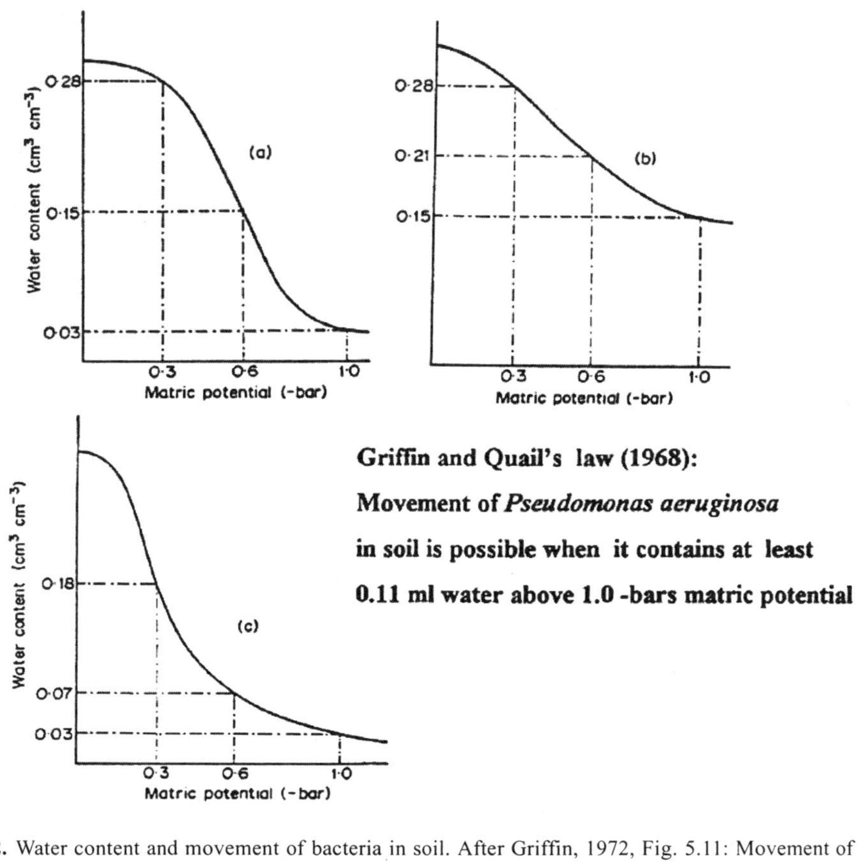

Figure 2. Water content and movement of bacteria in soil. After Griffin, 1972, Fig. 5.11: Movement of *Pseudomonas aeruginosa* in hypothetical soils: (a) loamy sand; (b) Clay loam; (c) sand. By permission of Chapman & Hall Ltd., London, UK.

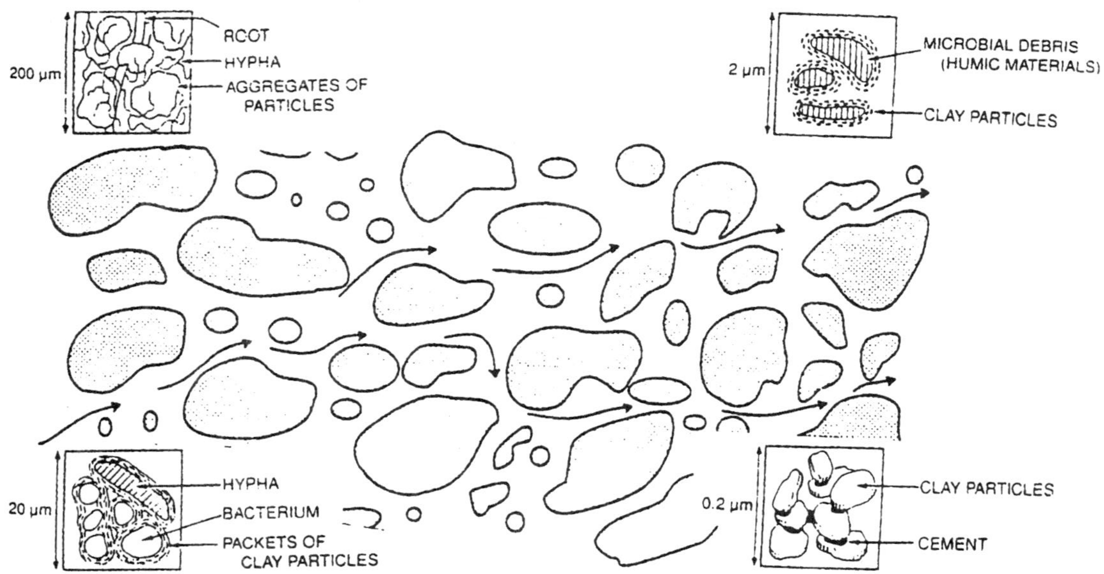

Figure 3. Indirect paths of contaminant and bacterial movement, and various microenvironments of different sizes in soil. After Alexander, 1994, Figs. 10.2 and 10.3. By permission of Academic Press, San Diego, CA.

Table 3. Microbial properties and bioavailability

1.	Morphology: cells (bacteria) *vs* mycelium (actinomycetes and fungi)
2.	Size: 0.2–1x 0.5–5 μm (bacteria) *vs* 3–10 μm (fungi), 10–20 μm (protozoa)
3.	Motility: active (gliding, flagella) *vs* passive (growth, streaming water)
4.	Growth pattern: surface (bacteria) *vs* penetration (fungi, actinomycetes)
5.	Dissolving capacity (detergents)
6.	Chemotaxis
7.	Predation: bacteriophages, Myxobacteria, *Bdellovibrio*, protozoa
8.	Coordination (simultaneous and successive): aerobic/anaerobic, metabolic consortia, cometabolism
9.	Acclimation: enrichment and buildup, mutation and selection, induction
10.	Growth kinetics: maintenance energy, μmax and Ks (autochthonous *vs* zymogenous)

readily degradable source of carbon and energy, such as acetate, especially when the contaminant is present at a level too low to support growth and enzyme synthesis (Bouwer, 1992).

In comparison to soil, the aquatic environment is much simpler. Yet, there are many physical and chemical factors which affect bioavailability in an aquatic environment. Solubility and specific density of the contaminant determine its availability and location-whether floating on the water, sinking to the bottom onto the benthic layer, adsorbed to suspended clay and organic matter particles, or taken up by the aquatic micro- and macro-flora and fauna. In the latter case, the uptake process often leads to biomagnification of the contaminant, resulting in the intoxication of floral and faunal populations (Mitchell, 1974). This process is common in both soil and aquatic environments. Another phenomenon, common to terrestrial and aquatic environments, is the response to contaminants of the microbial population known as Elton's principle: The higher the pollution, the lower is the diversity of the microbial population, pollution resulting in a drastic reduction in biological versatility, and accompanied by an enrichment of few organisms which take over, multiply and reach exceptionally high levels (Elton, 1958). A typical example of this phenomenon is the explosive growth of spore-forming bacilli following the addition of small amounts of peptone to soil (Thimann, 1966).

In both terrestrial and aquatic environments, physical conditions and chemical components that support growth and enzyme synthesis must exist to allow bioremediation. Essential elements are carbon, nitrogen, phosphorus, iron, magnesium, sulfur and microelements. An analysis of the contaminated environment for the presence of these components in the right proportions, the addition of missing factors and, If necessary, an energy source, to ensure optimal conditions for degradation of the contaminant, are recommended.

The relationships between the degrading microorganisms and the bioavailability of environmental contaminants are summarized in Table 3.

Cell size and shape of the microbial cell affect its capacity to reach and colonize microsites, especially in soil. Submicroscopic viruses and bacteria 1–2 microns in size and having the shape of a ball or a rod, move more readily in soil , either passively or actively, than eukaryotic microorganisms such as protozoa which range in size from a few to 200 microns. Some bacteria, such as actinomycetes, and filamentous fungi, produce a mycelium consisting of hyphae which grow and spread by branching and and become elongated at their hyphal tips. As a result, the growing hyphal tip exerts mechanical pressure on the substrate, allowing members of these groups actively to penetrate into solid substrate such as plant debris and to move through soil by active growth, like plant roots. Other means of movement of these groups include production of spores which spread by water or by air streams, along with dust. In contrast to mycelium-forming organisms, unicellular bacteria cannot penetrate solid substrate. This difference in growth behavior is readily observed on solidified media: Whereas actinomycetes and fungi produce a mycelium which penetrates into the agar medium, spreads on its surface and into the air as an aerial mycelium, the site of growth and spreading of unicellular bacteria and fungi (yeasts) is limited to the agar surface. This growth pattern enables fungi and actinomycetes partially to avoid competition for space and nutrients with bacteria, which occurs especially in aquatic environments and in flooded soil, in which the latter group usually takes over (Henis *et al.*, 1961).

The meeting between the pollutant and its degraders can occur as a result of a passive microbial movement toward the contaminant along with water flow in soil, by growing toward the pollutant, such as in the case of fungi and actinomycetes, or through active movement by means of flagella or by gliding. Active movement and growth may be affected by positive or negative chemotaxis (Chet and Mitchell, 1976).

As already mentioned, other factors which affect the potential meeting between a pollutant and its degraders are its solubility and diffusibility. Microorganisms capable of degrading poorly soluble hydrophobic substrates, often produce detergents which increase their solubility, and thus their availability to biodegradation (Gutnick and Rosenberg, 1977; Rosenberg and Ron, 1995).

Again, this means that conditions which allow growth and synthesis must prevail to enable biodegradation to take place. The degrading population, during its growth, may be subjected to predation by protozoa, or be attacked by bacteriophages or by myxobacteria. Although the role of predation in microbial activity in soil has been studied in detail (Alexander, 1981), only a few studies have been conducted on the effect of parasitism and predation on bioremediation (Wiggins and Alexander, 1988). On the other hand, the phenomena of simultaneous and successive coordination among different groups of microorganisms during biodegradation of various pollutants have been studied in detail. A typical example of a successive coordination is bioremediation of fuel-contaminated terrestrial and aquatic environments (Rosenberg *et al.*, 1992). Crude oil consists of hundreds of different aliphatic and aromatic compounds which differ in their biodegradability, and which are successively degraded by different microorganisms, according to their bioavailability. Some of these compounds may be toxic to degraders of other compounds, and their removal enables the successive degradation to continue. Another form of coordination among different microorganisms is the formation of consortia for the purpose of simultaneous degradation of substrates, including pollutants. A typical example of such a case is the biodegradation of chlorobenzoate by the dehalogenating bacterium *Desulphomonile tiedjii,* the benzoate fermenting rod BZ-2 and the methanogen *Methanospirillum* sp. (Mohn and Tiedje, 1992).

A third mechanism associated with bioremediation is cometabolism (Alexander, 1971; Horvath, 1972; Focht, 1987; Alexander, 1994). In this case, microorganisms degrade the contaminant in the presence of available substrates, without obtaining any building blocks or energy from the contaminant, which is usually present in concentrations too low to enable growth and synthesis.

The common denominator of all the mechanisms mentioned above is that a pollutant's biodegradation is possible only under conditions which allow multiplication and biosynthesis. In many cases a lag period is observed between the contamination of a specific site and the onset of the contaminant's biodegradation. This lag period, also termed *acclimation period,* may be defined as "the length of time between the addition or entry of a chemical into an environment and evidence of its detectable loss" (Alexander, 1994) . This period varies enormously, depending on the chemical structure, its concentration, and the environment. Several mechanisms may be involved in this phenomenon, including the multiplication of the degraders, which prior to contamination probably were present in the environment at levels too low to enable them to degrade the chemical at a detectable rate. Another possible mechanism of acclimation is enzyme induction, the period required for the synthesis of specific enzymes involved in the degradation and their increase to a level which permits the detection of the chemical's degradation. A third mechanism is the requirement for a time period for the occurrence of mutation and selection processes, resulting in the formation and enrichment of mutants which are better equipped than their ancestors for multiplying in the contaminated environment and for degrading the contaminant. Another factor which may be involved in acclimation, already mentioned above, is the time period required for the movement of the degrader(s) and the contaminant toward each other.

One common denominator of the factors which affect the acclimation period is the requirement for environmental conditions which allow for the occurrence of degraders' multiplication, and synthesis of enzymes and detergents. In most cases, the presence of the contaminant and the selective multiplication and enrichment of the better adapted degraders, results in a temporary decline in microbial diversity in the contaminated environment.

Finally, a major factor which determines the level and composition of the microflora that develop in the contaminated environment is growth kinetics (Alexander, 1994). Microorganisms differ in their capacity to utilize natural and synthetic substrates. These differences are reflected by different rates of multiplication and growth , and by different capacities to utilize substrates at different ranges of concentration, as expressed by the values of their maximum growth rate (μmax) and saturation constant (Ks), respectively. Whereas μmax is the relative amount of cell mass synthesized by the cell in a time unit, Ks is the concentration of the growth - limiting substrate which allows the cell to grow at a rate of one half of its μmax. Thus, bacteria like *Escherichia coli*, which live in the intestine of many animals, grow quite fast but require a relatively high level of nutrients. Microorganisms also differ in their requirements for a minimum level of carbon and energy sources to maintain their viability (maintenance energy). When reaching alien terrestrial and aquatic environments, they are unable successfully to compete with the indigenous microflora. On the other hand, indigenous oligotrophic organisms, which colonize soil and water, require a much lower substrate level in order to reach their maximum growth rate. They can not compete with eutrophs at high substrate levels, but the lower their Ks, the better is their competitive capacity at lower substrate levels.

CONCLUSIONS

Bioremediation of contaminated environments requires the presence of microorganisms capable of detoxification or mineralization of the contaminant, and environmental conditions that allow multiplication, selection, enrichment, growth and enzyme synthesis by the degrading microorganisms. Optimization of the bioremediation process can be achieved by prior analysis of the chemical and physical factors that affect microbial activity, amending them as much as possible and, if necessary, adding degrading cultures to the contaminated environment.

REFERENCES

Alexander, M., 1971. Microbial Ecology. John Wiley & Sons, Inc., New York, 511 pp.

Alexander, M., 1981. Why microbial predators and parasites do not eliminate their prey and hosts. *Annu. Rev. Microbiol.* **35:** 113–133.

Alexander, M., 1994. Biodegradation and Bioremediation. Academic Press, San Diego, CA, 302 pp.
Bollag, J-M., 1992. Decontaminating soil with enzymes. *Environ. Sci. Technol.* **26**: 1876–1881.
Bouwer, E.J., 1992. Bioremediation of organic contaminants in the subsurface, p. 287–318. In R. Mitchell, (ed.), Environmental Microbiology, John Wiley & Sons, New York, 411 pp.
Chet, I., and Mitchell, R., 1976. Biological aspects of microbial chemotactic behavior. *Annu. Rev. Microbiol.* **30:** 221–239.
Elton, C.S., 1958. The Ecology of Invasions by Animals and Plants. Methuen, London.
Focht, D.D., 1987. Ecological and evolutionary considerations on the metabolism of xenobiotic chemicals in soil, p. 157–166. In L.L. Boersma (ed.), Future Developments in Soil Science Research, Soil Science Society of America, Inc., Madison, WI.
Griffin, D.M., 1972. Ecology of Soil Fungi. Syracuse University Press, Syracuse, NY, 193 PP.
Griffin, D.M., and Quail, G., 1968. Movement of bacteria in moist, particulate systems. *Aust. J. Biol. Sci.* **21**: 579–582.
Gutnick, D.L., and Rosenberg, E., 1977. Oil tankers and pollution: A microbiological approach. *Annu. Rev. Microbiol.* **31**: 379–396.
Henis, Y., Keller, P., and Keynan, A., 1961. Inhibition of fungal growth by bacteria during cellulose decomposition. *Can. J. Microbiol.* **7**: 857–863.
Horvath, R.S., 1972. Microbial cometabolism and the degradation of organic cmpounds in nature. *Bacteriol. Rev.* **36**: 146–155.
Mitchell, R., 1974. Introduction to Environmental Microbiology. Prentice Hall International Inc., London. 355 pp.
Mohn, W. and Tiedje, J.M., 1992. Microbial reductive dehalogenation. *Microbiol. Rev.* **56**: 482–507.
Prince, R.C., 1992. Bioremediation of oil spills, with particular reference to the spill from the Exxon Valdez, p. 19–34. In J.C. Fry, G.M. Gedd, R. Herbert, C.W. Jones and I.A.
Watson-Craik (eds.), Microbial Control of Pollution. 48th Symposium of the Society for General Microbiology, Cambridge University Press, Cambridge, UK.
Rosenberg, E., Langmann, R., Kushmaro, A., Taube, R., Adler, E., and Ron, E., 1992. Petroleum bioremediation: a multiphase problem. *Biodegradation* **3**: 337–350.
Rosenberg, E., and Ron, E.,1996. Bioremediation of petroleum contamination, p. 100–124. In R.L. Crawford and D.L. Crawford (eds.), Bioremediation: Principles and Applications, Cambridge University Press, Cambridge, UK.
Thimann, K.V., 1966. The Life of Bacteria, 2nd Ed. Macmillan Co., New York, 909 pp.
Wiggins, B.A. and Alexander, M., 1988. Role of protozoa in microbial acclimation for mineralization of organic chemicals in sewage. *Can. J. Microbiol.* **34**: 661–666.

21

BIODEGRADATION OF ORGANIC POLLUTION INVOLVING SOIL IRON(III) SOLUBILIZED BY BACTERIAL SIDEROPHORES AS AN ELECTRON ACCEPTOR

Possibilities and Perspectives

Alexander A. Kamnev,[1] Lyudmila P. Antonyuk,[1] and Vladimir V. Ignatov[1,2]

[1]Laboratory of Biochemistry
Institute of Biochemistry and Physiology of Plants and Microorganisms
Russian Academy of Sciences, 410015 Saratov, Russia
[2]Department of Biochemistry and Biophysics, Faculty of Biology
Saratov State University, 410600 Saratov, Russia

SUMMARY

Microbial degradation of organic matter, like any oxidation process, requires an oxidizing agent. The role of the latter may be played by oxygen (the most common electron acceptor). However, under microaerobic conditions (ground water, subsurface soil) the supply of oxygen is limited, which, in its turn, may limit the biodegradation rate. An alternative possibility is the use of other electron acceptors, among which iron(III) is generally abundant in soil, yet being poorly soluble at physiological pH values. In this case its bioavailability can be essentially increased, e.g. by adding a chelating agent: NTA, EDTA, etc.

It should be noted that such an externally introduced chelator would increase also the bioavailability of a number of heavy metals which might then be easily accumulated in plants or otherwise affect the biota, including possible suppression of biodegrading microorganisms. This drawback can be avoided by using bacterial siderophores which selectively solubilize iron(III) and can be used by dissimilatory iron(III)-reducing bacteria that couple iron(III) reduction to oxidative degradation of organics. In this chapter, the possibilities are discussed for applying bacterial siderophores, as well as iron(III)-solubilizing bacterial cultures both "compatible" with the bacteria-biodegraders in sharing the resulting iron(III) chelate and viable in the presence of the organic matter to be degraded, as

Novel Approaches for Bioremediation of Organic Pollution, edited by
Fass *et al.* Kluwer Academic / Plenum Publishers, New York, 1999.

well as other possible contaminants. Plant-associated bacteria of the genus *Azospirillum* seem to be beneficial for the above applications as co-inoculants in concert with bacteria-biodegraders.

1. INTRODUCTION

Contamination of both surface and subsurface (deeper soil layers, ground waters) environments with diverse organic matter, largely represented by oil leakage and spills, constitutes a serious and acute problem of increasing environmental, as well as public health and economic significance (see, e.g. Holliger et al., 1997; Razo-Flores et al., 1997). At the same time it should be noted that, whereas e.g. heavy metals as contaminants have to be either physically removed from the environment or immobilized by (co)precipitation, adsorption, or an appropriate change in their chemical (e.g., oxidation) state ideally yielding much less or non-toxic species (see Lovley and Coates, 1997; White et al., 1997, and references therein), organic pollutants can in principle be essentially or, under appropriate conditions, even completely degraded in the course of chemical and/or biochemical (i.e., enzymatic) processes finally up to CO_2 and H_2O.

It seems important to emphasize that biological methods of remediation, which involve a wide range of diverse microorganisms capable of degrading toxic contaminants (bioremediation), in either ex-situ or in-situ variants, particularly for engineered bioremediation (see below), for a successful application require thorough knowledge of not only microbiological and biotechnological aspects of the processes occurring between the first treatment of the contaminated site and its complete cleanup, but also of their physical chemistry. Some noticeable underestimation of the latter aspect, on the background of an impressive progress in biological and biotechnological developments, was, for example, unambiguously revealed at the recent International Society for Environmental Biotechnology Meeting on Bioremediation (24-27 September 1997, Leipzig, Germany), as concluded by Professor M. Moo-Young who convened the vivid round-table discussions after the meeting.

In this chapter, the above aspects of the bioremediation processes indicated in the title are considered jointly, together with a survey of recent relevant literature.

2. GENERAL BIOREMEDIATION STRATEGIES FOR ORGANIC POLLUTION

Bioremediation of organic pollution with the help of bacteria and other microorganisms possessing degradative capacities is a highly attractive strategy. It may be realized ex situ (using, e.g. pump-and-treat methods for groundwater, soil excavation and treatment in bioreactors, etc.); however, it is hardly worth mentioning that this way, along with a number of its drawbacks (Holliger et al., 1997), is highly laborious and expensive (Vel'kov, 1995) and may even not be applicable at all to, e.g. vast territories of contaminated soil and subsurface aquifers. As an alternative, in situ bioremediation is in many cases possible; the corresponding processes may be be distinguished as intrinsic and engineered (Figure 1). The latter way has to be applied if the rate of in situ intrinsic (natural) biodegradation involving indigenous microflora is insufficient, or if the process is incomplete, which may be due to lack or rapid consumption of nutrients, electron acceptors (oxi-

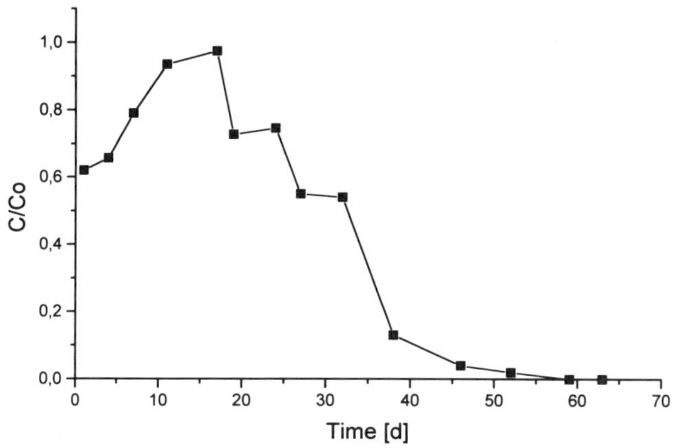

Figure 1. Scheme of ex situ and in situ bioremediation processes.

dizing agents), insufficiency or low activity of appropriate microorganisms-biodegraders, etc. (Golovlyova, 1992; Vel'kov, 1995).

Analysis of the literature on bioremediation (see, e.g. Golovlyova, 1992; Holliger et al., 1997; Ismailov, 1988; Mandelbaum et al., 1997; and references cited therein) shows that in oil contaminated aquifers a number of microbially mediated processes are generally stimulated (Table 1). This may in principle be due to an increased specific activity (i.e., per unit of biomass) of the corresponding microorganisms; however, considering a number of reports on suppression of some microbiological processes in oil contaminated environments (Ismailov, 1988; Kireeva, 1994; Kireeva et al., 1997), this may also be a result of increased population sizes of certain microorganisms overcompensating the suppression effect, if any.

3. BIOREMEDIATION OF ORGANIC POLLUTION AS A REDOX PROCESS

3.1. Involvement of Electron Acceptors

From the viewpoint of chemistry, biodegradation of organic pollution is an oxidation process which, as such, must evidently involve an electron acceptor. The role of the latter is commonly played by oxygen; however, in many cases which iclude the application of (or the use of indigenous) anaerobic microorganisms-biodegraders and/or poor aeration of the contaminated site (subsurface soil and groundwater with a limited supply of oxygen; oil-contaminated soil particles with a drastically decreased oxygen permeability) an alter-

Table 1. Influence of oil organics on microbial processes in contaminated soil

Stimulation	Inhibition
Denitrification	Nitrification
Ammonification	Cellulose degradation
Nitrogen fixation	
Carbohydrates oxidation	

native electron acceptor has to be utilized. The role of the latter may be played by a number of both inorganic (nitrate, nitrite, sulphate, multivalent metal ions in a higher oxidation state, etc.) and organic substrates, such as humic substances, (poly)phenolics, etc. (Barkovskii et al., 1994; Holliger et al., 1997; Lovley, 1991, 1997; Lovley et al., 1996). It is also essential that certain organics can only or largely be biodegraded under anaerobic conditions (Golovlyova, 1992; Holliger et al., 1997).

3.2. Microbial Transformations of Soil Iron(III)

3.2.1. Assimilatory and Dissimilatory Microbial Iron(III) Reduction. One of the most generally abundant potential electron acceptor in subsurface environments is iron(III). It should however be noted that, owing to its high hydrolytic capability, at pH values already over 4 ferric iron is extremely poorly soluble. Considering the very low solubility product of ferric hydroxide being of the order of 10^{-38} to 10^{-40} (Lindsay and Schwab, 1982; Lur'e, 1989) definitely depending on the degree of its crystallinity (the higher the latter, the lower the former, which is regulated by the thermodynamic stability of the dissolving phase), it may be easily calculated that at physiological (i.e., circumneutral) pH values the equilibrium concentration of free (aqueous) iron(III) ion would be many orders of magnitude lower than the minimum required for growth of almost any living cell (Sigel and Sigel, 1998). In order to overcome iron limitation, the general iron assimilation mechanisms in microorganisms involve the excretion of siderophores, specific chelates of a relatively low molecular weight which selectively bind iron(III) forming strong complexes with stability constants within the range 10^{23}-10^{35} (Braun, 1997), thus increasing its bioavailability (Briat, 1992; Sigel and Sigel, 1998). In order to sequester iron(III) from the complex after its solubilization and transfer to the cell, its assimilation scheme necessarily includes a reduction step (Figure 2). In this case assimilatory iron(III) reduction generally takes place either on the cell surface or inside the cell using physiological reducing agents; in certain cases an alternative in situ reduction mechanism may be realized (Kamnev, 1998; Kamnev and Kuzmann, 1997-a; Rioux et al., 1986).

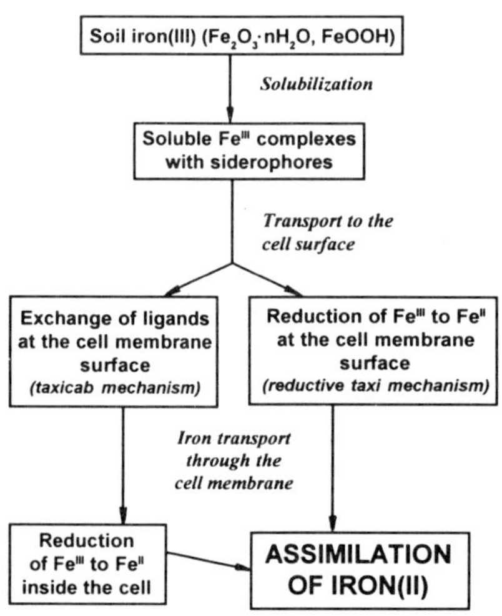

Figure 2. General assimilation scheme of soil iron.

Table 2. Dissimilatory iron(III)-reducing microorganisms

Genera	Species
Aeromonas	*A. hydrophila*
Bacillus	*B. infernus*
Clostridium	*C. pasteurianum*
Desulfobacter	*D. hydrogenophilus*
Desulfobacterium	*D. autotrophicum*
Desulfobulbus	*D. propionicus*
Desulfovibrio	*D. desulfuricans, D. vulgaris*
Desulfuromonas	*D. acetexigens, D. acetoxidans, D. palmitatis*
Desulfuromusa	*D. bakii, D. kysingii, D. succinoxidans*
Ferrimonas	*F. balearica*
Geobacter	*G. chapellei, G. humireducens, G. hydrogenophilus, G. metallireducens, G. sulfurreducens*
Geospirillum	*G. barnseii*
Geothrix	*G. fermentens*
Geovibrio	*G. ferrireducens*
Pelobacter	*P. acetylenicus, P. acidigallici, P. carbinolicus, P. propionicus, P. venetianus*
Rhodobacter	*R. capsulatus*
Shewanella	*S. alga, S. potomacii, S. putrefaciens*
Thiobacillus	*T. ferrooxidans, T. thiooxidans*
Wolinella	*W. succinogenes*

In contrast, iron(III)-reducing bacteria obtain energy by coupling the process of dissimilatory iron(III) reduction to oxidation of organic matter (Lovley and Phillips, 1988), including toxic pollutants (Lovley, 1991, 1997), leading to its biodegradation. Known microorganisms capable of dissimilatory iron(III) reduction are rather diverse (Table 2) and fall in a number of different phylogenetic groups; the most intensively studied genera Geobacter and Shewanella are often isolated from sedimentary environments (Lovley, 1997).

It is worth mentioning that the chemical composition, particle size and shape, stability, and surface physicochemical properties of naturally formed ferric (oxy)hydroxides may essentially differ from those of conventionally precipitated compounds (or otherwise synthesizes under laboratory conditions) often used in model experiments (see, e.g. Lovley and Woodward, 1996). This is evident from the fact that in natural waters or subsurface soils oxidation of iron(II) proceeds very slowly, thus lacking any local pH heterogeneity, in contrast to precipitation by adding a base or by a relatively rapid hydrolysis (Deng, 1997). This may be essential and should be kept in mind while interpreting the results of experiments on microbial reduction of artificially obtained ferric hydroxides.

3.2.2. Possible Mechanisms for Dissimilatory Microbial Iron(III) Reduction. There are three most general mechanisms for dissimilatory iron(III) reduction by microorganisms (Lovley, 1991, 1997) schematically illustrated in Figure 3.

i. Contact reduction is realized when the bacterium attaches directly to an iron(III)-bearing soil particle. This process was suggested (Caccavo et al., 1997) to involve certain cell surface proteins which, at a first stage, efficiently chelate iron(III) thus making a "store" of electron acceptor directly at the cell surface. It was also found that bacterial adhesion to the mineral may be transient and reversible, and the physical and chemical interactions between the bacterium and mineral particles, the mechanisms of which are still largely unclear, may be rather specific and complex.

ii. Iron(III) may be bound by a chelate and, in the resulting soluble form, be transported to the cell with subsequent dissimilatory iron(III) reduction occurring at the bacterial cell surface similar to the reduction step in the above case. It was experimentally shown (Lovley and Phillips, 1988; Munch and Ottow, 1983) that the iron(III)-reducing bacteria studied did not excrete any chelator which could solubilize iron(III). However, such a chelator added to the medium may greatly facilitate enzymatic dissimilatory iron(III) reduction (Lovley, 1991) and stimulate the coupled process of biodegradation, including (poly)aromatic hydrocarbons (Lovley et al., 1994; Lovley and Woodward, 1996; Trepte, 1997).

The above mechanisms (i) and (ii), being essentially different, nevertheless have a common feature consisting in the fact that iron(III) reduction in both cases occurs at the bacterial cell surface (see Figure 3).

iii. Iron(III) may be reduced by a specific compound (e.g., humic substances and their analogs) serving as an electron shuttle between the cell and the mineral particle (Lovley, 1997; Lovley et al., 1996). At a first step, the shuttle molecule accepts an electron from the cell and then transfers it to the mineral, abiotically reducing iron(III) contained therein thus being recovered to the initial state. This mechanism, although similar to (ii) in that the cell and the mineral (i.e., iron(III) source) need not be in contact, is however fundamentally different from (ii) in the mode of electron transfer to iron(III).

3.3. Possible Involvement of Iron(II), Formed by Dissimilatory Iron(III) Reduction, in Bioremediation

Ferrous iron released after the enzymatic microbial reduction of iron(III) is also known to be capable of abiotic (chemical) reduction of chlorinated organics (Bradley and Chapelle, 1996; Glass, 1972; Gorby et al., 1996) and nitroaromatic compounds (Heijman et al., 1995;

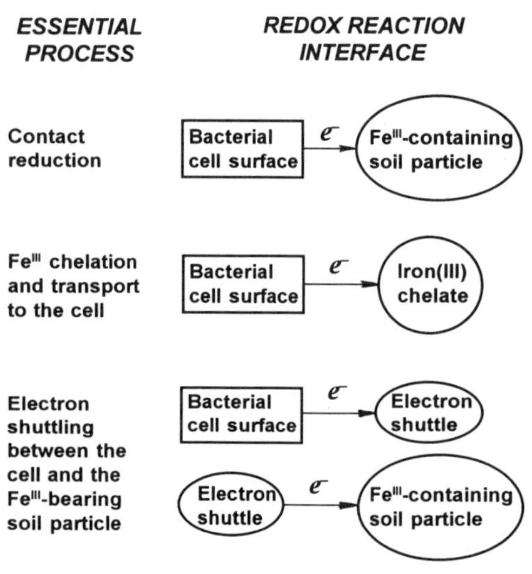

Figure 3. Scheme of general mechanisms of dissimilatory microbial iron(III) reduction.

Holliger et al., 1997), thus potentially contributing to a multifunctional further remediation of complex pollutants (e.g., organic contaminants containing chloro- and nitroaromatics) and returning the resulting recovered iron(III) into the redox bioremediation cycle.

However, the chemical reducing activity of iron(II) ions seems to be significantly better expressed in the adsorbed state, e.g. at the surface of iron(III)-containing mineral particles (Klausen et al., 1995). These conditions may be less favorable for microbial reduction of iron(III) involving iron(III) chelates, which results in the formation of iron(II) at a reduction site independent from the "parent" iron(III) source (viz., soil or mineral particles), and more favorable for the mechanisms of contact reduction and shuttle-mediated reduction of iron(III) (Lovley, 1997; Lovley et al., 1996), both obviously resulting in the formation of ferrous species directly adsorbed at or otherwise superficially attached to the "parent" mineral iron(III)-bearing particle.

It is interesting to note for comparison that chelator-stimulated microbial reduction of synthetic poorly crystalline ferric oxide was found to proceed faster when the latter had been adsorbed onto clay particles (Lovley and Woodward, 1996). This, together with the above mentioned finding (Klausen et al., 1995), indicates that some surface-related stimulation of the chemical activity of the adsorbed iron species might take place.

It should also be mentioned that, though the amino aromatic derivatives formed in the course of anaerobic reduction of nitroaromatics may generally be even more toxic than the latter parent contaminants (Holliger et al., 1997), they have been shown to irreversibly bind to certain soil components (e.g., humic substances, montmorillonitic clay) both under aerobic (Daun et al., 1995) and anaerobic conditions (Lenke et al., 1997).

4. THE APPLICATION OF EXTERNAL IRON(III) CHELATORS FOR STIMULATING BIODEGRADATION

4.1. Positive Effects of Iron(III) Chelators Added to the Contaminated Aquifer

As has been mentioned above, the stimulating effect of an externally introduced iron(III) chelating agent on the biodegradation rate of oil organics was observed with nitrilotriacetic acid (NTA) (Lovley et al., 1994). Further investigation of the mechanism of the stimulating effect showed it to consist in enhanced solubilization of iron(III) from the sediments caused by the chelator (Lovley and Woodward, 1996). At present, laboratory and in situ experiments with NTA and ethylenediaminetetraacetate (EDTA) are performed to test the possibility of biodegradation of a range of polyaromatic hydrocarbons (PAHs) most resistant to enzymatic oxidation (Trepte, 1997).

It should be noted that contaminated subsurface aquifers usually contain insufficient amounts of intrinsic iron(III) chelators, if at all, though some evidence for in situ intrinsic biodegradation under iron(III)-reducing conditions reported recently (Lovley, 1997) in the absence of intrinsic iron(III) chelators points to the possible involvement of natural electron shuttles.

4.2. Possible Negative Processes Accompanying the Addition of Strong Chelating Agents

Along with a well-documented positive effect on the biodegradation rate of organics coupled to dissimilatory iron(III) reduction, the addition of such general strong chelating

Table 1. Stability constants (K_s) of some ethylenediaminetetraacetate ($EDTA^{4-}$) complexes $[M\ EDTA]^{n-4}$ (M= metal ion, n=oxidation state) at 20 °C-30 °C and ionic strength 0.1 (if not indicated otherwise)

Metal	$logK_s$	Metal	$logK_s$	Metal	$logK_s$
Al^{3+}	16.5	Cu^{2+}	18.8	Pb^{2+}	18.04
Bi^{3+}	27.4	Fe^{2+}	14.2	Tl^{3+}	37.8 (I=1)
Cd^{2+}	16.46	Fe^{3+}	24.23 (I=0)	V^{3+}	25.9
Co^{2+}	16.31	Hg^{2+}	21.8	$(V^{IV}O)^{2+}$	18.0
Co^{3+}	40.6 (I=0.2)	Mn^{2+}	14.04	Zn^{2+}	16.26
Cr^{3+}	23.4	Mn^{3+}	24.9 (I=0.2)		

agents as NTA and EDTA may well have accompanying negative effects. In case of a complex pollution by, e.g. organics and heavy metals, or otherwise in the presence of the latter (even in an insoluble or immobilized form with a low toxicity), an essential increase in their solubility and solubilization rate may well be observed caused by a competitive complexation with the chelator.

It should be mentioned that a higher level of heavy metals in soil by itself does not necessarily correlate with their increased concentrations in plants growing thereupon, which may obviously be connected with different degrees of metal availability; rather, a combination of certain environmental factors or even a lower soil pH may facilitate the uptake of heavy metals by plants (Gondola and Kadar, 1994/1995).

Stability constants (K_s) for complexes of a number of heavy metals with, e.g. EDTA (Lur'e, 1989) are either comparable with that of iron(III) for which $logK_s = 24.2$ (for example, for chromium(III), mercury(II)) or even higher (for bismuth(III), cobalt(III), thallium(III), vanadium(III), manganese(III); Table 3). Thus, in the presence of these metals their solubilization and the corresponding increase in their bioavailability may result in their accumulation in plants or even have a poisoning effect on the further developing biota.

Moreover, it has been documented that microbial biodegradation of PAHs in the presence of heavy metals may be essentially retarded owing to their toxicity towards the microorganisms-biodegraders; some special measures which have to be taken to reduce the latter effect include the proposed application of surfactant-modified clay adsorbents (Malakul et al., 1997). The presence of some heavy metals was also shown to slow down bioremediation of other organic pollutants (Panchenko and Turkovskaya, 1996).

5. POSSIBILITIES FOR THE INVOLVEMENT OF BACTERIAL SIDEROPHORES AS IRON(III) CHELATORS FOR STIMULATING BIODEGRADATION

It has been mentioned above that under iron-limiting conditions many bacteria produce specific iron(III)-chelating ligands (siderophores) thus increasing its bioavailability. It is essential that in assimilation of iron from the resulting complex, specific bacterial outer membrane receptor proteins are involved which have a high specificity for the ferric siderophore used as iron donor for a particular strain (Braun, 1997). As a result, certain bacteria may use their siderophores to induce iron stress upon other microorganisms, including plant pathogens, thus suppressing their growth and enhancing plant growth (Emery, 1980; Kloepper et al., 1980).

In contrast, dissimilatory reduction of iron(III) by microorganisms, which use it solely as an electron acceptor, is obviously not particularly sensitive to the form of iron(III). This means that iron(III) solubilized by siderophores may in principle be utilized by dissimilatory iron(III) reducers which can couple its reduction to biodegradative oxidation of organics. In this case it is essential that after the reduction the resulting iron(II)-siderophore complex is essentially less stable and the ligand is easily released, further chelating iron(III) in the environment. It is clear that this mechanism of using another microbe's iron(III) in the chelate by dissimilatory iron(III) reducers is identical to the above mentioned chelator-mediated mechanism (ii), the only difference consisting in the source of the chelating agent produced in situ by (an)other organism(s).

6. POSSIBILITIES FOR APPLYING PLANT-ASSOCIATED DIAZOTROPHIC BACTERIA IN ANAEROBIC BIODEGRADATION UNDER IRON(III)-REDUCING CONDITIONS

Plant-associated nitrogen-fixing bacteria are well-documented inoculants used in agriculture (Bashan and Holguin, 1997; Okon, 1994; Umarov, 1986). Their positive influence on the overall soil fertility and plant-growth-promoting effects have been under intensive investigation for decades, although such complicated systems still leave much to be understood.

One of the main features of these microorganisms is their capability of fixing molecular nitrogen which can be used by the plant partner of the association (Dobereiner and Pedrosa, 1987). In general, the insufficiency of bound nitrogen often observed in soil despite the diversity of diazotrophic bacteria (Umarov, 1986) is due to the fact that the fraction of prokaryotes in the microbial biomass is about 2.5% only (Polyanskaya et al., 1995), while non-nitrogen-fixing fungi dominating in the rest part evidently consume the majority of the substrate. Thus, inoculation with diazotrophs improves the supply of bound nitrogen to the soil; nevertheless, a number of other beneficial functions of these bacteria were found essentially contributing to the promotion of plant growth, including biosynthesis and excretion of plant growth regulators (Prikryl et al., 1985; Costacurta and Vanderleyden, 1995).

Considering the above mentioned processes occurring during bioremediation of organic pollution, the application of plant-associated diazotrophic bacteria in concert with dissimilatory iron(III) reducing bacteria which biodegrade organics seems promising. Among the prospective co-inoculants from the large family of rhizobacteria, bacteria of the genus *Azospirillum* may in many cases be suitable owing to a number of its beneficial features briefly discussed below. It is also helpful that azospirilla have been for two decades under thorough investigation with regard to their microbiology, genetics, biochemistry, and ecology (Fendrik et al. 1995).

It should be emphasized that, considering the overall enhancement of the bioremediation process in the rhizosphere of plants (Anderson and Coats, 1994), the plant-growth-promoting effect of *Azospirillum* spp. is highly beneficial. This microaerophilic bacterium is capable of fixing atmospheric dinitrogen even in the presence of bound inorganic nitrogen in the medium in the form of ammonium or nitrate (Ruppel and Merbach, 1997). Also, its ability to produce phytohormones is well documented (Costacurta and Vanderleyden, 1995; Iosipenko and Ignatov, 1995).

Another very essential feature of *Azospirillum* is the production of siderophores (Bacchawat and Ghosh, 1987; Mori et al., 1995) which, as has been discussed above, may participate in dissimilatory iron(III) reduction by biodegrading bacteria as an iron(III) chelator. Under anaerobic conditions, *Azospirillum* spp. was shown to be viable and capable of using nitrate and, possibly, nitrite (but not iron(III)!) (Bothe et al., 1981).

Many of *Azospirillum* strains have been shown to biodegrade (poly)phenolics and other aromatics under oxygen-limiting conditions (Barkovskii et al., 1995-a, 1995-b; Barkovskii et al., 1994). In addition, *Azospirillum lipoferum* was shown to reduce chloronitroaromatics both under aerobic and unaerobic conditions using nitrate as a nitrogen source (Russel and Muszynski, 1995). Thus, these bacteria may also directly contribute to bioremediation catabolizing intermediate products.

Among other beneficial features of *Azospirillum*, its high motility both in vertical and horizontal directions detected in field experiments (Bashan and Levanony, 1987) should be mentioned which, being connected with root growth of the host plant, is regulated by both redox taxis and chemotaxis of the bacterium (Zhulin et al., 1996; Zhulin and Taylor, 1995).

In addition, *Azospirillum brasilense* was found to be resistant to high concentrations of Al, Mn, and Zn (Gowri and Srivastava, 1996-a) which, in the case of zinc, is based on a reduced metal uptake (Gowri and Srivastava, 1996-b); Cu was found to be still less toxic for *Azospirillum* strains than Zn (Biro et al., 1995). This might be essential in case of complex pollution including heavy metals. There were also indications about acid tolerance of some *Azospirillum* species (Magalhaes et al., 1983). All the above considered features of azospirilla indicate that they are very promising as co-inoculants which could contribute to bioremediation either indirectly (increasing the overall fertility of the soil, facilitating plant growth by supplying plant roots with nutrients and growth promotors, excreting siderophores) and, in certain cases, directly, catabolizing some organics and/or intermediate partly oxidized biodegradation products.

ACKNOWLEDGMENTS

This work was supported in part by the EU (INTAS Project No. 96-1015). A.A.K. also acknowledges support from the Organizing Committee of the 42nd OHOLO Conference "Novel Approaches for Bioremediation of Organic Pollution" (Eilat, Israel, May 3-7, 1998) and the Russian Foundation for Basic Research (RFBR travel grant 98-04-58564) for participation in the Conference where this material was presented.

REFERENCES

Anderson, T.A., and Coats, J.R. (eds.) (1994) Bioremediation through Rhizosphere Technology, ACS Symposium Series 563, American Chemical Society, Washington, D.C. 249 pp.

Bacchawat, A.K., and Ghosh, S. (1987) Iron transport in *Azospirillum brasilense*: role of siderophore spirilobactin. J. Gen. Microbiol. 133:1759-1765.

Barkovskii, A.L., Bouillant, M.-L., and Balandreau, J. (1994) Polyphenolic compounds respired by bacteria, p. 28-42. In T.A. Anderson and J.R. Coats (eds.), Bioremediation through Rhizosphere Technology, ACS Symposium Series 563, American Chemical Society, Washington, D.C.

Barkovskii, A.L., Bouillant, M.-L., Monrozier, L.J., and Balandreau, J. (1995-a) *Azospirillum* strains use phenolic compounds as intermediates for electron transfer under oxygen-limiting conditions. Microb. Ecol. 29:99-114.

Barkovskii, A.L., Korshunova, V.E., and Pozdnyacova, L.I. (1995-b) Catabolism of phenol and benzoate by *Azospirillum* strains. Appl. Soil Ecol. 2:17-24.

Bashan, Y., Holguin, G. (1997) *Azospirillum*-plant relationships - environmental and physiological advances (1990-1996). Can. J. Microbiol. 43(2):103-121.

Bashan, Y., and Levanony, H. (1987) Horizontal and vertical movement of *Azospirillum brasilense* Cd in the soil and along the rhizosphere of wheat and weeds in controlled and field experiments. J. Gen. Microbiol. 133:3473-3480.

Biro, B., Bayoumi Hamuda, H.E.A.F., and Kecskes, M. (1995) Associative, symbiotic N_2-fixers and scavenger strains, affected by Cu^{2+} and Zn^{2+} in vitro, p. 495-500. In I. Fendrik, M. del Gallo, J. Vanderleyden, and M. de Zamaroczy (eds.), Azospirillum VI and Related Microorganisms: Genetics, Physiology, Ecology, Springer, Berlin.

Bradley, P., and Chapelle, F.H. (1996) Anaerobic mineralization of vinyl chloride in Fe(III)-reducing aquifer systems. Environ. Sci. Technol. 30:2084-2086.

Braun, V. (1997) Avoidance of iron toxicity through regulation of bacterial iron transport. Biol. Chem. 378:779-786.

Briat, J.-F. (1992) Iron assimilation and storage in prokaryotes. J. Gen. Microbiol. 138:2475-2483.

Caccavo, F., Jr., Schamberger, P.C., Keiding, K., and Nielsen, P.H. (1997) Role of hydrophobicity in adhesion of the dissimilatory Fe(III)-reducing bacterium *Shewanella alga* to amorphous Fe(III) oxide. Appl. Envir. Microbiol. 63:3837-3843.

Costacurta, A., and Vanderleyden, J. (1995) Synthesis of phytohormones by plant-associated bacteria. Crit. Rev. Microbiol. 21:1-18.

Daun, G., Lenke, H., Desiere, F., Stolpmann, H., Warrelmann, J., Reuss, M., and Knackmuss, H.-J. (1995) Biological treatment of TNT-contaminated soil by a two-stage anaerobic/aerobic process, p. 337-346. In W.J. van den Brink, R. Bosman, and F. Arendt (eds.), Contaminated Soil'95, Kluwer, Dordrecht.

Deng, Y. (1997) Formation of iron(III) hydroxides from homogeneous solutions. Water Res. 31:1347-1354.

Dobereiner, J., and Pedrosa, F. (1987) Nitrogen-fixing Bacteria in Nonleguminous Crop Plants. Springer, Berlin.

Emery, J. (1980) Iron deprivation as a biological defence mechanism. Nature (London) 287:537.

Fendrik, I., del Gallo, M., Vanderleyden, J., and de Zamaroczy, M. (eds.) (1995) Azospirillum VI and Related Microorganisms: Genetics, Physiology, Ecology, Springer, Berlin.

Glass, B.L. (1972) Relation between the degradation of DDT and the iron redox system in soils. J. Agric. Food Chem. 20:324-327.

Golovlyova, L.A. (1992) Microbial methods for decontamination of soil and ground water. Biotekhnologiya (No. 5):60-64.

Gondola, I., and Kadar, I. (1994, Pub. 1995) Heavy metal content of flue-cured tobacco leaf in different growing regions of Hungary. Acta Agron. Hung. 43(3-4):243-251.

Gorby, Y., Kennedy, D., and Workman, D. (1996) Reductive dechlorination of carbon tetrachloride by biogenic ferrous iron, p. 107. In 1996 International Symposium on Subsurface Microbiology, 15-21 September 1996, Davos, Switzerland.

Gowri, P.M., and Srivastava, S. (1996-a) Zinc resistance in *Azospirillum brasilense* Sp7 and its applications, p. 135-143. In M. Moo-Young, W.A. Anderson, and A.M. Chakrabarty (eds.), Environmental Biotechnology, Kluwer, Dordrecht.

Gowri, P.M., and Srivastava, S. (1996-b) Reduced uptake based zinc resistance in *Azospirillum brasilense* Sp7. Curr. Sci. 71:139-142.

Heijman, C., Grieder, E., Holliger, C., and Schwarzenbach, R.P. (1995) Abiotic reduction of nitroaromatic compounds coupled to microbial iron reduction in laboratory aquifer columns. Environ. Sci. Technol. 29:775-783.

Holliger, C., Gaspard, S., Glod, G., Heijman, C., Schumacher, W., Schwarzenbach, R.P., and Vazquez, F. (1997) Contaminated environments in the subsurface and bioremediation: organic contaminants. FEMS Microbiol. Rev. 20:517-523.

Iosipenko, A., and Ignatov, V. (1995) Physiological aspects of phytohormone production by *Azospirillum brasilense* Sp7, p. 307-312. In I. Fendrik, M. del Gallo, J. Vanderleyden, and M. de Zamaroczy (eds.), Azospirillum VI and Related Microorganisms: Genetics, Physiology, Ecology, Springer, Berlin.

Ismailov, N.M. (1988) Microbiology and enzymatic activity of oil-contaminated soils, p. 42-56. In M.A. Glazovskaya (ed.), Rehabilitation of oil-contaminated soil ecosystems, Nauka, Moscow (in Russian).

Kamnev, A.A. (1998) On the possibility of reductive solubilization of iron(III) by some plant and microbial metabolites as an alternative to siderophore secretion. Dokl. Akad. Nauk (Moscow) 359 (No. 5): in press.

Kamnev, A.A., and Kuzmann, E. (1997-a) Some rhizobacterial metabolites of non-siderophore nature as possible solubilizing agents for soil ferric species, p. 85-86. In P. Carmona, R. Navarro and A. Hernanz (eds.), Spectroscopy of Biological Molecules: Modern Trends. Annex, UNED, Madrid.

Kireeva, N.A. (1994) Microbiological Processes in Oil Contaminated Soils, Bashkir State University Press, Ufa, 171 pp. (in Russian).
Kireeva, N.A., Novoselova, E.I., and Khaziev, F.Kh. (1997) Enzymes of nitrogen metabolism in soil polluted with oil. Izv. Ross. Akad. Nauk. Ser. biol. (No. 6):755-759.
Klausen, J., Tröber, S.P., Haderlein, S.B., and Schwarzenbach, R.P. (1995) Reduction of substituted nitrobenzenes by Fe(II) in aqueous mineral suspensions. Environ. Sci. Technol. 29:2396-2404.
Kloepper, J.W., Leong, J., Teintze, M., and Schroth, M.N. (1980) Enhanced plant growth by siderophores produced by plant growth-promoting rhizobacteria. Nature (London) 286:885-886.
Lenke, H., Warrelmann, J., Daun, G., Walter, U., Sieglen, U., and Knackmuss, H.-J. (1997) Bioremediation of TNT-contaminated soils, p. 15. In T. Münker (ed.), International Society for Environmental Biotechnology ISEB'97 Meeting on Bioremediation, 24-27 September 1997, Leipzig, Germany. Abstracts, UFZ Umweltforschungszentrum Leipzig-Halle GmbH, Leipzig.
Lindsay, W.L., and Schwab, A.P. (1982) The chemistry of iron in soils and its availability to plants. J. Plant Nutr. 5:821-840.
Lovley, D.R. (1991) Dissimilatory Fe(III) and Mn(IV) reduction. Microbiol. Rev. 55:259-287.
Lovley, D.R. (1997) Microbial Fe(III) reduction in subsurface environments. FEMS Microbiol. Rev. 20:305-313.
Lovley, D.R., and Coates, J.D. (1997) Bioremediation of metal contamination. Curr. Opinion Biotechnol. 8:285-289.
Lovley, D.R., Coates, J.D., Blunt-Harris, E.L., Phillips, E.J.P., and Woodward, J.C. (1996) Humic substances as electron acceptors for microbial respiration. Nature (London) 382:445-448.
Lovley, D.R., and Phillips, E.J.P. (1988) Novel mode of microbial energy metabolism: organic carbon oxidation coupled to dissimilatory reduction of iron or manganese. Appl. Environ. Microbiol. 54:1472-1480.
Lovley, D.R., and Woodward, J.C. (1996) Mechanisms for chelator stimulation of microbial Fe(III)-oxide reduction. Chem. Geol. 132:19-24.
Lovley, D.R., Woodward, J.C., and Chapelle, F.H. (1994) Stimulated anoxic biodegradation of aromatic hydrocarbons using Fe(III) ligands. Nature (London) 370:128-131.
Lur'e, Yu.Yu. (1989) Handbook of Analytical Chemistry, 6th edn., Khimiya Publishers, Moscow (in Russian).
Magalhaes, F.M.S., Baldani, J,I., Souto, S.M., Kuykendall, J.R., and Dobereiner, J. (1983) A new acid tolerant *Azospirillum* species. Ann. Acad. Bras. Sci. 55:417-430.
Malakul, P., Brouhard, G., Srinivasan, K., and Wang, H.Y. (1997) Metal toxicity reduction in PAH biodegradation using metal chelating adsorbents, p. 145. In T. Münker (ed.), International Society for Environmental Biotechnology ISEB'97 Meeting on Bioremediation, 24-27 September 1997, Leipzig, Germany. Abstracts, UFZ Umweltforschungszentrum Leipzig-Halle GmbH, Leipzig.
Mori, E., Fani, R., and Fulchieri, M. (1995) Involvement of lon protease of *Azospirillum brasilense* in iron uptake, p. 137-141. In I. Fendrik, M. del Gallo, J. Vanderleyden, and M. de Zamaroczy (eds.), Azospirillum VI and Related Microorganisms: Genetics, Physiology, Ecology, Springer, Berlin.
Munch, J.C., and Ottow, J.C.G. (1983) Reductive transformation mechanism of ferric oxides in hydromorphic soils. Environ. Biogeochem. Ecol. Bull. (Stockholm) 35:383-394.
Okon, Y. (ed.) (1994) *Azospirillum*/Plant Associations, CRC Press, Boca Raton.
Panchenko, L.V., and Turkovskaya, O.V. (1996) Surfactant waste water treatment in the presence of accompanying pollutants, p. 359. In K.G. Skryabin (ed.), Bayev Memorial Conference, May 20-22, 1996. Abstracts, "Bioengineering" Centre, Russian Academy of Sciences, Moscow.
Polyanskaya, L.M., Golovchenko, A.V., and Zvyagintsev, D.G. (1995) Microbial biomass in soils. Dokl. Akad. Nauk (Moscow) 344:846-848.
Prikryl, Z., Vancura, V., and Wurst, M. (1985) Auxin formation by rhizospheric bacteria as a factor of root growth. Biol. Plant. 27:159-163.
Razo-Flores, E., Donlon, B., Lettinga, G., and Field, J.A. (1997) Biotransformation and biodegradation of N-substituted aromatics in methanogenic granular sludge. FEMS Microbiol. Rev. 20:525-538.
Ruppel, S., and Merbach, W. (1997) Effect of ammonium and nitrate on $^{15}N_2$-fixation of *Azospirillum* spp. and *Pantoea agglomerans* in association with wheat plants. Microbiol. Res. 152:377-383.
Russel, S., and Muszynski, S. (1995) Reduction of 4-chloronitrobenzene by *Azospirillum lipoferum*, p. 369-375. In I. Fendrik, M. del Gallo, J. Vanderleyden, and M. de Zamaroczy (eds.), Azospirillum VI and Related Microorganisms: Genetics, Physiology, Ecology, Springer, Berlin.
Sigel, A., and Sigel, H., eds. (1998) Metal Ions in Biological Systems. Vol. 35. Iron Transport and Storage in Microorganisms, Plants, and Animals, Marcel Dekker, New York. 824 pp.
Trepte, B. (1997) Fe(III) ligands as oxidants to enhance bioremediation, p. 139. In T. Münker (ed.), International Society for Environmental Biotechnology ISEB'97 Meeting on Bioremediation, 24-27 September 1997, Leipzig, Germany. Abstracts, UFZ Umweltforschungszentrum Leipzig-Halle GmbH, Leipzig.
Umarov, M.M. (1986) Associative Nitrogen Fixation, Moscow University Press, Moscow (in Russian).

Vel'kov, V.V. (1995) Bioremediation: principles, problems, approaches. Biotekhnologiya (No. 3-4):20-27.

White, C., Sayer, J.A., and Gadd, G.M. (1997) Microbial solubilization and immobilization of toxic metals: key biogeochemical processes for treatment of contamination. FEMS Microbiol. Rev. 20:503-516.

Zhulin, I.B., Bespalov, V.A., Johnson, M.S., and Taylor, B.L. (1996) Oxygen taxis and proton motive force in *Azospirillum brasilense*. J. Bacteriol. 178:5199-5204.

Zhulin, I.B., and Taylor, B.L. (1995) Chemotaxis in plant-associated bacteria: the search for the ecological niche, p. 451-459. In I. Fendrik, M. del Gallo, J. Vanderleyden, and M. de Zamaroczy (eds.), Azospirillum VI and Related Microorganisms: Genetics, Physiology, Ecology, Springer, Berlin.

22

$^{13}C/^{12}C$ STABLE ISOTOPE FRACTIONATION OF TOLUENE BY ANAEROBIC DEGRADATION

A New Method to Monitor Biological Degradation *in Situ*?

Rainer U. Meckenstock,[1,*] Eva Annweiler,[2] Rolf Warthmann,[1] Bernhard Schink,[1] Walter Michaelis,[2] and Hans H. Richnow[2]

[1]University of Konstanz, Department of Microbial Ecology
Universitätsstr. 10, 78457 Konstanz, Germany
[2]University of Hamburg, Institute of Biogeochemistry and Marine Chemistry
Bundesstrasse 55, 20146 Hamburg, Germany

ABSTRACT

In microbial degradation experiments with soil percolation columns and a pure bacterial culture the consumption and the $^{13}C/^{12}C$ stable carbon isotope composition of aromatic hydrocarbons were monitored in order to evaluate if isotopic fractionation is applicable as a parameter to evaluate biodegradation processes in field studies.

A soil percolation column was filled with aquifer material from a coal-tar contaminated site and run under anoxic conditions. After a lag phase of 20 days degradation of toluene was observed concomitant with sulfate reduction. Strain TRM1, a new sulfate-reducing bacterium that grew with toluene as the sole carbon and energy source was isolated from the column.

In batch degradation experiments with strain TRM1 the isotopic composition of the residual toluene was recorded by IRM-GC-MS analysis. The observed δ ^{13}C-shift of 6 ‰ (-29 to -23 ‰ PDB) upon the degradation of 300 μM to 10 μM toluene indicated a considerable isotopic fractionation. The kinetic isotope fractionation factor was $\alpha C = 1.0017$ under closed system conditions.

Isotopic fractionation by sulfate dependent toluene degradation was also measured in a soil percolation column used as a model system for aquifer conditions. Upon degradation a stable toluene gradient of 145 μM to < 6 μM built up along the column and was used as a quasi closed system to determine the *in situ* kinetic isotope fractionation factor.

* To whom correspondence should be addressed: Email: Rainer.Meckenstock@Uni-Konstanz.de

Novel Approaches for Bioremediation of Organic Pollution, edited by
Fass *et al.* Kluwer Academic / Plenum Publishers, New York, 1999.

The toluene δ ^{13}C changed from -29 ‰ at the inlet to -26 ‰ at the outlet with a kinetic isotope fractionation factor of $\alpha C = 1.0014$ under closed system conditions.

Our laboratory observations suggest that the determination of isotopic fractionation of organic pollutants might become a tool to elucidate and quantify biological degradation in aquifers.

1. INTRODUCTION

Microbial activity is suggested to be the major process to reduce organic pollutants in aquifers. Biological degradation is usually recorded in controlled laboratory experiments which simulate subsurface conditions, or by monitoring the disappearance of the organic pollutants in the field [Kuhn et al., 1985; Hutchins, 1991; Albertsen and Christensen, 1994].

Decreasing concentrations of contaminants in aquifers could also be due to several non-biological processes such as evaporation, dilution, or adsorption to soil constituents. In the context of risk assessment a clear characterisation of *in situ* biodegradation is extremely important to predict the fate of organic pollutants in contaminated aquifers. Therefore, methods are needed to distinguish between *in situ* biodegradation and non-biological processes. Currently, there is no method available to characterise the *in situ* biological degradation of individual compounds in an aquifer system unambiguously.

The $^{13}C/^{12}C$ stable carbon isotope chemistry of naturally occurring ^{13}C (1,11 %) may provide clues to assign biodegradation processes in the environment [Peterson and Fry, 1987; Aggarwal et al., 1997]. For example, during biodegradation organic contaminants are oxidised to CO_2 which may alter the isotopic composition of dissolved carbonates in aquifer systems. Commonly, the isotopic composition of mineral oil constituents is in the range of the biological precursor material between -25 ‰ and -35 ‰ and during biodegradation the isotopically light CO_2 is released into the aquifer [Peters and Moldowan, 1993]. This changes the isotopic signature of dissolved carbonates which could be used to evaluate biological degradation of petroleum derived compounds in the aquifer [Höhener et al., 1998].

The difference in the isotopic composition of organic matter and inorganic carbonates originates from the biological isotopic fractionation of $^{13}CO_2$ which occurs during primary production (CO_2-fixation) in algae and plants [O'Leary, 1981]. A similarly significant isotopic fractionation is observed in CO_2 reduction during methanogenesis where methane is strongly depleted in ^{13}C by -32 ‰ to -79 ‰ [Games et al., 1978; Botz et al., 1996].

In many biological transformation processes the lighter substrate isotopes are preferentially used whereas an enrichment of heavy isotopes is observed in residual substrates. For example, during microbial methane oxidation a characteristic enrichment of ^{13}C in the residual methane fraction occurs indicating that the lighter methane is preferentially consumed [Whiticar and Faber, 1985; Blair and Aller, 1995]. Similarly, an isotopic shift of about +12 ‰ in the residual acetate fraction has been demonstrated in degradation experiments with aceticlastic methanogenic bacteria [Gelwicks et al., 1994]. Moreover, the H_2-dependent CO_2 reduction by methanogenic bacteria leads to an enrichment of ^{13}C in the residual CO_2 [Fuchs et al., 1979; Krzycki et al., 1987].

$^{13}C/^{12}C$ isotopic fractionation of crude oil constituents and chlorinated hydrocarbons during biodegradation has been investigated [Stahl 1980; Lebedew et al., 1969; Ertl et al., 1996] but most studies concerning isotopic fractionation of organic substrates during microbial degradation were confined to low molecular weight compounds and gases.

Our approach focuses on the $^{13}C/^{12}C$ isotopic effects of residual aromatic hydrocarbons occuring during anaerobic degradation by sulfate-reducing bacteria.

2. MATERIALS AND METHODS

2.1. Soil Column Experiments

Sediment cores from the anoxic zone of a coal-tar contaminated aquifer near Stuttgart, Germany, were sampled avoiding contact with air. Large stones were picked form the sample and the remaining gravel was used to fill soil percolation columns without further treatment.

Aquifer conditions were simulated in glass columns (4.5 cm x 40 cm) equipped with 5 sampling ports at the inlet and at 2.5 cm, 7 cm, 16 cm, 26 cm, and 36 cm (port 1–5) distance from the inlet. The flow (bottom to top) was adjusted with peristaltic pumps (Minipuls3, Gilson, Middleton, WI, USA) to one bed volume per day. All tubings except the tygon tubes of peristaltic pump were made of gas-tight Viton (MULTIMED, Kirchheim, Germany). Toluene (Fluka, Buchs, Switzerland) was added using 1000 μl syringes (Unimetrics Corp, Shorwood, IL) in connection with a SPP22i syringe pump (WPI, Berlin, Germany) and directly injected into a 100 ml mixing chamber (150 μM final concentration). The carbonate-buffered freshwater mineral medium [Widdel and Bak, 1982] was reduced with 3 mM $Fe^{II}SO_4$ instead of 1 mM Na_2S. Soil columns were run in the dark at a constant temperature of 16 °C.

2.2. Isolation of Strain TRM1 and Batch Culture Experiment

Strain TRM1 was enriched on the same bi-carbonate-buffered medium supplemented with 10 mM Na_2SO_4 and 3 mM $FeSO_4$ and with 5 μl toluene directly added to 50 ml medium in 100 ml serum bottles sealed with Viton rubber stoppers (70 shore) (MAAG TECHNIK, Dübendorf, Switzerland). The inoculum was taken from the toluene degrading soil column.

After repeated transfers the strain was isolated by serial agar dilution according to [Widdel and Bak, 1992].

2.3. Determination of Hydrocarbon Concentrations

250 μl-samples were withdrawn from the soil column ports with syringes, diluted in 1 ml ethanol (p.a.) and centrifuged. HPLC analysis (BECKMAN, System-Gold) of the hydrocarbons was performed with a silica reverse phase column (Grom-SIL ODS %-ST 5 μM 150X5 mm, GROM, Herrenberg, Germany) by isocratic elution with 70 % acetonitrile in 30 % 100 mM ammonium phosphate buffer (pH 2.6). Elution profiles were recorded with a Beckman UV-detector 166 at 206 nm and quantifications were performed by comparison with an external toluene standard.

2.4. Isotope Ratio Monitoring Gas Chromatography-Mass Spectrometry

Isotope ratios of hydrocarbons were determined by an isotope ratio monitoring gas chromatography-mass spectrometry system (IRM-GC-MS) consisting of a GC-unit, interfaced to a Finnigan MAT combustion device with a water removal assembly and coupled to a Finnigan MAT 252 mass spectrometer. The effluent stream from the GC was combusted to CO_2 which was transferred continuously to the mass spectrometer to determine the $^{12}CO_2/^{13}CO_2$ ratios. The IrmGC-MS system was controlled by the ISODAT software (Finnigan MAT). Methodical details are described elsewhere [Hayes et al., 1990; Merrit et al., 1994]. Isotope ratios are reported in δ-notation as standard differences to the Pee Dee Belemnite standard in parts per thousand ($\delta\ ^{13}C$ ‰). The system was calibrated with an external CO_2 standard and the performance of the instrument was checked daily with a standard solution of hydrocarbons with known $\delta^{13}C$ values. All samples were measured in triplicates with standard deviations usually lower than 0.8 ‰.

Calculations:

$\delta\ ^{13}C$ values (PDB, PeeDee Belemnite standard) were calculated with equation (1).

$$\delta^{13}C[‰] = \frac{\left(^{13}C/^{12}C \text{ sample - } 1\right)}{^{12}C/^{13}C \text{ standard}} \tag{1}$$

Calculation of the kinetic isotope fractionation factor αC was performed with the Raleigh equation for a closed system (2) [Rayleigh, 1896] as modified by Hoefs (1997) and Whiticar and Faber (1985).

$$\delta\ ^{13}C_{tm} = \delta\ ^{13}C_i + 1000\ (1/\alpha C - 1)\ \ln f \tag{2}$$

$\delta\ ^{13}C_{tm}$ is the actual isotopic signature at a given time t, $\delta\ ^{13}C_i$ is the initial isotopic signature of the substrate and $f = (C_{tm} / C_i)$ is the fraction of substrate remaining in the sample at any time t_m.

The equation can be solved for the kinetic isotope fractionation factor αC. If ($\delta\ ^{13}C_{tm}$ - $\delta\ ^{13}C_i$) /1000 is plotted over ln f, for all time points t_m the slope of the linear regression curve of

$$\frac{\delta^{13}C_{tm} - \delta^{13}C_i / \ln f}{1000} = 1/\alpha C - 1 \tag{3}$$

gives the kinetic isotope fractionation factor αC.

3. RESULTS

For the initial degradation experiments soil columns were filled with contaminated aquifer material and continuously supplied with toluene-enriched medium. Toluene recovery at the outlet of the column was only 60 % in the starting phase due to sorption to the soil material but increased continuously to 100 % when the column material got saturated with toluene (fig. 1).

After operation for 20 days an efficient degradation of toluene was observed (fig. 1). The toluene concentration at the effluent of the column decreased from 145 μM to less than 6 μM within the 45 days. Sampling of the ports along the column showed that a toluene gradient from the inlet to the outlet was generated mainly in the first 15 cm of the column (fig. 2). The bulk of the degradation took place in the first 2–5 cm of the column. Therefore, a quasi stationary system along the flow direction was established. An electron balance which was calculated by the consumption of sulfate showed a recovery of about

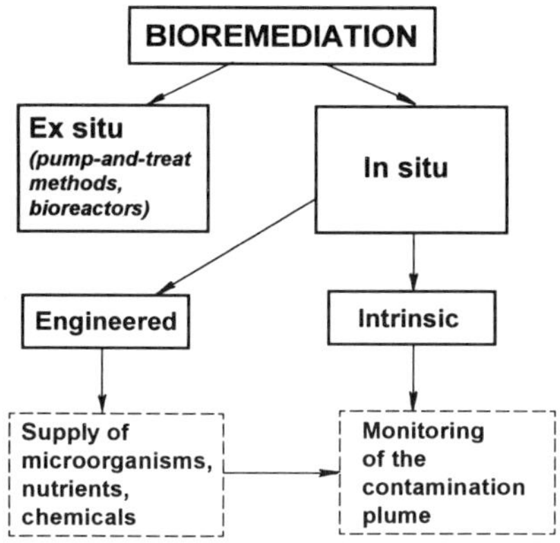

Figure 1. Degradation of toluene in the percolation column as a function of time C/Co depicts the toluene concentration at the outlet divided by the inlet concentration.

80 % of the electrons from toluene in the consumed sulfate which was most likely reduced to sulfide (data not shown).

A sulfate-reducing bacterium, strain TRM1, was isolated from the soil column with toluene as the sole carbon and energy source by enrichment in batch cultures and subsequent agar dilution series. The strain grew anaerobically on toluene with a doubling time of 36 hours at 30° C (Fig. 3 a). The electron balance showed a recovery of about 80 % of the electrons from toluene in the produced sulfide which suggests a complete oxidation of the toluene to CO_2. No other products such as organic acids or alcohols could be detected with HPLC analysis.

TRM1 was grown on toluene as sole carbon source in batch cultures. In order to determine the isotopic composition of the residual toluene fraction by IRM-GC-MS batch cultures were sampled periodically (Fig. 3 a) and the δ ^{13}C-values were plotted over the toluene concentration (Fig. 3 b). The isotopic composition of the applied toluene was - 29 ‰ (PDB) which is common for industrial mineral oil derived products. With decreasing toluene concentration in the batch increasing $^{13}C/^{12}C$ ratios of the remaining toluene were observed. Consequently, ^{13}C-toluene was enriched in the residual fraction. The increase of the δ ^{13}C-signature by 6 ‰ is significant compared to the standard deviation of 0.5 ‰.

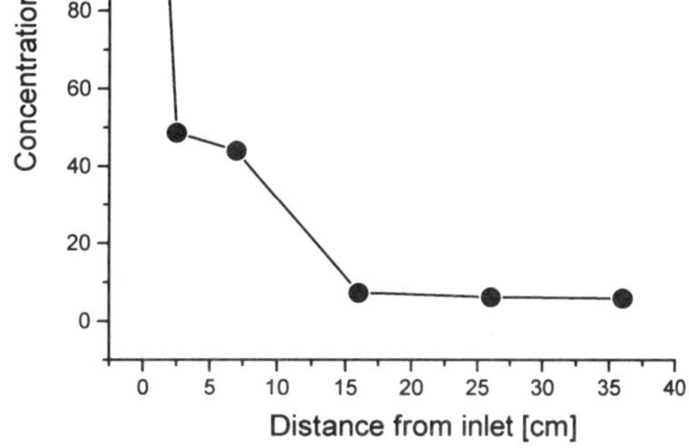

Figure 2. Toluene degradation gradient in the soil percolation column at day 50 from the inlet through port 5.

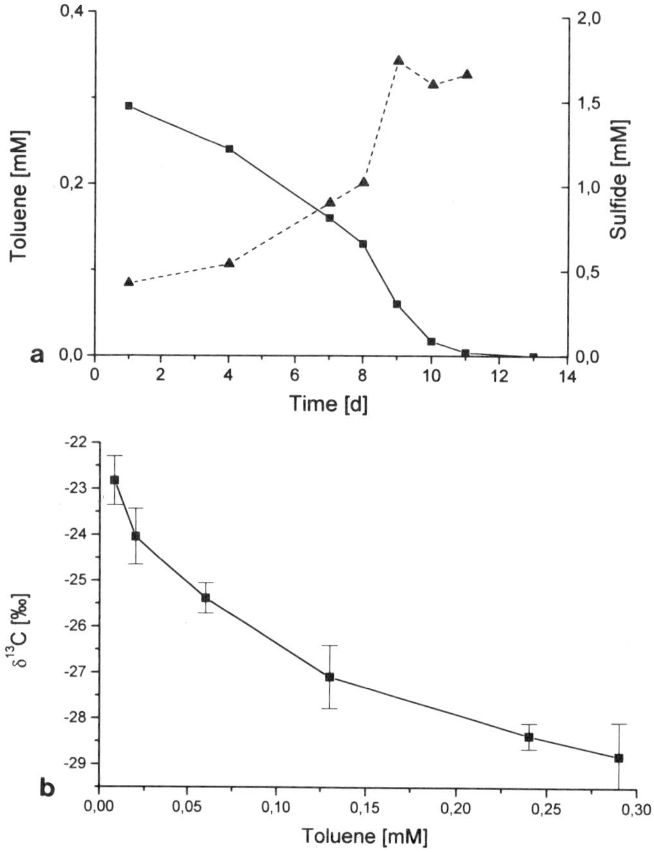

Figure 3. Growth curve of strain TRM1 on sulfate and on toluene as the sole carbon and energy source. a) Solid line, toluene concentration; dashed line, sulfide concentration. b) $^{13}C/^{12}C$ isotopic fractionation of residual toluene given in ‰ (PDB).

Presuming the batch experiment is a closed system, the kinetic isotopic fractionation factor αC can be calculated using the Raleigh equation [Rayleigh, 1896; Hoefs, 1997; Whiticar and Faber, 1985]. Based on the δ-values of the degradation curve (Fig. 3 b) the fractionation factor for toluene was $\alpha C = 1.0017$ at 30° C.

The soil percolation column represents a more complex biological system and may simulate aquifer conditions more realistically than a batch culture. Thus, the isotope fractionation of toluene during biodegradation in the soil column was analysed by sampling at the different ports and analysing for toluene concentration and isotopic composition. Plotting the δ-values versus the toluene concentration, an isotopic fractionation of 3 ‰ could be seen upon degradation of 140 mM toluene at 16° C. The fractionation factor $\alpha C = 1.0014$ was calculated using the Rayleigh equation as before.

4. DISCUSSION

Microbial degradation is suggested to be the dominant factor leading to the reduction of organic compounds in the plume of contaminated aquifers. Nevertheless, it is extremely difficult to quantify the biological contribution to the degradation process *in situ*. This is partly due to the fact that samples which have been removed from the aquifer and transferred to the lab in order to investigate biodegradation differ from field conditions. Degradation rates which are monitored in the laboratory study often do not occur in the field because the conditions in the aquifer might not be favourable. However, apart from laboratory studies there is to date no reliable method available which can monitor biological degradation of specific substrates in a contaminated aquifer *in situ*.

We isolated a new sulfate reducing strain which could grow with toluene as the sole carbon and energy source and differed from earlier discribed toluene degrading sulfate reducing bacteria [Beller et al., 1996; Rabus et al., 1993; Heider and Fuchs, 1997]. The anaerobic degradation of toluene by strain TRM1 revealed considerable carbon isotope fractionation of the residual substrate with a kinetic fractionation factor $\alpha C = 1.0017$.

In order to elucidate fractionation processes in a more complex system we studied the carbon isotopic fractionation of toluene in a soil percolation column as a model for an aquifer system. Strong degradation of toluene was observed in the first 2–5 cm of the percolation column and a stable gradient of the toluene concentration was generated. With respect to the stable toluene gradient the soil column behaved as a quasi-closed system in which the decreasing toluene concentration along the column profile was comparable to decreasing toluene concentrations over time in a batch experiment. The conditions of a quasi closed system are as well realistic for the plume of a contaminated aquifer assuming that the evaporation into the gas phase of the unsaturated zone is negligible.

In the percolation column study a fractionation factor $\alpha C = 1.0014 \pm 0.0002$ was observed. The difference to the batch culture fractionation factor of $\alpha C = 1.0017$ is rather low and in the range of the measurement error.

As could be shown by our experiments, the isotopic fractionation of a specific microbial community is comparable to the isotopic fractionation by pure cultures in laboratory experiments. This concept may be applied to evaluate biodegradation in natural aquifers considering a closed or quasi-closed system with relatively stable environmental conditions. Both presumptions are typical for many aquifers since temperature and physico-chemical conditions are stable over time. In this case isotopic composition and concentration of pollutants in the aquifer can be obtained from water samples and isotopic fractionation factors from laboratory experiments. The three parameters are required to estimate the microbial *in situ* degradation process, a factor which is very important to predict the future behavior of a contamination plume.

Important factors which in addition could influence the extent of isotopic fractionation are: temperature, electron donors and different types of microbial communities or degradation pathways. Nevertheless, the biologically induced $^{13}C/^{12}C$ isotopic fractionation might become a tool to investigate the fate of substrates in the environment and to predict biodegradation in contaminated aquifers.

ACKNOWLEDGMENT

We are grateful to Dr. Pierre Albrecht and Patrick Wehrung for providing facilities and technical assistance for the IRM-GC-MS measurements. This paper represents publication no. 44 of the priority program 546 "geochemical processes with long-term effects in anthropogenically affected seepage- and groundwater" with financial support by the Deutsche Forschungsgemeinschaft (grants Mi 157-12 and Schi 180/7-1).

REFERENCES

Aggarwal, P. K., M. E. Fuller, M. M. Gurgas, J. F. Manning, and M. A. Dillon (1997). Use of stable oxygen and carbon isotope analyses for monitoring the pathways and rates of intrinsic and enhanced *in situ* biodegradation. Environ. Sci. Technol. **31**: 590–596.

Albrechtsen, H.-J. and T. Christensen (1994). Evidence for microbial iron reduction in a landfill leachate-polluted aquifer (Vejen, Denmark). Appl. Environ. Microbiol. **60**: 3920–3925.

Beller, H. R., A. M. Spormann, P. K. Sharma, J. R. Cole, and M. Reinhard (1996). Isolation and characterization of a novel toluene-degrading, sulfate-reducing bacterium. Appl. Environ. Microbiol. **62**: 1188–1196.

Blair, N. E. and R. C. Aller (1995). Anaerobic methane oxidation on the Amazon shelf. Geochim. Cosmochim. Acta **59**: 3707–3715.

Botz, R., H.-D. Pokojski, M. Schmitt, and M. Thomm (1996). Carbon isotope fractionation during bacterial methanogenesis by CO_2 reduction. Org. Geochem. **25**: 255–262.

Ertl, S., F. Seibel, L. Eichinger, F. H. Frimmel, and A. Kettrup (1996). Determination of the 13C/12C isotope ratio of organic compounds for the biological degradation of tetrachloroethene (PCE) and trichloroethene (TCE). Acta hydrochim. hydrobiol. **24**: 16–21.

Fuchs, G., R. Thauer, H. Ziegler, and W.Stichler (1979). Carbon isotope fractionation by *Methanobacterium thermoautotrophicum*. Arch. Microbiol. **120**: 135–139.

Games, L. M., J. M. Hayes, and R. P. Gunsalus (1978). Methane producing bacteria: natural fractionation of stable carbon isotopes. Geochim. Cosmochim. Acta **42**: 1295–1297.

Gelwicks, J. T., J. B. Risatti, and J, M. Hayes (1994). Carbon isotope effects associated with aceticlastic methanogenesis. Appl. Environ. Microbiol. **60**: 467–472.

Hayes J. M., K. H. Freeman, B. N. Po, and C. H. Hoham (1990). Compound-specific isotope analysis. A novel tool for reconstruction of ancient biogeochemical processes. In: Advances in Organic Geochemistry 1989 (Durand B. & Behar F. eds.). Org. Geochem. **16:** 1115–1128.

Heider, J. and G. Fuchs (1997). Anaerobic metabolism of aromatic compounds. Eur.J. Biochem. **243**: 577–596.

Hoefs, J. (1997). Stable isotope geochemistry (4.th edition). Springer Verlag, Berlin, Germany.

Höhener, P., D. Hunkeler, A. Hess, T. Bregnard, and J. Zeyer (1998). Methodology for the evaluation of engineered *in situ* bioremediation: lessons from a case study. J. Microbiol. Methods **32**: 179–192.

Hutchins, S. R. (1991). Biodegradation of monoaromatic hydrocarbons by aquifer microorganisms using oxygen, nitrate, or nitrous oxide as the terminal electron acceptor. Appl. Environ. Microbiol. **57:** 2403–2407.

Kryzcki, J. A., W. R. Kenealy, M. J. DeNiro, and J. G. Zeikus (1987). Stable isotope fractionation by *Methanosarcina barceri* during methanogenesis from acetate, methanol, or carbon dioxide-hydrogen. Appl. Environ. Microbiol. **53**: 2597–2599.

Kuhn, E. P., P. J. Colberg, J. L. Schnoor, O. Wanner, A. J. B. Zehnder, and R. P. Schwarzenbach (1985). Microbial transformation of substituted benzenes during infiltration of river water to groundwater: laboratory column studies. Environ. Sci. Technol. **19**: 961–968.

Lebedew, W. C., W. M. Owsjannikow, G. A. Mogilewskij, and W. M. Bogdanow (1969). Fraktionierung der Kohlenstoffisotope durch mikrobiologische Prozesse in der biochemischen Zone. Angewandte Geologie **15**: 621–624.

O'Leary, M. H. (1984). Carbon isotope fractionation in plants. Phytochemistry **20**: 553–567.

Merritt D.A., Brand W.A., and Hayes J.M. (1994) Isotope ratio-monitoring gas chromatography mass spectrometry -Methods for isotopic calibration. Org. Geochem. **21**, 573–583.

Peters K.E. and Moldowan J.M. (1993) The biomarker guide. Prentice Hall, Englewood Chliffs, New Jersey.

Peterson, B. J. and B. Fry (1987). Stable isotopes in ecosystem studies. Ann. Rev. Ecol. Syst. **18**: 293–320.

Rayleigh, J. W. S. (1896). Theoretical considerations respecting the separation of gases by diffusion and similar processes. Philos. Mag. **42**: 493.

Rabus, R., R. Nordhaus, W. Ludwig, and F. Widdel (1993). Complete oxidation of toluene under strictly anoxic conditions by a new sulfate-reducing bacterium. Appl. Environ. Microbiol. **59**: 1444–1451.

Stahl, W. J. (1980). Compositional changes and 13C/12C fractionations during the degradation of hydrocarbons by bacteria. Geochim. Cosmochim. Acta **44**: 1903–1907.

Whiticar, M. J. and E. Faber (1985). Methane oxidation in sediment and water column environments-Isotope evidence. Advances Organic. Geochem. **10**: 759–768.

Widdel, F. and F. Bak (1992). Gram-negative mesophilic sulfate-reducing bacteria, Vol. **4.** Edition 2. P. 3352–3378. In A. Balows, H. G. Trüper, M. Dworkin, W. Harder and K. H. Schleifer (eds.), The Procaryotes, New York, Springer Verlag.

23

DEGRADATION OF ATRAZINE IN CONTAMINATED SUB-SOIL AND WASTEWATER BY WHOLE CELLS OF *Pseudomonas* sp. STRAIN ADP AND BY THE ENZYME ATRAZINE CHLOROHYDROLASE

Raphi T. Mandelbaum, Nir Shapir, and Carl Kauffmann

The Institute of Soils, Water, and Environmental Sciences
Volcani Research Center, Bet Dagan, 50250 Israel

1. INTRODUCTION

The outstanding performance of atrazine with respect to selective herbicidal effects and crop tolerance is coupled with relative recalcitrance in the environment and tendency to leach to groundwater. Nevertheless, over 83 million pounds of the three major *s*-triazine herbicides (atrazine, cyanazine, and simazine) were used in the cultivation of corn in the United States alone in 1990 (National agricultural statistics service, 1991). This large scale usage inevitably resulted in contamination of environmental sites such as—agricultural fields, lakes, rivers, sediments and ground water which are subtle living ecosystems and await remediation. The seasonal contamination of the Mississippi river and the Great Lakes Basins in May-July can best illustrate the magnitude of the contamination problem. Contamination of atrazine in these surface waters were reported to sometimes exceed the Maximum Contaminant Levels (MCLs) of 3 μg/L (Nelson and Jones, 1994). It should however be appropriate to mention that the environmental risks associated with these contaminations are by far less dramatic than those associated with chemicals such as DDT, dioxins or PCB's.

Nevertheless, the permitted MCL in drinking water (as low as 0.1 ppb in Europe) posed atrazine among the pesticides that have ground water advisory statements on their labels. Moreover, incidental spills of atrazine or the discharge of atrazine manufacturing wastewater containing high levels of residual atrazine pose a significant contamination risk to soils and water.

Cook (1987) and Erickson and Lee (1989) reviewed the examination of steps, necessary to biodegrade high concentrations of s-triazines. In their reviews, bioremediation of

Novel Approaches for Bioremediation of Organic Pollution, edited by
Fass *et al.* Kluwer Academic / Plenum Publishers, New York, 1999.

atrazine from contaminated soil or water was not considered feasible due to the absence of efficient atrazine degrading cultures.

In the 1980's the general microbial biodegradation pathway of *s*-triazine herbicides became better understood and emerged to consist of a series of hydrolytic cleavage of chloro, amino and alkylamino groups from the *s*-triazine ring (Erickson and Lee, 1989) and/or oxidative dealkylation of the side chains. Pathways generally converge to cyanuric acid, which is further subject to hydrolytic ring cleavage to CO_2 and NH_4^+ via hydrolysis to biguanidin, biuret, and urea.

The isolation of *Pseuodomonas* sp. strain ADP (*P.* ADP) (Mandelbaum et al., 1995) and the characterization of the gene and protein involved in the dechlorination of atrazine (De souza et al., 1995) in recent years, enabled the development of biotechnological tools to remove atrazine from soils and water. Several factors are been thought to influence the competitiveness of indigenous and applied microorganisms in soils, aquifers surface waters and in wastewater. These can be divided into biotic and abiotic factors which include: nutrient limitation and specificity, moisture requirement, electron acceptors, pH, temperature, soils texture and porosity, organic matter content, antibiotic and bacteriocin production, solute types and concentrations, numbers and type of other indigenous microorganisms, selective predation by protozoa, residence time in soil, and ability to form resting structures. All of these factors, either singly or in combination, play major roles in the selective activity of various members of the microbial community. Depending on the microbial population that occupies the soil, biodegradation of *s*-triazines varies in rate and yield various metabolites with variable environmental consequences.

The present report describes the potential use of *P.* ADP and its atrazine chlorohydrolase enzyme to decontaminate soils and water contaminated with atrazine via dechlorination of the regulated compound atrazine to its non-regulated metabolite hydroxyatrazine.

2. ATRAZINE DEGRADATION IN SUB-SOILS

Soil is the major arena for herbicide application, action and eventual dissipation. Even with post-emergence applications when the plant foliage primarily receives the herbicide, the soil surface is the end recipient. The vadose zone beneath the cultivated horizon is the last barrier that can detoxify, adsorb or immobilize leaching herbicides before they contaminate the aquifer. The destruction of herbicides applied to soil is of great practical importance and essential to their sustained use, yet different soils in different depths exhibit differences in their ability to degrade herbicides such as atrazine (Sparling and Aislabie, 1996). Therefore, understanding of the activity of atrazine degrading bacteria in various soil strata is important for proper agricultural use and moreover so when bioremediation of contaminated soil or aquifer sediment is considered.

2.1. Non Saturated Sub-Soils

To date, most of the studies on atrazine degradation focused on the root zone and on dissipation and leaching into the subsurface, and only a few have considered the degradation in the unsaturated zone (Ritter,1990; Linn et al., 1993; Bottoni et al., 1996; Ma and Selim, 1996). However, since even small amounts of pesticides can seriously contaminate groundwater (Kordel et al., 1992), information on microbial degradation in the unsaturated zone is important for the understanding of the overall fate of atrazine and have a significant implication on bioremediation of atrazine spills.

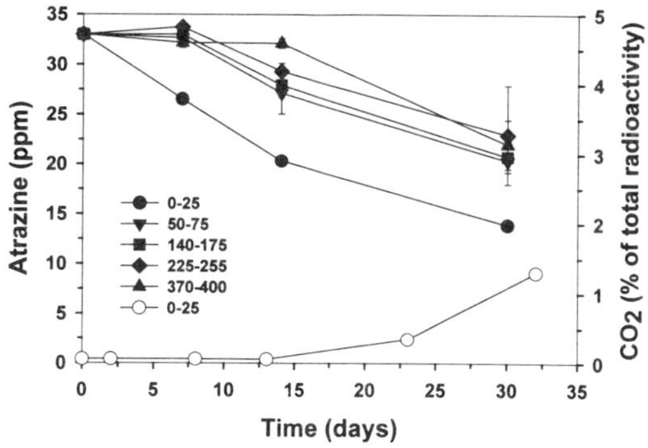

Figure 1. Atrazine degradation (solid signs) and mineralization (empty signs) in non-augmented soil samples. Numbers in the graph refer to soil depths in cm.

To elucidate the fate of atrazine and the activity of the atrazine degrading bacterium *P.* ADP in a non saturated soil, a soil profile was sampled to a depth of 4 m. Soil analysis data indicated spatial heterogeneity with clayey soil sandwiched between two sandy layers. The surface sandy layer (0–25 cm) was underlayed by a more clayey soil (50–255 cm), which was followed by another sandy layer (Shapir and Mandelbaum, 1997). In the surface layer, higher organic matter content and Kd values were measured. These values declined with increasing soil depth, and soil microbial biomass (Biomass-N) and activity (hydrolase and dehydrogenase) had already fallen below the experimental detection limits at a soil depth of 2.5 m (data not shown). Generally, the decrease in microbial activity along the soil profiles was not linear, and there were steady levels of microbial activity and biomass in the 75–175 cm horizon. The highest microbial activity and biomass were associated with the upper soil layer, that contained more than 1% organic matter. The zone between 75 and 175 cm, where the activity was relatively constant had an organic matter content of 0.6–1%, and in the zone with no activity, soil organic matter content was below 0.4%.

Atrazine degradation was measured in various soil depths along the soil profile. Significant atrazine degradation was detected in samples taken from the upper soil level. More than 50% of the applied atrazine (33 ppm) was degraded in 30 days of incubation (Figure 1). However, in soil samples taken from lower parts of the profile, significantly less atrazine degradation occurred. Less than 33% degradation was recorded by the end of 30 days in all soil samples from deeper than 50 cm. The results of our experiments are in line with previous results which indicated lower degradation rates in subsurface soils because of their lower microbial populations (Fomsgaard, 1995; Kordel et al., 1995; Veeh et al., 1996).

Atrazine mineralization in the top layer with most activity, occurred to a much lesser extent. Thus, indicating that only partial degradation of atrazine in the soils took place. After 30 days, about 1% of the initial atrazine amount in the upper soil layer had been mineralized (Figure 1), while in soil samples from deeper parts of the profile no atrazine mineralization had taken place (data not shown).

When *P.* ADP was inoculated in soils amended with [U-ring-^{14}C]atrazine, 90–100% of the label was recovered as $^{14}CO_2$ in 15 days after inoculation (Figure 2). However, marked differences in the initial mineralization rates of atrazine were recorded.

Samples from depths of 0–75 cm and 370–400 cm showed enhanced rates of mineralization, whereas mineralization in the strata between these levels lagged. In these experiments citrate was added as an external carbon source to support the activity of *P.* ADP.

From these data we could conclude that the potential for atrazine degradation exists in all soil strata. The most important factor that governs atrazine mineralization in the

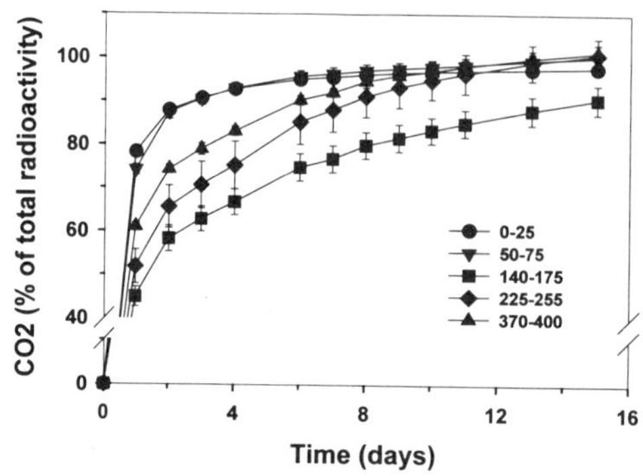

Figure 2. Atrazine mineralization in soil samples taken from soil profile and inoculated with *P.* ADP. Numbers refers to soil depths in cm.

soils is the presence of an atrazine-mineralizing population. From previous pesticide leaching data collected in the same location (data not shown) and in accordance with previously published data (Walker et al., 1996; Azevedo et al., 1997) pesticide distribution 1 year after application was mostly in the 10–100 cm layer.

It is suggested that bioremediation of soils in the upper level or in deeper strata is feasible by amendment with the bacterium *P.* ADP.

2.2. Saturated Sub-Soils

Significant amounts of atrazine leach beneath the aerobic soil layer into lower soil strata and into groundwater where oxygen limited conditions prevail (Ma and Selim, 1996; Sorenson et al., 1994). The finding of *s*-triazine herbicides and their metabolites in subsurfaces (Fomsgaard, 1994; Ng et al., 1995) and groundwater (Fomsgaard, 1994; Kolpin et al., 1996), has increased the importance of elucidating their degradation pathways and rates in the subsurface environments (Aislabie and Lloydjones, 1995). Once *s*-triazines reach groundwater they become less amenable to biodegradation (Wood et al, 1991). Even when atrazine metabolism was observed under natural aquifer conditions, the rates are slow. In fact in all studies that reported on atrazine dissipation in aquifers, very slow or not existing biological degradation rates has been documented (Goswami et al., 1971; Mcmahon and Chapelle, 1992; Wehtje et al., 1983). The relatively slow degradation rates of atrazine has been contributed to several factors, including application history, soil nutrient conditions, temperature, oxygen content, redox potential and the presence of atrazine degrading microorganisms.

To study the ability of *P.*ADP to degrade atrazine in sediments from a contaminated aquifer, soil above the aquifer and sediments from 30 cm below the water table, were collected from a shallow aquifer at Drengsted, Jutland, Denmark. The aquifer underlined a cornfield continuously cultivated and sprayed with either atrazine (until banned in 1995) or terbuthylazine (after 1995). Atrazine mineralization in 20 g soil samples was measured in biometric flasks. Twenty-four h after atrazine application, the samples were amended with 7.5 ml phosphate buffer, containing sodium nitrate and *P.* ADP cells, to yield 3.5 mM of nitrate (including the pre-existing nitrate), and 1×10^6 bacterial cells g^{-1} sediment. A non-inoculated control was added with nitrate only. Half of the sediment samples were incubated under anaerobic conditions and amended with citrate (as a carbon source) to yield a final concentration of 6.8 µmole sodium citrate g^{-1} sediment.

Anaerobism was achieved by placing the samples in an anaerobic hood in which the atmosphere was flushed with high purity nitrogen gas. Atrazine mineralization was recorded in sediments contaminated with 0.01 ppm atrazine and inoculated with *P.* ADP cells under both aerobic and anaerobic conditions (Table 1). 15 days after inoculation over

Table 1. Atrazine mineralization in aquifer sediments

Treatment	Atrazine concentration			
	0.1 ppm		10 ppm	
	% from initial	Total amount mineralized (ng/g)	% from initial	Total amount mineralized (μg/g)
Anaerobic with citrate	76.0	7.6	80.0	8.0
Anaerobic without citrate	78.3	7.8	1.5	0.15
Aerobic without citrate	79.3	7.9	3.4	0.34

70% of the atrazine was mineralized in the sediment not amended with citrate. Citrate amendment resulted in a significant increase in the denitrification rate (Data not shown), but reduced the rate of atrazine mineralization. Only 55% of the atrazine were mineralized after 4 days from inoculation. However, after 15 days of incubation nearly 80% (8 ng g^{-1} dw soil) mineralization was recorded irrespectively of the incubation conditions. When higher atrazine concentrations were present (10 ppm), citrate amendment in anaerobic conditions significantly increased atrazine degradation rates relative to atrazine degradation under both aerobic and anaerobic conditions without citrate. After 15 days almost 80% of the amended atrazine was mineralized in the citrate amended treatment, while only 1% mineralization was recorded in treatments not receiving citrate (Table, 1). Interestingly, citrate amendment largely improved atrazine degradation when atrazine was present in high concentration, but not in the low concentration treatment. It has been previously suggested that atrazine serves as a N source for *Pseudomonas* sp. strain ADP (Mandelbaum et al., 1995) and therefore it is assumed that C:N ratio is an important factor determining the mineralization rate of atrazine.

It should be noted however, that although the percentage atrazine mineralized in the 10 ppm atrazine treatment not receiving citrate is lower than in the 0.01 ppm, the total mineralized amount was much higher. After 15 days of incubation 160 and 365 ng g^{-1} dw soil of atrazine were mineralized under anaerobic and aerobic conditions, respectively without citrate addition. The highest quantity of atrazine was mineralized under anaerobic conditions with citrate e.g. 8 μg g^{-1} dw soil.

The present data prove that the absence of an efficient atrazine mineralizing bacteria in the sediments may be the major reason for the persistence of atrazine in the aquifer. Previously, Splichal and Schwab (1991) suggested that the prolonged persistence of atrazine under aquifer conditions ($t_{½}$ = 209–648 d) was an indication of less microbial activity or the absence of atrazine degraders. Our results indicate that inoculation with *P.*ADP can result in a rapid mineralization of atrazine in contaminated aquifer sediments.

3. ATRAZINE DEGRADATION IN WASTEWATER

Wastewater from atrazine manufacturing plants contains large amounts of residual atrazine and atrazine synthesis products, which must be removed before disposal. One of the obstacles to biological treatment of these wastewaters is their high salt content, e.g., up to 4% (w/v). To enable biological treatment we have adopted *P.* ADP to degrade atrazine in the presence of relevant salt concentrations (e.g. ~ 3% NaCl). We also considered the ability of using enzyme based systems for the dechlorination of atrazine, in order to overcome limitations imposed by whole cell live systems. The use of enzymes to detoxify industrial wastewater did not attract much attention due to the high cost of enzyme-based systems. However, the increased cost of toxic wastewater treatment coupled with the use of new, relatively inexpensive and efficient enzyme immobilization methods may increase the interest in enzymatic wastewater treatment.

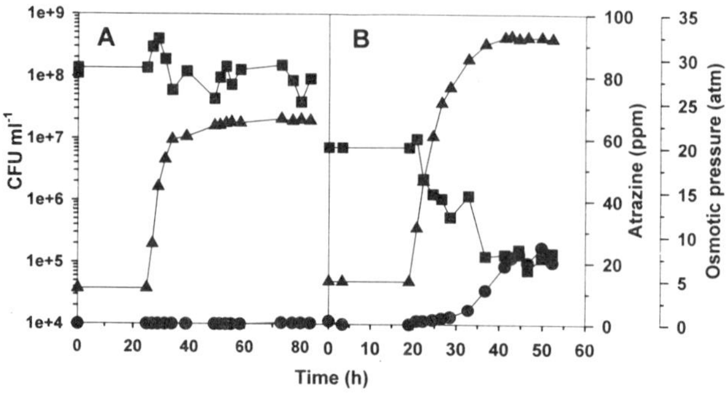

Figure 3. *P.* ADP growth and its atrazine degradation activity as affected by increasing medium salinity. A = 3% NaCl, B= 5% NaCl. (■) - Bacterial concentration (CFU ml^{-1}); (▲) - osmotic pressure (atm); (●) - Atrazine concentration (ppm).

Here, we report on the adaptation of *P.* ADP to tolerate high salt concentrations, the use of sol-gel entrapped atrazine chlorohydrolase (atzA) and the construction of chimeric enzymes for the removal of atrazine residues from wastewater.

3.1. Adaptation of *P.* ADP in Continuous Cultures Exposed to Increasing Osmotic Pressure

To grow in an osmotically stressful environment, a bacterial cell must establish and maintain its internal pressure above that of its surrounding medium (Abee et al. 1990), which it commonly achieved through the accumulation of osmotically active solutes (osmolytes) in the cytosol (Csonka, 1989; Le Rudulier et al., 1984). *P.* ADP was grown in a chemostat. The dilution rate was 0.167 h^{-1} with double strength atrazine medium previously described by Mandelbaum et al. (1993). Atrazine was the limiting growth factor and served as the sole nitrogen source. After the culture had reached a steady state when fed from a reservoir containing 0.1% NaCl, the feeding solution was replaced with a 3% or 5% NaCl in double strength atrazine medium. Samples for bacterial counts and for determination of the residual atrazine concentration (HPLC), were aseptically removed at selected time intervals. Complete degradation of atrazine during the first 20–22 h (when NaCl content was low), indicated a physiologically healthy culture. (Figure 3A and Figure 3B, respectively). The bacterial counts during the first 22 h reached 1×10^8 cells ml^{-1} and 1×10^7 cells ml^{-1} (Figure 3A and Figure 3B, respectively). When the NaCl concentration in the reservoir medium was gradually raised to 3.0%, *P.* ADP could maintain its atrazine degradation activity (Figure 3A). Accumulation of the disaccharide trehalose and a small amount of an unknown compound, suspected to be the unique dipeptide *N*-acetylglutaminylglutamine amide (NAGGN), were detected using NMR analysis (Shapir et al., 1998). The amounts of trehalose and NAGGN within late log phase cells, grown on 3% NaCl, were 82.2 mg g^{-1} protein and 25.4 mg g^{-1} protein, respectively.

3.2. Atrazine Degradation by *P.* ADP in Authentic Wastewater from an Atrazine Manufacturing Plant

These results indicate that under a gradual, slow increase in salt concentration *P.* ADP can synthesize enough osmolyte to protect itself from drying out. When the reservoir medium NaCl concentration was raised to 5.0% (Figure 3B) and the osmotic pressure rose above 20 atm the bacterium could not degrade atrazine effectively and cell counts dropped two orders.

Industrial effluents were collected from a wastewater biological treatment facility in a herbicide factory. Atrazine degradation was measured in wastewater or saline (0.1% NaCl) (control) supplemented with 25 ppm atrazine. Sodium citrate was added as a carbon source, to give a final concentration of 2% (w/v). Media were inoculated with either a non-adapted or 3% NaCl-adapted *P.* ADP suspension to yield a final concentration of 5×10^7 cells ml^{-1}. At regular intervals, samples were removed, to evaluate atrazine degradation using a gas chromatograph. The wastewater osmotic pressure was 22 atm (3% NaCl), and its pH was neutral (6.9); it contained high concentrations of nitrogenous compounds originating from the synthesis. Only treatment with adopted bacteria significantly improved the degradation of atrazine in the wastewater (Figure 4). This result proves the ability of salt adopted *P.*ADP to successfully compete with the indigenous bacterial population of the effluents and degrade atrazine in the presence of other nitrogenous compounds.

The addition of a carbon source (sodium citrate) further improved the degradation rate and complete degradation was observed after 4 days. The hydrocarbons in the wastewater could provide some carbon and energy sources for *P.* ADP, as indicated by the similar initial atrazine degradation rates in the citrate-amended and non-amended wastewater. However, after 24 h, the rate of degradation in the non-amended medium leveled, while the citrate-amended culture completely degraded atrazine.

The results of the present study suggest that *P.* ADP can be used for atrazine degradation in salt-containing wastewater. Further studies on the ability of *P.* ADP to utilize carbon sources available in authentic wastewater are underway.

3.3. Entrapment of Atrazine Chlorohydrolase in Sol-Gel Matrix

The sol-gel process is collectively referring to several processes in which a solution or sol undergoes a sol-gel transition. In its broadest sense the term sol-gel refers to the preparation of inorganic oxides by wet chemical methods, irrespective of the final form of the product, monolith, crystalline or amorphous (Buckley and Greenblatt, 1994; Hench and West, 1990). Using sol-gel materials for mechanical entrapment of enzymes permitted stabilization of the proteins tertiary structure due to the tight gel network (Avnir et al., 1994). Moreover,

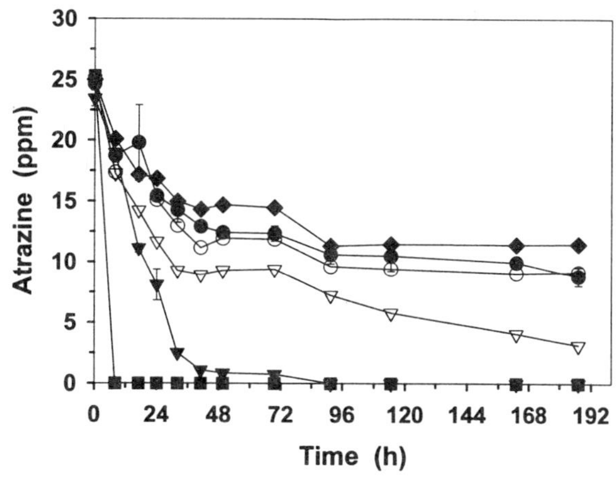

Figure 4. Atrazine degradation in authentic wastewater by non-adapted (●) and 3.0% NaCl adapted (▲) *P.* ADP, with or without the addition of citrate (solid and empty symbols, respectively). (◆)—Authentic wastewater without addition of bacteria; Square—Saline water. Bars represent the standard error.

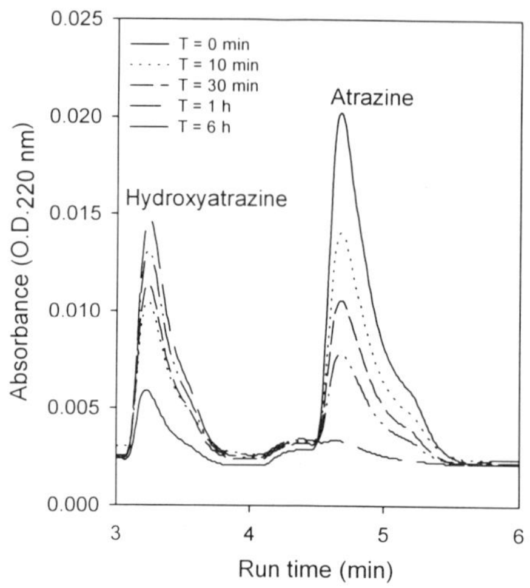

Figure 5. HPLC chromatogram from a typical activity experiment. The overlapping graphs show the concentration of atrazine and hydroxyatrazine after 0, 10 min, 30 min, 1 hour and 6 hours reaction time. The peak of atrazine decreased until totally degraded after 6 hours, while the hydroxyatrazine peak concomitantly increased.

the easy insertion of substituent groups into a silicate matrix may provide the entrapped enzymes with beneficial micro environments (Reetz et al., 1995; Reetz et al., 1996).

AtzA was purified from crude cell extracts of wild-type *P.* ADP and immobilized in sol-gel glass matrices according to Kauffmann and Mandelbaum (1998). The immobilized enzyme completely transformed atrazine to hydroxyatrazine during 5 reaction cycles carried out over 3 weeks. In all reaction cycles all the atrazine was converted to hydroxyatrazine. The kinetics were as illustrated in Figure 5, which shows an example of the conversion of atrazine to hydroxyatrazine during the forth reaction cycle. Mass balance calculations evidenced stochiometric transformation of atrazine to hydroxyatrazine (data not shown).

AtzA activity was significantly improved by increasing the ratio of hydrophobic to hydrophilic sol-gel components (Figure 6). Under the best conditions, the sol-gel immobilized atzA maintained more than 40% of the measured activity of the non-immobilized protein.

In previous results from our laboratory (Kauffmann and Mandelbaum, 1996) less than 1% of the activity of the free protein preparation was maintained upon immobilization in sol-gel. These results are now largely improved by incorporating the hydrophobic methyltrimethoxysilane (MTrMOS) into the otherwise hydrophilic tetramethoxysilane

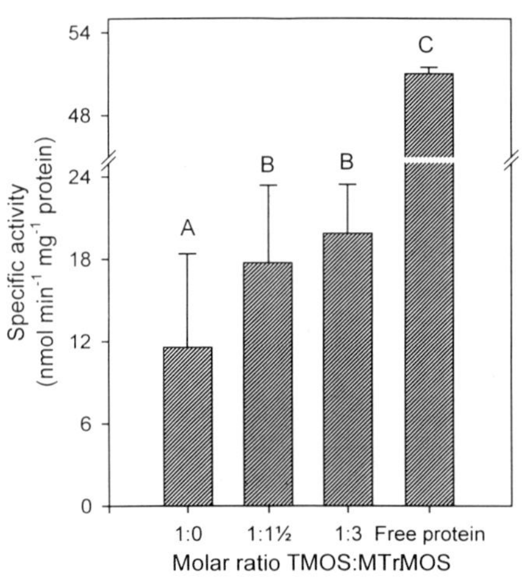

Figure 6. Effect of sol-gel hydrophobisity on atzA activity (one unit = 1 nmol atrazine converted by 1 mg atzA in one min). A higher ratio of MTrMOS to TMOS results in a more hydrophobic sol-gel matrix. Error bars represent standard deviation values. One way analysis of variance (Bonferroni's method) showed that the results marked A rs. B rs. C were significantly different from each other ($P < 0.05$).

(TMOS) sol-gel matrix. Further improvement of the activity might be achieved with a faster turn-over enzyme. The Km and Vmax values of free recombinant atzA for atrazine has been reported to be 149 μM (32 ppm) and 2.6 μmol of hydroxyatrazine per min per mg of protein, respectively (De Souza et al., 1996). Site directed mutagenesis could additionally improve the activity (Larry Wackett, personal communication). Therefore, the given rates can probably still be improved.

It is concluded that entrapment of active atzA in sol-gels was possible and that high activity was maintained over at least 3 weeks. The more hydrophobic sol-gel performed better, probably due to better affinity with the substrate (atrazine is a hydrophobic herbicide) and/or possible improved interaction with the protein.

3.4. CBD atzA Fusion Protein Cloning

Cellulose binding domains (CBDs) from several bacterial species provide a new class of affinity tags with appealing biotechnological attributes, including: (i) Effective immobilization of CBD fusion proteins to cellulose-based matrices without the need for covalent crosslinking; and (ii) A readily available, low cost affinity matrix (cellulose) with inherent low non-specific protein binding characteristics. Moreover, after the construct has been overexpressed in a bacterial or fungal system, the construction of a bio-filter is achieved in one easy step, e.g. breaking the cells (or harvesting an excreted soluble enzyme) and reacting the protein with the cellulose which will later be used as a bio-filter to remove atrazine from wastewater (Figure 7).

3.4.1. Bacterial Strains, Plasmids, Cloning of CBD-AtzA and Activity Tests. *E. coli* XL1-Blue was used for all cloning experiments. *E. coli* BL21 was used for all the expression experiments. The expression vector pET-CBD-180 (Oded Shoseyov, personal communication), were used for cloning, and pMD4 (De Souza et al., 1995) was used as PCR template.

AtzA was PCR amplified using pMD4 as a template. The primers were constructed as follows: N-terminal primer containg a NcoI restriction site:

5′AAAACCATGGCGCAAACGCTCAGCATCC 3′.

C-terminal primer containg a stop codon followed by a BamHI restriction site:

5′ AAAAGGATCCTAGAGGCTGCGCCAAg 3′.

The primers were designed and synthesized in a way that enabled forced cloning of the *atzA* gene (NcoI/BamHI) into the pET-CBD-180 plasmids. PCR conditions were optimised and the PCR amplified product was digested with NcoI and BamHI. The expected 1.4 Kb DNA fragment was isolated from a 0.6% agarose gel and the NcoI/BamHI fragment was ligated using T4 Ligase into the pET-CBD-180 plasmid predigested with the same enzymes. The ligation mixture was used to transform *E. coli* XL1-Blue competent cells and transformed colonies were selected on LB agar plates containing 100 g ml^{-1} ampicillin. Successful construct containing the DNA insert was designated pET-CBDclos/atzA (Figure 8) and transformed into *E. coli* BL21.

Overexpression was carried out using the pET system (Anonymous, 1995), where expression is driven by bacteriophage T7 DNA polymerase. The host strain used was *E. coli* BL21. The expression was induced by isopropyl-beta-D-thiogalactopyranoside (IPTG). For protein purification, transformed cells were grown overnight in M9 minimal medium containing 50 μg ml^{-1} ampicillin. After dilution 1:80 in TB medium containing 100 μg ml^{-1} ampicillin, cells were grown to 1.5 O.D. 600nm at 37 ^{0}C, at this point IPTG

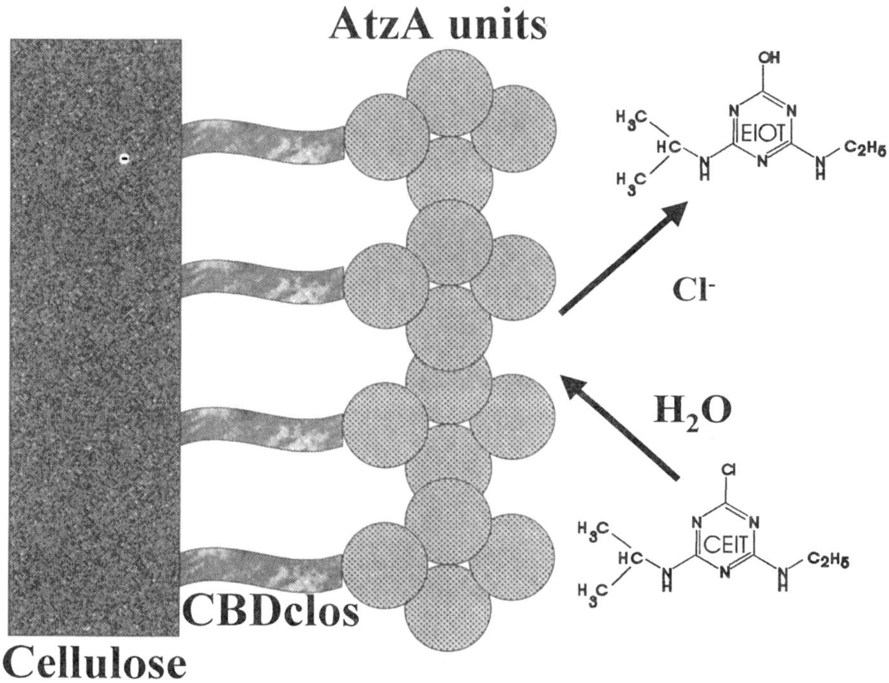

Figure 7. Illustration of a CBD atzA fusion protein immobilized on cellulose.

was added to a final concentration of 0.5 mM. After 4 h the cells were harvested by centrifugation and resuspended in tris buffer pH 7. The cells were lysed by three passages through a homogenizer applying an internal pressure of 10,000 psi. Inclusion bodies were isolated by centrifugation at 9,000 x g, washed and analyzed by gel electrophoresis.

Approximately 70 mg of washed inclusion bodies (wet weight) were suspended in 1 ml of phosphate buffer (pH 7). The activity test (500 μl) was carried out in phosphate buffer at 30 ^{0}C. Inclusion bodies (14 mg ml^{-1}), CoSO4 (1 mM) and [U-^{14}C]atrazine were incubated 22 h, and extracted with ethyl acetate (200 μl). Aliquots (80 μl) from the aqueous and from the organic phases were mixed with Ultima Gold, scintillation cocktail and counted. The specific activity of the non-soluble part of the broken cells from 6 clones

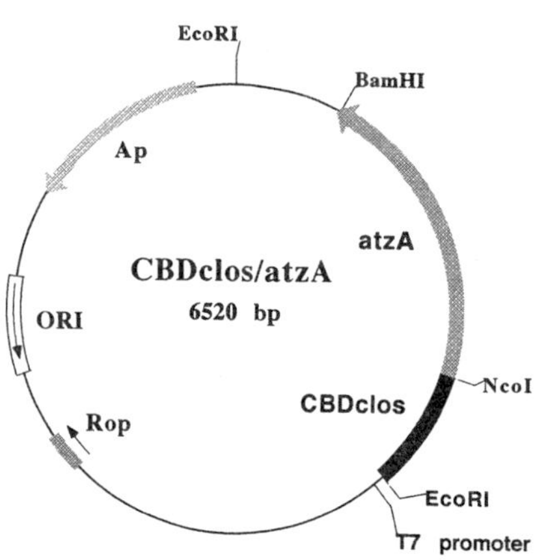

Figure 8. pET-CBDclos-atzA vector carrying ampicillin resistance. The *atzA* gene were inserted between the NcoI and the BamHI restriction sites at the N-terminal of the CBD_{clos}.

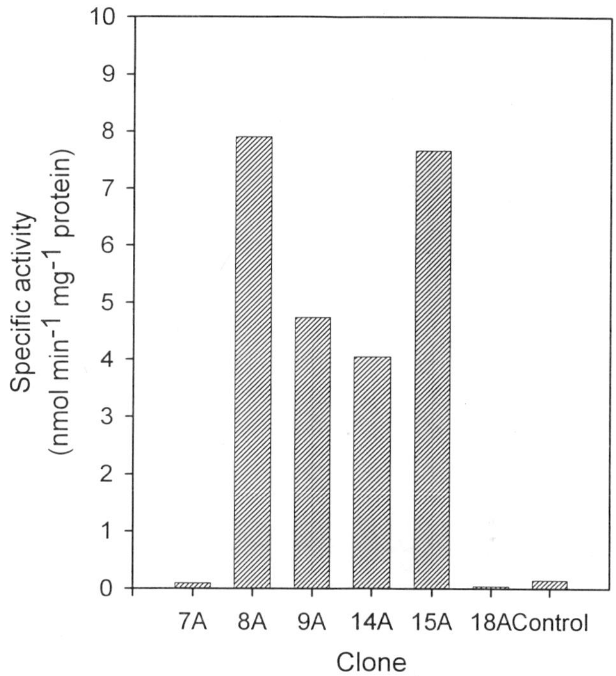

Figure 9. Specific activity of the non-soluble part of broken cells from 6 selected clones of *E.* coli BL21 transformed with the pET-atzA or the pET-CBDclos-atzA plasmid. 7A-9A: pET_{atzA} clones. 14A-18A: $pET_{CBDclos\ atzA}$ clones.

was determined. Clones 14A and 15A (atzA-CBD) showed the highest atrazine dechlorination activity (nmol product min^{-1} mg^{-1} protein) (Figure 9), similar to the activity of clones 9A and 8A that overexpressed recombinant atzA without CBD.

4. A CONCLUDING REMARK

The isolation of *Pseudomonas* sp. strain ADP, and the characterization of the genes and enzymes involved in the atrazine mineralization pathway made a breakthrough in our ability to bioremediate excess atrazine from soils and water. The sequencing of the relevant genes specifically enables the use of novel treatment methods such as the CBD fusion protein and using novel tools to track the fate of whole cells in the environment.

ACKNOWLEDGMENT

We thank Janis MacFarland of Novartis Corporation for providing the s-triazine standards. This work was supported, in part, by a grant from the Israeli-American BARD program, by a grant from Novartis Crop Protection Inc., and by a grant no. 8867197 for strategic research in bioremediation from the Israeli Ministry of Sciences. Also, Mr. Kauffmann received a grant from the Danish National Science Research Council for which we thank.

REFERENCES

Abee, T. R., Palmen, K. J, Klaas, and Konings, W. N. (1990). Osmoregulation in *Rhodobacter sphaeroides*. J Bacteriol. **172**:149–154.

Anonymous. (1995). pET System Manual. Novagen, Madison, WI.

Aislabie, J., and Lloydjones, G. (1995). A review of bacterial degradation of pesticides. Aust. J. Soil Res. **33**:925–942.

Avnir, D., Braun, S., Lev, O., and Ottolenghi, M. (1994). Enzymes and other proteins entrapped in sol-gel materials. Chem. Mat. **6**:1605–1614.

Azevedo, A. S., Kanwer, R. S., Singh, P., Ahuja, L., and Pereira, L. S. (1997). Simulation of atrazine transport using root zone water quality model for Iowa soil profiles. J. Environ. Qual. **26:**153–164.

Bottoni, P., Keizer, J., and Funari, E. (1996). Leaching indices of some major triazine metabolites. Chemosphere **32:**1401–1411.

Buckley, A. M., and Greenblatt, M. (1994). The sol-gel preparation of silica gels. J. Chem. Educ. **71:**599–602.

Cook, A. M. (1987). Biodegradation of s-triazine xenobiotics. FEMS Microbiol. Rev. **46:** 93–116.

Csonka, L. N. (1989). Physiological and genetic responses of bacteria to osmotic stress. Microbiol Rev. **53**:121–147.

De Souza, M. L., Sadowsky, M. J., and Wackett, L. P. (1996). Atrazine chlorohydrolase from Pseudomonas sp. strain ADP: gene sequence, enzyme purification and protein characterization. J. Bacteriol. **178:**4894–4900.

De Souza, M. L., Wackett, L. P., Boundymills, K. L., Mandelbaum, R. T., and Sadowsky, M. J. (1995). Cloning, characterization, and expression of a gene region from *Pseudomonas* sp strain ADP involved in the dechlorination of atrazine. Appl. Environ. Microbiol. **61**:3373–3378.

Erickson, L.E., Lee, K. H. (1989). Degradation of atrazine and related s-triazines. Crit. Rev. Environ. Cont. **19**:1–14.

Fomsgaard, I. S. (1995). Degradation of pesticides in subsurface soils, unsaturated zone—a review of methods and results. Int. J. Environ. Anal. Chem. **58:** 231–245.

Goswami, K. P., and Green, R. E. (1971). Microbial degradation of the herbicide atrazine and its 2-hydroxy analog in submerged soils. Environ. Sci. Technol. **5**:426–429.

Hench, L. L., and West, J. K. (1990). The sol-gel process. Chem. Rev. **90:**33–72.

Kauffmann, C., and Mandelbaum, R. T. (1998). Entrapment of atrazine chlorohydrolase in sol-gel matrix. J. Biotechnology In Press.

Kauffmann, C. G., and Mandelbaum, R. T. (1996). Entrapment of atrazine degrading enzymes in sol-gel glass. J. Biotechnology **51:**219–225.

Kolpin, D. W., Thurman, E. M., and Goolsby, D. A. (1996). Occurrence of selected pesticides and their metabolites in near- surface aquifers of the midwestern united states. Environ. Sci. Technol. **30:**335–340.

Kordel, W., Herrchen, M., Klein, M., Traub-Eberhard, U., Kloppel, H., and Klein, W. (1992). Determination of the fate of pesticides in outdoor lysimeter experiments. Sci. Total Environ. **123/124**:421–434.

Kordel, W., Wahle, U., Knoche, H., and Hund, K. (1995) Degradation capacities of chlorotoluron and simazine in subsoil horizons. Sci. Total Environ. **171:**43–50.

Le Rudulier, D., Strøm, R., Dandekar, A. M., Smith, L. T., and Valentine, R. C. (1984). Molecular biology of osmoregulation. Science **224**:1064–1068.

Linn, D. M., Carski, T. H., Brusseau, M. L., and Chang, H.-F. (1993) Sorption and degradation of pesticides and organic chemicals in soil. SSSA Special Publication No. 32. Soil Science Society of America, Inc.: Madison, WI.

Ma, L., and Selim, H. M. (1996). Atrazine retention and transport in soils. In Reviews of Environmental Contamination and Toxicity. Ware, G. W., Ed.; Springer-Verlag: New York, p 129–173.

Mandelbaum, R. T., Allan, D. L., and Wackett, L. P. (1995). Isolation and characterization of a *Pseudomonas* sp that mineralizes the s-triazine herbicide atrazine. Appl. Environ. Microbiol. **61**:1451–1457.

Mandelbaum, R. T., Wackett, L. P., and Allen, D. L. (1993). Mineralization of the s-Triazine ring of atrazine by stable bacterial mixed culture. Appl. Environ Microbiol. **59**:1695–1701.

Mcmahon, P. B., Chapelle, F. H., and Jagucki, M. L. (1992). Atrazine mineralization potential of alluvial-aquifer sediments under aerobic conditions. Environ. Sci. Technol. **26**:1556–1559.

National Agricultural Statistics Service. Agricultural Chemical Usage: 1990 Field crop survey. 1991.

Nelson, H., and Jones, R. (1994). Potential regulatory problems associated with atrazine, cyanazine, and alachlor in surface water source drinking water. Weed Technol. **8:**852–861.

Ng, H.Y. F., Gaynor, J. D., Tan, C. S., and Drury, C. F. (1995). Dissipation and loss of atrazine and metolachlor in surface and subsurface drain water: a case study. Water. Res. **29:**2309–2317.

Reetz, M. T., Zonta, A., and Simpelkamp, J. (1995). Efficient heterogeneous biocatalyst by entrapment of lipases in hydrophobic sol-gel materials. *Angew. Chem.* Int. Ed. Engl. **34:**301–303.

Reetz, M. T., Zonta, A., and Simpelkamp, J. (1996). Efficient immobilization of lipases by entrapment in hydrophobic sol-gel materials. Biotechnol. Bioeng., **49:**527–534.

Ritter, W. F. (1990). Pesticide contamination of groundwater in the United States—a review. J. Envirom. Sci. Health **B25:**1–29.

Shapir, N., and Mandelbaum, R. T. (1997). Atrazine degradation in subsurface soil by indigenous and introduced microorganisms. J. Agri. Food Chem. **45**:4481–4486.

Shapir, N., Mandelbaum, R. T., and Gottlieb, H. (1998). Atrazine degradation in salin waste water by *Pseudomonas* sp. strain ADP. J. Indust. Microbiol. Biotechnol. In press.

Sorenson, B. A., Koskinen, W. C., Buhler, D. D., Wyse, D. L., Lueschen, W. E., and Jorgenson, M. D. (1994). Formation and movement of C^{14} atrazine degradation products in a clay loam soil in the field. Weed Sci. **42:**618–624.

Sparling, G.P., and Aislabie, J. (1996). Soil type affects the rate of atrazine decomposition and potential for aquifer contamination. The NZ Society for Horticultural Science; Hamilton, NZ. p. 11.

Splichal, P. A., and Schwab, A. P. (1991). Atrazine and alachlor degradation in shallow Kansas aquifers. Nonpoint source pollution conference Tacoma, Washington. March.20.21.

Veeh, R. H., Inskeep, W. P., and Camper, A. K. (1996). Soil depth and temperature effect on microbial degradation of 2,4-D. J. Environ. Qual. **25:**5–12.

Walker, A., Welch, S. J., Melacini, A., and Moon, Y. H. (1996). Evaluation of three pesticides leaching models with experimental data for alachlor, atrazine and metribuzine. Weed Res. **36:**37–47.

Wehtje, G. R., Spalding, R. F., Burnside, O. C., Lowry, S. R., and Leavitt, J. R. C. (1983). Biological significance and fate of atrazine under aquifer conditions. Weed Sci. **31:**610–618.

Wood, M., Harold, J., and Johnson, A. (1991). The potential for atrazine degradation in aquifer sediments. Monograph British Crop Protection Council.

24

FIELD APPLICATION OF A GENETICALLY ENGINEERED MICROORGANISM FOR POLYCYCLIC AROMATIC HYDROCARBON BIOREMEDIATION PROCESS MONITORING AND CONTROL

Gary S. Sayler,[1,2] Chris D. Cox,[1,3] Robert Burlage,[4] Steven Ripp,[1] David E. Nivens,[1] Claudia Werner,[1] Yeonghee Ahn,[1] and Udayakumar Matrubutham[1]

[1]Center for Environmental Biotechnology
[2]Department of Microbiology
[3]Department of Civil and Environmental Engineering
University of Tennessee, Knoxville, Tennessee 37996
[4]Environmental Sciences Division
Oak Ridge National Laboratory, Oak Ridge, Tennessee 37831

1. INTRODUCTION

On October 30, 1996, the U.S. Environmental Protection Agency (EPA) commenced the first test release of genetically engineered microorganisms (GEMs) for use in bioremediation. This study was initiated by investigators at the University of Tennessee Center for Environmental Biotechnology (CEB) in collaboration with the Environmental Sciences Division (ESD) of Oak Ridge National Laboratory (ORNL). This test was conducted under the auspices of the U.S. Department of Energy's (DOE) Natural and Accelerated Bioremediation Research Program (NABIR). The specific objectives of the investigation were multifaceted and include 1) testing the hypothesis that a GEM can be successfully introduced and maintained in a bioremediation process, 2) testing the concept of using, at the field scale, reporter organisms for direct bioremediation process monitoring and control, and 3) acquiring data that can be used in risk assessment decision making and protocol development for future field release applications of GEMs.

The genetically engineered strain under investigation is *Pseudomonas fluorescens* strain HK44 (King *et al.*, 1990). The original *P. fluorescens* parent strain was isolated from polycyclic aromatic hydrocarbon (PAH) contaminated manufactured gas plant soil (MGP). *P. fluorescens* HK44 was created by the incorporation of a plasmid containing a salicylate

Novel Approaches for Bioremediation of Organic Pollution, edited by Fass *et al.* Kluwer Academic / Plenum Publishers, New York, 1999.

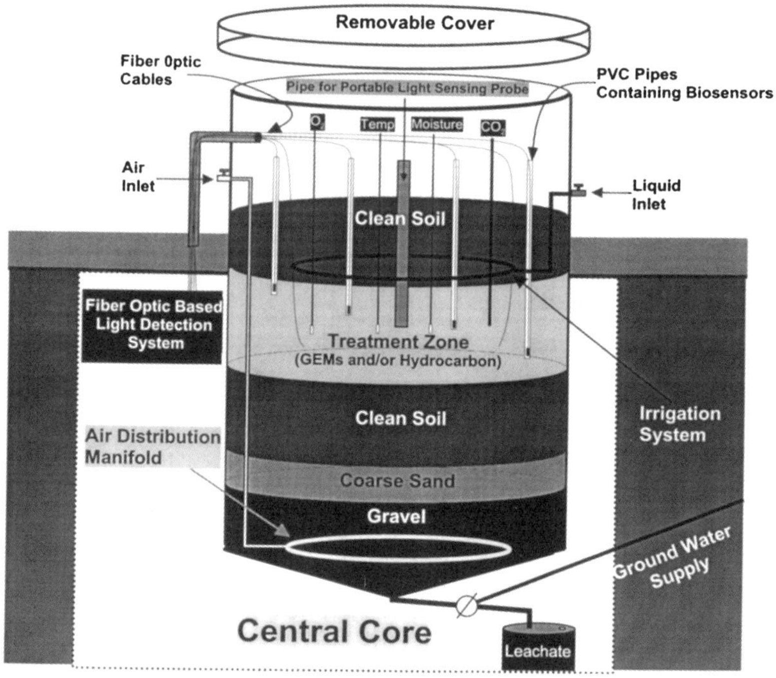

Figure 1. Schematic representation of a lysimeter containing a stratified bed and *in situ* sensors.

inducible operon and the gene cassette for bacterial bioluminescence (*lux*) from *Vibrio fischerii*. When naphthalene is metabolized to salicylate, the *lux* transcriptional gene fusion is induced and expressed to produce enzymes responsible for a bioluminescent signal at an approximate wavelength of 490 nm. Thus, this bacterium is able to biodegrade naphthalene (as well as other substituted naphthalenes and other PAHs) and is able to function as a living bioluminescent reporter for the presence of naphthalene contamination, its bioavailability, and the functional process of biodegradation.

A unique component of this field investigation was the availability of an array of large subsurface soil lysimeters at ORNL (Figure 1). These lysimeters were originally constructed for quantifying and characterizing leachate derived from low-level radioactive wastes but were never used for this purpose. They were instead extensively modified for the GEM release experiments. The lysimeters permitted a semi-contained, controllable field test environment. In addition, they allowed accurate field replication and provision of control tests. The sheer size of each lysimeter (4 m deep by 2.5 m in diameter) allowed for extensive sampling, the deployment of both well-established and prototypic *in situ* sensors, and localized testing of treatments prior to global changes. Large sample cores permitted numerous assays to be performed, including procedures that determine cell enumeration data, genotypic information, hydrocarbon and naphthalene concentrations, bioluminescence, and soil dry weights. *In situ* sensors monitored temperature, bioluminescence, vapor phase volatile hydrocarbon contamination, oxygen concentration, soil moisture content, and carbon dioxide concentration. Furthermore, the lysimeters are exposed to environmental conditions over a two year period, thus providing a wealth of information on *in situ* survivability of GEMs. This article describe our experience associated with the release of a genetically modified microorganism, the lysimeter facility and its associated instrumentation, as well as representative data collected during the first eighteen months of operation.

2. METHODOLOGY

2.1. Lysimeter Design and Set-Up

The lysimeter array consists of seven large lysimeters (4 m deep by 2.5 m in diameter) positioned in a circular array around a central core (6 m deep by 6 m in diameter) that houses instrumentation and plumbing for each of the lysimeters. The lysimeters and the central core were constructed of galvanized steel pipe set into concrete foundations and covered with removable stainless steel lids. Six of the lysimeters were filled with stratified beds of soil (Figure 1) and the seventh contained groundwater and was used to adjust the water level in the other lysimeters. The stratified bed consisted of a 31 cm layer of graded gravel, a 61 cm layer of coarse sand, a 92 cm layer of uncontaminated soil, a 92 cm treatment layer and a 61 cm cap of clean soil placed on top of the bed (Figure 1). The uncontaminated soil was a Huntington loam topsoil consisting of 42% sand, 40% silt, 18% clay, and 1.3% organic carbon.

Originally, a portion of this topsoil was artificially contaminated with naphthalene, anthracene and phenanthrene. A concrete mixer truck was used to combine 23 m^3 of soil with 23 kg of naphthalene and 0.23 kg each of anthracene and phenanthrene, producing approximate final average concentrations of 1000 mg/kg for naphthalene and 10 mg/kg for anthracene and phenanthrene. These concentration are representative of a PAH mixture from a typical MGP site.

The treatment zone in Lysimeters 1, 2, and 4 received the contaminated soil inoculated with HK44, Lysimeters 3 and 5 received uncontaminated soil inoculated with HK44, and Lysimeter 6 received a layer of uninoculated, contaminated soil (Table 1). Soil layers were added in approximate 10 cm increments, each of which was packed to a dry density of 1300 kg/m^3. In Lysimeters 1–5, each 10 cm increment of soil within the 92 cm treatment zone was sprayed with approximately 4 L of cell suspension using a backpack spray tank equipped with an extended nozzle. Anderson Air samplers and gravity plates were used to monitor dispersion rates of the GEMs (Ford *et al.*, 1998). One air sampler was placed 1 m from the outer edge of the lysimeter being sprayed and another was placed the

Table 1. Experimental treatments of each lysimeter

Lysimeter	Initial concentration (CFU/g soil)[1]	Chemically contaminated[1]	Nutrient amendment[2]	Aeration[3]
1	1.81 (± 1.02) x 10^6	+	+	+
2	8.31 (± 3.93) x 10^5	+	+	+
3	1.87 (± 0.09) x 10^6	-	-	-
4	1.90 (± 0.30) x 10^6	+	+	+
5	1.33 (± 0.99) x 10^6	-	-	-
6	—	+	+	+

[1]Into each lysimeter was pumped 200 L of Exxon Univolt 60 transformer oil containing 24 g/L naphthalene and 2.4 g/L anthracene

[2]Nutrient amendment consisted of 150 L minimal salts medium per lysimeter. (minimal salts contains in g/L: $NaNO_3$, 2; KH_2PO_4, 0.75; $FeCl_3$, 0.003; $MgSO_4$, 0.1; $CaCl_2$, 0.005; Na_2HPO_4, 0.25)

[3]Lysimeters were aerated at 2 SCFM for 4 h twice weekly.

same distance directly opposite on the other side. Plates were placed in concentric rings at distances of 1, 2, and 4 m from the lysimeter that was being sprayed with the HK44 culture. A total of 260 Anderson and gravity plates were used.

A plumbing system was installed to allow for liquid delivery either from the top of the treatment zone or bottom of the lysimeter (Figure 1). The upper irrigation system used a coil of drip irrigation tubing placed at the top of the treatment zone and was used to deliver supplemental nutrient treatments. The lower plumbing arrangement was designed to collect leachate in 55-gallon drums located beneath each lysimeter. As required by the EPA, any leachate percolating out of the lysimeters was collected and tested for the presence of *P. fluorescens* HK44. When full, drums were drained into large polytanks and disinfected with hypochlorite solution. After adequate sampling to ensure no survivability of *P. fluorescens* HK44 cells, the polytanks were transported to an ORNL waste treatment plant for disposal. In addition to leachate collection, the lower plumbing system was used to adjust the water level in each lysimeter. This was accomplished by gravity feeding groundwater contained within the seventh lysimeter into the lysimeter of choice. In addition, an air distribution manifold was placed in the bottom of each lysimeter to provide aeration.

2.2. Preparation of the Original Inoculum

The *P. fluorescens* HK44 inoculum was prepared in a 500 L New Brunswick fermenter (Model IF-100) containing a medium consisting of 160 g yeast extract, 1600 g peptone, 160 g ammonium nitrate and 800 mL glycerol in 400 L water. After a 22 h growth period the culture was harvested in a continuous centrifuge and the resulting cell paste resuspended in 60 gallons of saline. This suspension was transported to the site in two 30 gallon drums.

2.3. *In Situ* Treatment Zone Sensors

Various types of sensors were deployed to monitor parameters such as oxygen concentration, temperature, carbon dioxide concentrations, naphthalene vapor, moisture levels, and reporter bioluminescence in the treatment zone.

Oxygen and temperature sensors and data collection/storage devices (Datawrite Research Corp., Visalia, CA) were used to measure, record, and store oxygen data in Lysimeters 1, 2, 3 and 5 and temperature data in Lysimeter 2. Data were typically measured in thirty-minute intervals for a 10 day period, after which the data were retrieved with a laptop computer.

A novel biosensor was used to detect naphthalene vapors in the treatment zone of Lysimeter 1, 2 and 3. This biosensor utilizes alginate-encapsulated *P. fluorescens* HK44 cells (Stormo and Crawford, 1992) and a multi-channel fiber-optic based detection system. Encapsulated cells are placed in a well supported by a subassembly that fixes the distance between a fiber optic cable (1 mm diameter, spc1000/1150b Fiberguide Industries, Stirling, NJ) and the alginate encapsulated cells. The subassembly is then sealed within a light-tight housing constructed from a porous stainless steel tube, two end caps and a standard bulk-head tubing connector. The biosensor is lowered into one of four polyvinyl chloride pipes (PVC) buried at different depths within the treatment zones of Lysimeter 1, 2, and 3 (Figure 1). If naphthalene is present, light emitted from HK44 cells is transferred through the cable within a light-tight box. The fiber optic cable terminates into a 28-position circular fiber holder. A liquid light guide (Oriel Corp., Model 77566) attached to a rotational stage (Newport Corp., Motion-Master 2000) is positioned directly in front of the terminal end of the fiber-optic cable and collects light for 5 sec (Sayler *et al.*, 1995). The recorded counts are an average of 50 counting events, each lasting 0.1 sec. Photons collected by the light guide are detected with a photomultiplier tube (PMT) (Hammamatsu Photonics, Model R-4632) mounted in a thermoelectrically-cooled housing regulated at -35 °C. The pulses from the PMT are processed by a photon counting module (Hammamatsu Photonics, Model C3866) and counted with a computer counter/timer board. Custom DOS-based software caches the photon count data and controls the motion of the rotational stage. Once the data is stored, the computer activates a programmed motion until light is collected from each fiber-optic probe, and then continuously repeats the cycle.

Determination of bioluminescence from soil GEMs was also monitored with the multi-channel fiber optic-based detection system. Thirteen additional fiber optic cables were buried directly into the treatment zone in Lysimeters 1, 2, and 3. Bioluminescence from soil microorganisms directly below the fiber optic cable was quantified as described above.

In addition, a portable field deployable PMT-based probe was designed and used as a process monitoring and control tool. This tool offers increased sensitivity over the fiber optic based detection system. The probe consisted of a unique light-tight and water-tight housing and a PMT module equipped with an embedded microcontroller integrating all components necessary for photon counting and a RS-232-C interface (HC135-01, Hamamatsu, Middlesex, NJ). The probe housing was attached to a 1.5-inch pipe and lowered down 3-inch internal diameter pipes (some equipped with Plexiglass windows) to examine soil at the bottom of the pipes. A light-tight interface was designed to block out extraneous light. Light data were collected and stored on a laptop computer.

2.4. Soil Sampling

Soil cores were removed from each lysimeter using a Cole-Parmer soil sampler (Cole-Parmer Industries, Vernon Hills, IL). Zero-contamination sampling tubes (2.3 cm diameter and 61 cm in length) were used to prevent cross-contamination of soil cores. Typically, a site was sequentially sampled with 3 cores taken at different depths totaling 185 cm in length. The first core traversed the clean cap and was discarded. The second and third cores traversed through the 92 cm deep treatment zone and the 30 cm of soil directly beneath. Cores were stored at 4°C for transport. Once in the lab, cores were divided into 30.5 cm sections, each section thoroughly mixed, and stored separately in sterile Whirlpak bags (Fort Atkinson, WI).

2.5. Microbiological Analyses

One gram samples were removed from each Whirlpak bag and vortexed in 9 mL 0.1% sodium pyrophosphate for 1 min. Dilutions were then prepared in saline solution (0.85% NaCl) and plated in triplicate on YEPG agar plates (0.2 g yeast extract, 2.0 g polypeptone, 1.0 g glucose, 0.2 g NH_4NO_3, 17 g agar per liter, pH 7.0) to obtain total bacterial counts and on YEPSS agar plates (0.2 g NH_4NO_3, 0.2 g yeast extract, 0.5 g sodium salicylate, 2.7 g sodium succinate, 1 g polypeptone, 17 g agar, per liter pH 7.0) containing tetracycline at 14 mg/L to obtain *P. fluorescens* HK44 counts. In both cases cycloheximide was added at 50 mg/L to inhibit fungal growth. Plates were incubated in the dark at room temperature for four days. To verify the presence of the genetically engineered plasmid, colony hybridizations were performed on presumptive *P. fluorescens* HK44 colonies growing on YEPSS plates containing tetracycline. Colonies were transferred to 82 mm diameter nylon membranes (Millipore Corp., Bedford, MA) and hybridized with a 0.3 kb *luxA* gene probe under high stringency conditions (Sambrook et al., 1989).

2.6. Large-Scale Treatments Events

To provide an additional source of inducing substrate, naphthalene and anthracene were mixed with transformer oil (Exxon Univolt 60) and added to Lysimeters 1, 2, 4, and 6. Twenty-four kilograms of naphthalene crystals and 2 kg anthracene were dissolved in 220 gallons oil that proved to be an adequate solvent while still meeting regulatory requirements that it not contain RCRA (Resource Conservation and Recovery Act) hazardous constituents. Each of Lysimeters 1, 2, 4, and 6 received 55 gallons through the irrigation tubing located directly above the treatment zone. Theoretically, this was sufficient to produce an average loading capacity of 1000 mg/kg throughout the 92 cm contamination zone. Before adding the oil, the water table in each lysimeter was raised to the sand/clean soil interface. After all the oil was added, the water table was rapidly lowered to create a slight vacuum to induce the flow of oil into the contamination zone. This process was successful in preliminary laboratory experiments.

Nutrients in the form of minimal salts were added to Lysimeters 1, 2, 4, and 6. Fifty gallons of a minimal salts (2 g $NaNO_3$, 0.75 g KH_2PO_4, 0.003 g $FeCl_3$, 0.1 g $MgSO_4$, 0.005 g $CaCl_2$, and 0.25 g Na_2HPO_4 per liter) solution were added through the irrigation tubing. In microcosm studies performed in the laboratory, the nutrient solution and the supplemental addition of oil and naphthalene stimulated the growth of HK44 in lysimeter soil.

An air-pumping regime using an Inersoll-Rand 100 air compressor (Seattle, WA) began on April 11, 1997. The air-delivery scheme typically required pumping air twice a week for 4 hours at a rate of 2 SCFM.

3. RESULTS AND DISCUSSION

3.1. Risk Assessment

In July 1995, a premanufacture notification (PMN) was submitted to the EPA for proposed release of the GEM in the planned field test (Sayer, 1997). In March 1996 the University of Tennessee received an EPA Consent Order authorizing the field test under strict regulatory guidelines contained in Work Plans, Waste Management Plans, and Health and Safety Plans developed by ORNL and the CEB. Provisions in the consent order were to provide the EPA with monitoring data as well as a review of any technical changes in the work plan. This became necessary when the PMN was modified to incorporate a change in the method of soil inoculation from direct mixing to a spray application. The consent order itself was not an outright approval of field use, but rather a cautious tentative agreement of field use due to existing risk concerns. This risk assessment conclusion was made for several reasons: the recombinant genes are on a mobile genetic element, the genetic construct contains an antibiotic resistance marker, and some biovars of the species have the capacity to act as opportunistic plant pathogens, although HK44 shows no such potential based on limited analysis. Following additional DOE Environmental Health and Safety oversights and Readiness Reviews, the field test ultimately proceeded on October 30, 1996.

3.2. Soil Storage and Initial PAH Concentrations

Prior to the loading of soil into the lysimeters, the contaminated soil and an additional 46 m^3 of uncontaminated soil were stored in separate piles on a concrete pad under plastic tarps for approximately 80 days. A moderate aging and weathering period was desired to better simulate soils from contaminated sites since PAHs have been shown to strongly sorb to soils over time (Dzombak *et al.*, 1984). Unfortunately, a delay in receiving the numerous approvals required for releasing the GEMs resulted in a total of 180 days of contaminant soil aging. Consequently, approximately 95% of the naphthalene was lost through volatilization and natural biodegradation processes. A decision was made to proceed with the loading of the lysimeters since obtaining authorization to recontaminate the soil would result in significant additional delays. Consequently, a second contamination event occurred at a later date (Day 135 of the experiment).

3.3. Inoculation of the Soil

Cell density of the initial inoculum using plate counts was determined to be 1.42 x 10^7 cfu/mL. Inadvertent release of GEMs during spraying was inevitable but allowable under EPA guidelines as long as adequate precautions were taken and the degree of release was documented. A total of 260 Anderson Air sample and gravity plates were used. Of these, 36 were positive for HK44 growth, usually with less than 5 colonies per plate. No plates placed beyond a 3.5 m radius contained *P. fluorescens* HK44 colonies. Initial soil concentrations are shown in Table 1.

3.4. Population Dynamics

Two types of media were used to monitor the microbial populations in the lysimeters, YEPG and YEPSS (Figure 2). The native heterotrophic culturable population is defined by microorganisms that can grow aerobically on a medium containing nutrients that can be utilized by a wide variety of microorganisms (YEPG). Although it is known that only a small percentage of bacteria in soil are culturable under any condition, this population contains bacteria that can compete with *P. fluorescens* HK44 for nutrients. The second medium is more selective for *P. fluorescens* HK44 (YEPSS). This medium contains the antibiotic tetracycline and sodium salicylate, a carbon source for *P. fluorescens* HK44. Ideally, *P. fluorescens* HK44 would be the only bacterium to grow on this selected medium. However, soils contain a great diversity of bacteria, and in the control uninoculated but chemically contaminated lysimeter (Lysimeter 6), a background bacterial population capable of growth on the selective medium was found to pre-exist at an order of magnitude lower cell density (Figure 2). To accurately determine the percentage of *P. fluorescens* HK44 present on the plates, a colony hybridization assay was performed (Figure 3). All microbiological assays were performed to monitor the population dynamics occurring within each lysimeter before and after any treatment event and for no period of time greater than 4 months. This sampling regimen was used to study not only the treatment effect but also the treatment effect under changing environmental conditions. For example, soil temperature during this 18 month period had winter minimums of 6.75°C on January 26, 1997 (Day 88) and 11°C on February 9, 1998 (Day 467) and a summer maximum of 28.5°C on July 21, 1997 (Day 264). Temperature can have a profound effect on bioluminescence, since the enzymes responsible for bioluminescence are not stable at temperatures above 30°C (Meighen, 1991).

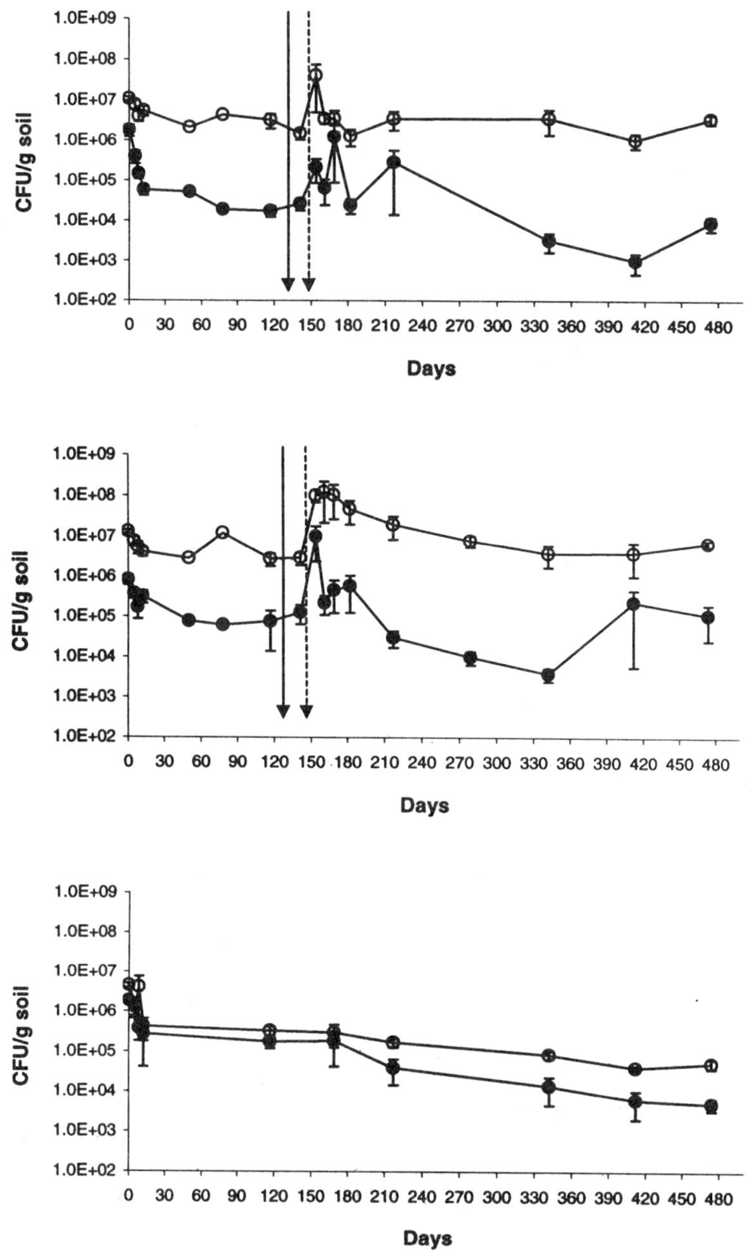

Figure 2. Population dynamics of *Pseudomonas fluorescens* HK44 and response to nutrient amendments following lysimeter inoculation. Total viable counts (o) and HK44 concentrations determined by selective plating on tetracycline (●) in each of the lysimeters. Open boxes (□) represent the number of indigenous tetracycline-resistant colony forming units in Lysimeter 6. Solid arrows indicate time of oil addition; dashed arrows indicate time of minimal medium addition. The heavy dashed arrow in Lysimeter 5 represents the time at which water was added in an attempt to lower soil temperatures.

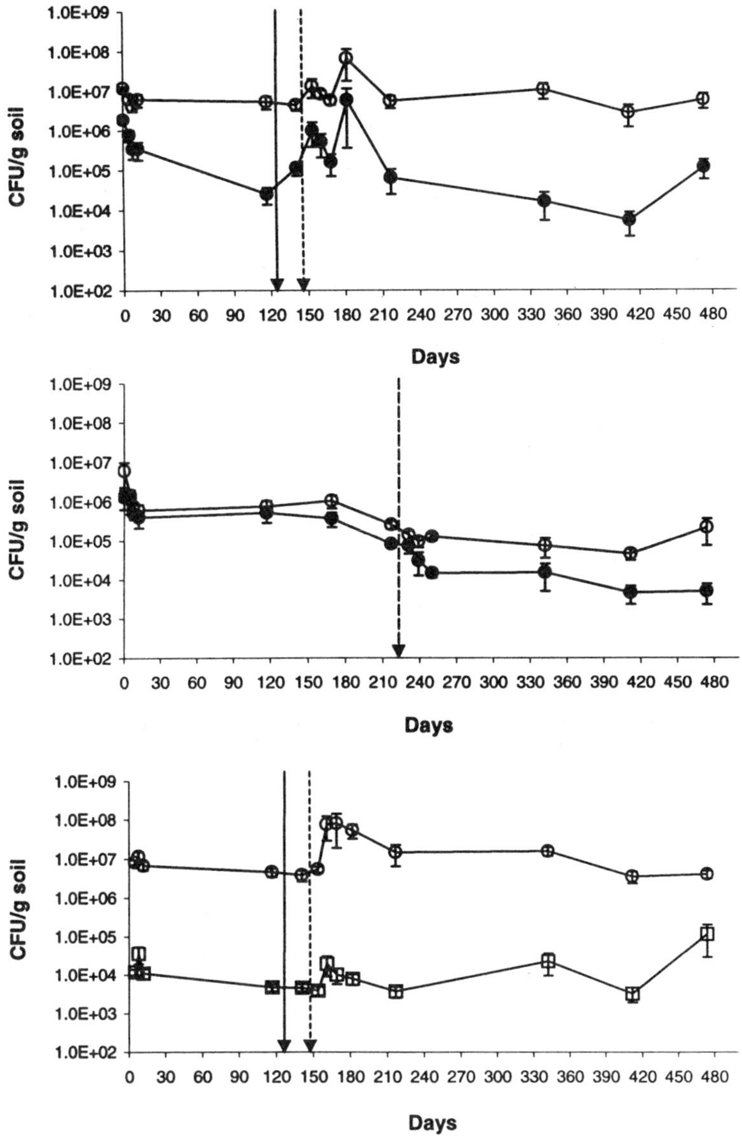

Figure 2. *Continued.*

Several trends are apparent relative to the long-term survival of *P. fluorescens* HK44 (Figure 2). After inoculation, *P. fluorescens* HK44 colony forming units per gram dry weight decreases in all lysimeters. For those chemically contaminated lysimeters inoculated with *P. fluorescens* HK44 (Lysimeters 1, 2, and 4), there is a high level of reproducibility in population maintenance and response. For the first 135 days of the experiment, both the native total culturable heterotrophic bacterial populations and GEMs declined reproducibly. Specifically, the *P. fluorescens* HK44 populations in Lysimeter 1, 2, and 4 decreased from the initial mean value of 1.5 (± 0.4 standard error) x 10^6 to 4.0 (± 1.9) x 10^4 cfu/g of soil on Day 117. After the addition of the transformer oil with naphthalene and minimal salts, the mean soil population

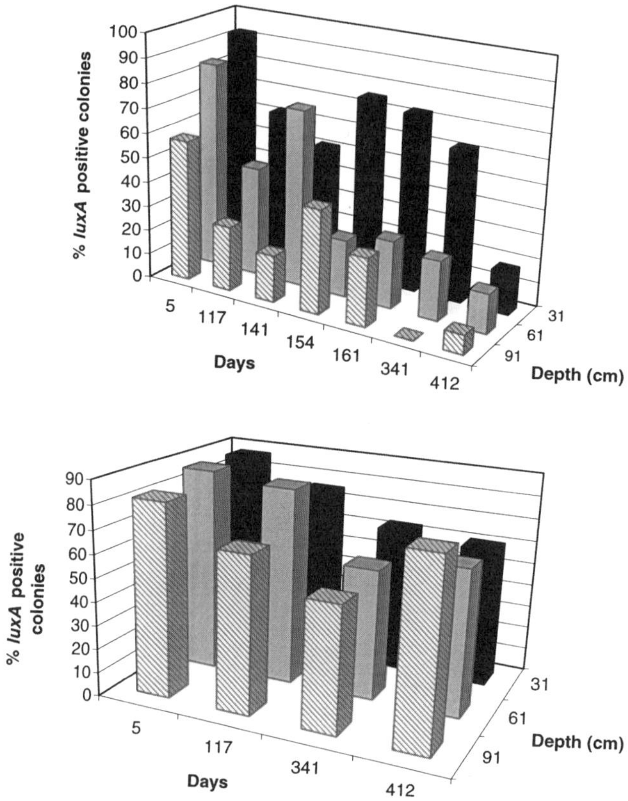

Figure 3. Colony hybridization data from Lysimeters 1 and 3. Results from Lysimeter 1 are typical of trends observed in Lysimeters 2 and 4 as well. The highest concentration of *P. fluorescens* HK44 cells is present in the first 31 cm of the contamination zone and gradually declines with depth. A large HK44 population was seen at the lowest depth when oil was added.

of GEMs increased to approximately 3.7 (± 3.1) x 10^6, then decreased to 1.4 (± 0.6) x 10^5 cfu/g of soil probably due to an insufficient air supply prior to Day 154. On this date, O_2 concentration was determined to be 2%. After modification of the air supply equipment, oxygen levels increased to approximately 20%, and the GEM populations rebounded, but again declined to 7.9 (±_3.4) x 10^4 cfu/g of soil on Day 474. Over the duration of the experiment, the percentage of tetracycline resistant bacteria in the total heterotrophic community decreased from 13% to approximately 1% on Day 474. However, these percentages overestimate the HK44 population, since the number of colonies detected on the selective medium probing positive for *lux* declined with time (Figure 3). Two plausible explanations for this finding exist. First, the nutrient augmentations used to stimulate HK44 also stimulated the native tetracycline population resistant (Figure 2). Second, the plasmid containing the *lux* genes may be unstable over many generations.

Within lysimeters containing GEMs without chemical contamination (Lysimeters 3 and 5), the populations have declined from 1.8 x 10^6 to 5 x 10^3 cfu/g of dry soil in Lysimeter 3 and from 1.3 x 10^6 to 5.1 x 10^3 cfu/g of soil in Lysimeter 5, with most of the decrease detected in the first 30 days. Furthermore, the percent of HK44 cells to total culturable heterotrophic populations has decreased from 41 to 10% in Lysimeter 3 and from 29 to 2.2% in Lysimeter 5. The only difference between the treatments for these lysimeters was a groundwater flooding event that occurred on Day 233. Throughout this study, the percentage of the colonies on the selective YEPSS medium that probed positive for the *luxA* gene ranged from 57 to 96% (for both Lysimeter 3 and 5), indicating that neither the tetracycline resistant bacteria nor plasmid instability is a significant problem in these lysimeters. These data, presented in Figures 2 and 3, demonstrate that a GEM can be introduced and maintained in contaminated soils, however nutrient augmentations can stimulate a competitive population.

3.5. Detection of Bioluminescence from HK44 Cells

Another important goal of this research is the detection of bioluminescence from GEMs. Bioluminescence from HK44 was used in alginate bead biosensors to monitor volatile PAHs. Initially, the bioluminescence from microorganisms within the biosensors was minimal due to the low concentration of volatile PAHs present in the treatment zone. Light production from the biosensors dramatically increased after the addition of the naphthalene in transformer oil, indicating that naphthalene vapor was present in the vapor phase of the treatment zone (Figure 4). The increase in the signal (Figure 4A) probably re-

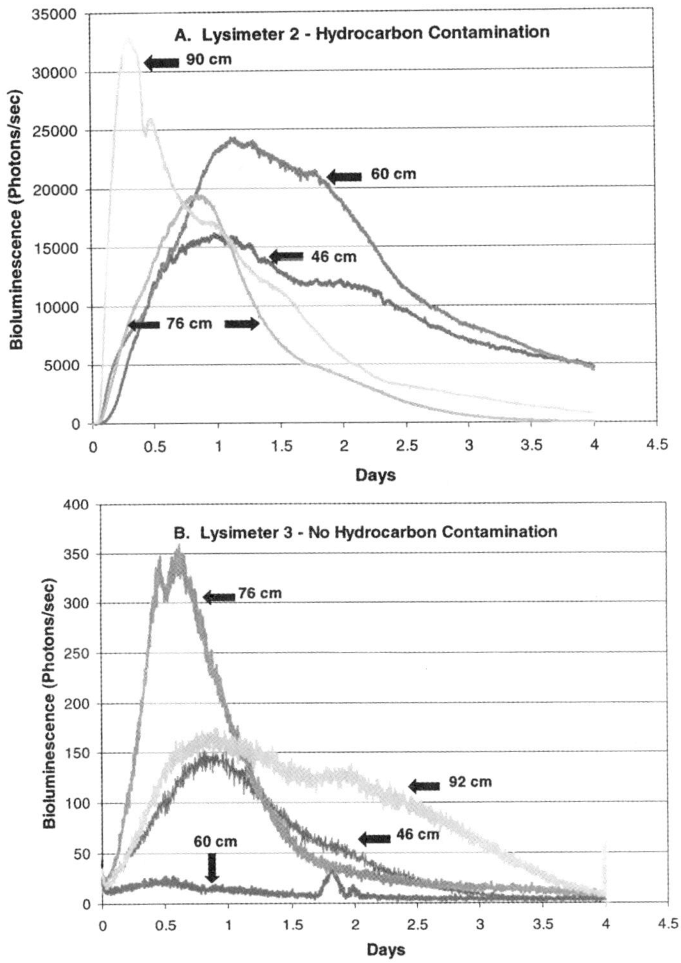

Figure 4. Bioluminescent response from encapsulated *P. fluorescens* HK44 cells in fiber optic biosensors at various depths in lysimeters 2 (A) and 3 (B). Lysimeter 3 contains no hydrocarbons and the low light response seen represents background levels of induction due to HK44 growth on exogenous sources of carbon in the alginate encapsulation medium. Depths originate from surface of treatment zone. Note the differences in the scale of the y-axis.

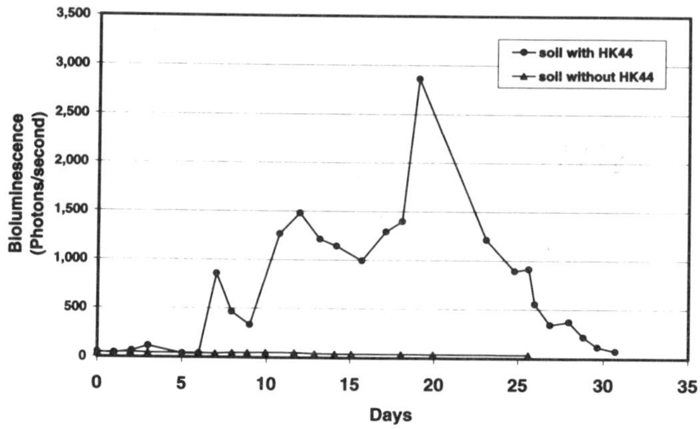

Figure 5. Bioluminescence from GEMs in lysimeter soil after the soil was challenged with naphthalene and a minimal salts solution. For comparison, dark counts from a soil that did not contain GEMs are presented (Lysimeter 6 soil).

sults from the induction and subsequent expression of the genes responsible for bioluminescence. In the presence of naphthalene vapor, the bioluminescent bacteria can continue to utilize nutrients and multiply, increasing the number of cells that in turn can increase the bioluminescence signal. After a few days, the growth of cells on or in the alginate beads may become limited by the lack of an essential nutrient causing a decrease in the signal. In some situations, the decrease in the bioluminescent signal may also be caused by dehydration of the alginate matrix. In Figure 4B, the low level of bioluminescence detected when no naphthalene is present may result from the basal level of expression of the enzymes responsible for bioluminescence in HK44. Qualitatively, the peak height or peak area of the signal decreased with concentration. However, quantitative assessment of the signal will require information about other parameters (including oxygen concentration, temperature, and relative humidity) that can either directly or indirectly affect the bioluminescence signal. Experiments to determine the best method to quantify the bioluminescent data are on-going. Yet even with these present limitations, these biosensors have proved to be useful tools for the real-time detection of volatile PAH contamination in soil.

GEMs in the soil were also used to monitor for the presence of naphthalene contamination, its bioavailability, and the functional process of biodegradation. Although the fiber-optic based system was also used to examine directly the lysimeter soil, detection of bioluminescence directly from HK44 cells in the soil has been unsuccessful, probably due to a lower than expected HK44 biomass. To increase sensitivity, a portable photon counting PMT-based system was designed to examine soil under windows located at the distal end of 3 inch pipes that terminate in the treatment zone. The probe was initially used in January of 1998, and has been used to show that much of the treatment zone is depleted of naphthalene. Localized areas examined by this detection system showed that bioluminescence from *P. fluorescens* HK44 can be stimulated if the minimal salts solution and naphthalene crystals were added to the treatment zone (Figure 5). For example, when 100 mg of naphthalene and 100 mL of minimal salts (time zero) were added to soil underneath the window, bioluminescence continued for approximately one month. Initially, the existing number of HK44 cells was probably too few to produce a measurable bioluminescent signal. After a 5 day lag, the population was large enough to produce a bioluminescent signal that continued for 26 days. Information gathered with this detection system was used to design the final perturbation involving a minimal salts addition and a water table adjustment to disperse the remaining naphthalene and oil within the treatment zone. This final test should allow for the detection of HK44 during a large scale perturbation.

4. SUMMARY

This research has demonstrated that GEMs can be introduced and maintained in environmental soils for at least 18 months and that the population dynamics can be affected by nutrient augmentation. For example, the presence of naphthalene can be used to enrich for *P. fluorescens* HK44 in soils and the lack of naphthalene causes the GEM population to decline. Furthermore, this study has provided evidence that bioluminescence from GEMs can be used as a tool for monitoring and controlling the bioremediation process. A GEM can bioluminesce and report conditions that are favorable for bioremediation. In the case where a GEM is present and not bioluminescing, unfavorable environmental or physiological conditions are revealed. If unfavorable conditions are detected, analytical and microbiological assays as well as localized treatments can be used to diagnose and correct the existing problem. This strategy can lead to rapid responses and eliminate the need for expensive microcosm studies that may or may not be amenable to scale-up.

5. FUTURE PLANS FOR THE LYSIMETER FACILITY

Upon completion of this two-year study, several other experiments utilizing the lysimeter arrays may be proposed to further enhance our understanding of the efficacy of bioremediative strategies in natural environments. Newly engineered bacterial strains that degrade PAHs more efficiently, generate greater light levels, and are more genetically stable are now available. Such activities can be assessed as a comparison to HK44. The use of surfactants is also being studied as a means of overcoming sorption between PAHs and soil particles, allowing the GEM to function more productively in its remediative efforts and increasing overall degradation efficiency. Other applications with engineered strains for cometabolic bioremediation of problematic contaminants such as DNAPL (Dense Non-Aqueous Phase Liquids; trichloroethylene (TCE) is a common example) are also being proposed for field research testing in the lysimeter array. Additionally we also have plans to fill another on-site lysimeter array with soil contaminated with uranium, radionuclides, heavy metals and mixed waste. We are especially interested in collaborations with other scientists who could benefit from studies in such a facility.

The use of large-scale field sites provides excellent scale-up for studying bacterial interactions under realistic conditions without direct release of organisms into natural ecosystems. This may be of special concern when considering the undefined risks associated with the dissemination of engineered genotypes to indigenous populations. The lysimeters provide ideal experimental sites for GEM release. They maintain a semi-contained environment with adequate safety backups and provisions for monitoring surrounding groundwater and soil to evaluate accidental GEM discharges. They are also equipped with sufficient instrumentation to monitor all relevant environmental parameters that may affect the GEM. Most importantly, though, is that they are large enough in size, both horizontally and vertically, to permit the high volume of sampling necessary to adequately assess the specific microbial interactions occurring. Results obtained from these experiments will provide valuable and essential insights into the bioremediative capabilities of GEMs, allowing for an economical and efficient means for their use in future commercial applications.

ACKNOWLEDGMENTS

Research sponsored by the Office of Biological and Environmental Research, U.S. Department of Energy. Oak Ridge National Laboratory is managed by Lockheed Martin Energy Research Corp. for the U.S. Department of Energy under contract number DE-AC05-96OR22464. S. Ripp was supported by an appointment to the Alexander Hollaender Distinguished Postdoctoral Fellowship Program sponsored by the U.S. Department of Energy, Office of Health and Environmental Research, and administered by the Oak Ridge Institute for Science and Education. Excellent technical and scientific support was provided by Bruce Applegate, Anthony Hay, Staci Kehrmeyer, Charles Steward, John Jarrell, Phil Jardine, Jim Easter, and Christopher Ford.

REFERENCES

Dzombak, D.A. and Luthy, R.G. (1984). Estimating adsorption of polycyclic aromatic hydrocarbons on soils. *Soil Sci.* **137**:292–308.

Ford, C.Z., Sayler G.S., and Burlage, R.S. (1998). *Appl. Microbiol. Biotechnol.*, (submitted).

Heitzer, A., Webb, O.F., Thonnard, J.E., and Sayler, G.S. (1992). Specific and quantitative assessment of naphthalene and salicylate bioavailability by using a bioluminescent catabolic reporter bacterium. *Appl. Environ. Microbiol.* **58**:1839–1846.

King, J.M.H., DiGrazia, P.M., Applegate, B., Burlage, R., Sanseverino, J., Dunbar, P., Larimer, F., and Sayler, G.S. (1990). Rapid, sensitive bioluminescent reporter technology for naphthalene exposure and biodegradation. *Science* **249**:778–781.

Meighen, E.A. (1991). Molecular biology of bacterial bioluminescence. *Microbiol. Rev.* **55**:123–142.

Sambrook, J., Fritsch E.F., and Maniatis T. (1989). Molecular cloning: A Laboratory Manual, Cold Spring Harbor Laboratory Press, Cold Spring Harbor, NY.

Sayer, P. (1997). Risk assessment for a recombinant biosensor, p. 269–279. In G.S. Sayler, J. Sanseverino, and K.L. Davis, (eds.), *Biotechnology in the Sustainable Environment,* Plenum Press, New York.

Sayler, G.S., Matrubutham, U., Palmer, R., and Kelly, C. (1995). Application of molecular biology to real time monitoring in bioremediation. Organization for Economic Co-operation and Development, Amsterdam Workshop, p. 323–334.

Stormo, K.E., and Crawford, L. (1992). Preparation of encapsulated microbial cells for environmental applications. *Appl. Environ. Microbiol.* **58**:727–730.

Webb, O.F., Bienkowski, P.R., Matrubutham, U., Evans, F.A., Heitzer, A., and Sayler, G.S. (1997). Kinetics and response of a *Pseudomonas fluorescens* HK44 biosensor. *Biotech. Bioeng.* **54**:491–502.

25

BIOSTABILIZATION TECHNOLOGY FOR TREATING PAH- AND PCP-IMPACTED SOIL TO ENVIRONMENTALLY ACCEPTABLE ENDPOINTS

Ten-Year Field Assessment

Jim Mueller,[1,*] Hap Pritchard,[2] Michael Tischuk,[3] Mitchell Brourman,[3]
Peter Swallow,[4] Margaret Tabe,[4] and John Smith[4]

[1]Dames & Moore
One Continental Towers Suite 1000
1701 Golf Road, Rolling Meadows, Illinois 60008
[2]Naval Research Laboratory
Code 6115, 4555 Overlook Ave., Washington D.C. 20375
[3]Beazer East, Inc.
One Oxford Centre Suite 3000, Pittsburgh, Pennsylvania, 15219
[4]ALCOA Environmental Science and Technology Development Group
100 Technical Drive, ALCOA Center, Pennsylvania, 15069-0001

ABSTRACT

Data from a 10-year field demonstration at a former wood treating site in Nashua, NH (USA) showed that soil heavily impacted by polycyclic aromatic hydrocarbons (PAH) from coal tar creosote and pentachlorophenol (PCP) was effectively managed using biostabilization. The remedial approach was initiated in 1986/1987 with active land treatment of heavily impacted soil in an engineered, solid-phase biotreatment system. This reduced an initial soil total PAH (tPAH) concentration of *ca.* 2,500 mg/kg and PCP concentration of *ca.* 75 mg/kg by >50% and >70%, respectively, within a two- to four-month period. Soil organic chemistry data showed that the soil was effectively stabilized in that there was a significant reduction in the concentration of the more leachable,

* (847) 228–0707 extension 131/fax: -1328, email CHIJGM@DAMES.COM

Novel Approaches for Bioremediation of Organic Pollution, edited by
Fass *et al.* Kluwer Academic / Plenum Publishers, New York, 1999.

bioavailable and potentially mobile compounds. Statistically significant data collected over the following six years (1987–1993) suggested a soil half-life for PCP under passive remediation conditions of 1,415 days. The soil half-life for tPAHs was 963 days. First-order decay rates for individual PAHs varied from 856 to 1,733 days. No vertical migration of PAHs or PCP was detected. Data collected in 1997 (*ca.* 10-years after active treatment) showed an additional half-life reduction in tPAHs from *ca.* 200 to 100 mg/kg soil given an additional 1,100 days of passive remediation; PCP was more recalcitrant. Nevertheless, passive bioremediation and natural attenuation processes effectively managed residual, less leachable and less mobile compounds as they slowly became bioavailable. Data from inorganic nutrient analyses, compound specific microbial counts, and TCLP analyses also support the position that biostabilization can provide a technically valid, safe and cost-efficient remediation strategy.

1. INTRODUCTION

Industrial use of fossil fuels and their derivatives (*e.g.*, refined petroleum products, coal-tar creosote) has, in some cases, led to elevated concentrations of organic constituents of interest (COI) in soil and groundwater. By nature, these COI have been around for eons (Blumer, 1976) hence the soil microflora has often adapted whereby naturally occurring microorganisms are capable of catabolizing these compounds. The rate and extent of PAH biodegradation is thus a function of the COI bioavailability and the physicochemical environment (Heitkamp and Cerniglia, 1987). Conversely, the biodegradation of pentachlorophenol (PCP) and other highly chlorinated, anthropogenic COI is often a slower process. This recalcitrance is due, in part, to the fact that indigenous microflora have been exposed to these compounds for a much shorter period of time (*ca.* since the 1950's). Nevertheless, many species of soil microbes have been identified with the ability to biodegrade PCP and related compounds (EPRI, 1993).

Bioremediation is referred to herein as the managed or engineered use of naturally occurring, catabolic biochemical processes to reduce the concentrations of COI in impacted media to acceptable levels in a timely manner. Over the years, many methods for implementing bioremediation *per se* have been devised (Mueller *et al*, 1993, 1996). Our experience with active bioremediation processes for many soils containing elevated levels of organic COI is that they tend to systematically reduce the amount of recoverable compound over time (Mueller *et al.,* 1991; Smith *et al.*, 1995), eventually reaching an apparent plateau. But continued observation shows that the COI concentrations often continue to decrease steadily over time, even under passive bioremediation conditions (Brown *et al.,* 1995; Pritchard *et al,* 1993; Smith *et al.*, 1995). Thus, it was considered that a combined active/passive biotreatment strategy could provide an effective means of managing materials impacted by organic COI through a series of processes termed "biostabilization" (Smith *et al.,* 1995).

Biostabilization as defined herein consists of two-phases. The first phase employs active bioremediation processes to affect significant reductions in leaching potential of the treated material. This is accomplished through a relatively rapid removal of the more mobile, more bioavailable organic phase (*e.g.,* not associated with soil). In the subsequent passive treatment phase, natural attenuation processes serve to ensure that no COI migrate away from the actively treated soil (*i.e.*, destruction/removal exceeds desorption). During the passive phase of treatment, however, residual COI concentrations may exceed conventional soil clean-up standards. Thus, acceptance of Alternative Relevant and Appropriate

Requirements (ARARs), or Environmentally Acceptable Endpoints (EAE), is essential to effective implementation of biostabilization (ALCOA, 1997; Smith *et al.*, 1996).

The objective of the studies described herein was to demonstrate, through long-term field studies, that biostabilization can be a safe and effective means of treating soils heavily impacted by organic wood preservatives and related COI. Following an active bioremediation phase to biostabilize the soil, 10- years of field data demonstrate that residual PAH and PCP concentrations continued to slowly reduce to concentrations near natural soil background levels. The data presented herein focus on demonstrating effective bioremediation, with emphasis given to biodegradation kinetics and constituent dissolution as they relate to the concept of biostabilization.

2. MATERIALS AND METHODS

2.1. Site Background

Wood treating operations at the Nashua New Hampshire Site were conducted from 1923 to 1983 by a number of companies that used large volumes of creosote and/or PCP to treat railroad ties, poles, *etcetera*. Standard industry practices for the time resulted in soil and groundwater COI impacts. The site has been characterized thoroughly, with much information reported elsewhere (Kopper's, 1987).

2.2. Remedial System Design and Implementation

In 1986, Kopper's Industries, Inc., contracted Keystone Environmental Resources, Inc. to install an Engineered Biodegradation System^{SM} (U.S. Patent No. 5,007,771) to actively treat soil containing elevated concentrations of organic COI; namely polycyclic aromatic hydrocarbons (PAHs) and pentachlorophenol (PCP) (Keystone, 1989). The biotreatment system (*ca.* 65′ wide and 200′ long) was unlined prepared soil sub-grade sloped to permit drainage of surface water to the adjacent storm water retention pond (Figure 1). The water in the storm water retention pond was used for moisture control in conjunction with a sprinkler system. A smaller control plot, approximately 65-feet wide by 30-feet long, was constructed and operated similarly to the biotreatment bed except that it was loaded with clean regional soil. The control plot provided non-impacted soil for purposes of analytical sampling and QA/QC. Earthen berms were constructed around the biotreatment bed and control plot to prevent surface water run-off and run-on. Pipe lysimeters were hydraulically placed beneath both the biotreatment bed and control plots. The pipe Lysimeters were installed approximately four feet below the bottom of the plots and were 12-inches in diameter.

2.3. Bioremediation Phases

Active biotreatment took place from May 1986 through December 1987 and consisted of four separate soil loadings or lifts (8- to 12-inch thick; see Figure 1) of COI impacted material onto the biotreatment bed, loading uncontaminated soil onto the control plot, and routine maintenance of the plot (see below). During each loading, soils were mixed with manure (*ca.* 2% by weight) and inorganic fertilizers (N and P) and applied to the biotreatment plot using a backhoe.

The control plot contained a single load of non-amended, non-impacted soil placed in 1986. During active biotreatment, the single 1986 loading was actively treated for a four-month period (May–September 1986) and then left fallow for the winter months. The three subsequent 1987 layers were actively treated for two months each. Routine maintenance op-

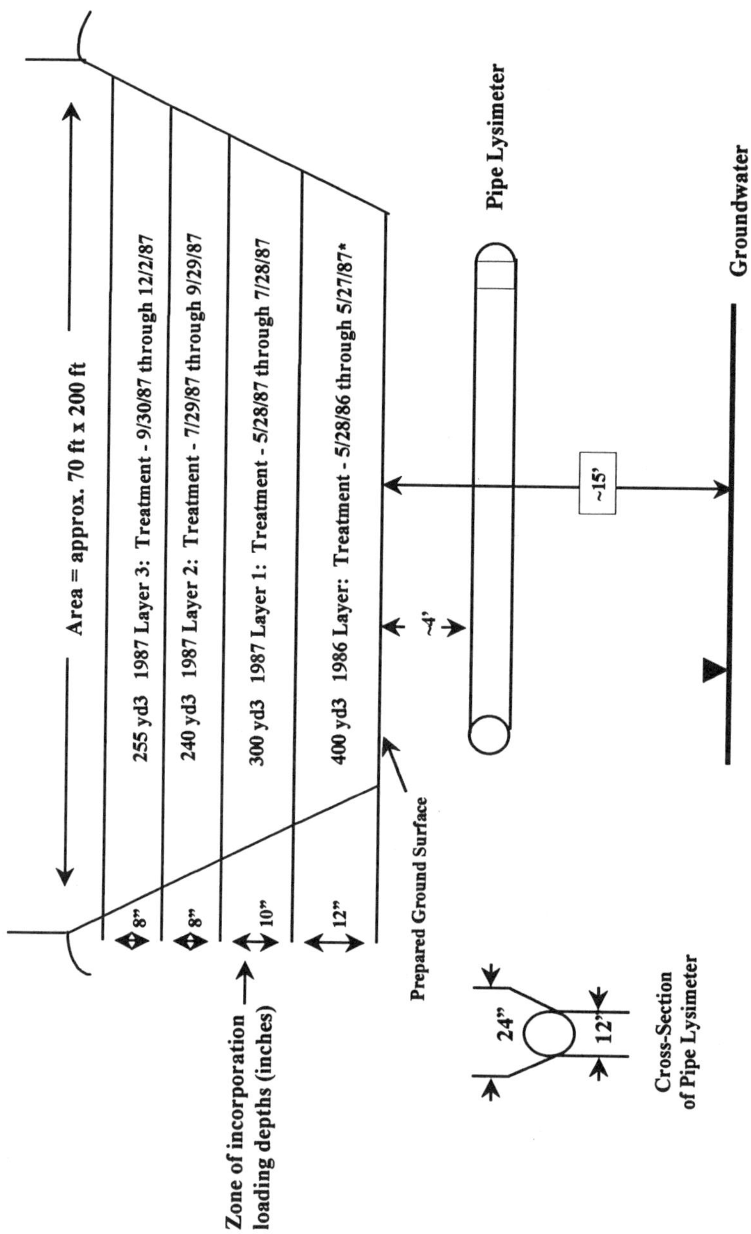

Figure 1. 1986–1987 soil loading profile.

erations included weekly tilling of the soil (rototiller), moisture control, pH adjustments and inorganic nutrient additions. The control plot was tilled only during 1986. The operating conditions maintained in the active biotreatment zone (*i.e.,* zone of incorporation – see Figure 1) were a pH between 6 to 7, moisture level ranging from 10 to 20% (weight) and a COI/N/P ratio of 50/2/1. Following active biotreatment of the final load in 1987, the material was left in place under "passive treatment" conditions (*i.e.*, no tilling, nutrient addition or artificial watering) and seeded with vegetation (sunflowers followed by regional grasses).

2.4. Performance Monitoring

To determine the performance of the bioremediation system, sampling events were conducted for untreated/treated soil, the prepared sub-grade soil beneath the biotreatment bed, groundwater up-gradient and down-gradient of the biotreatment bed, and pipe lysimeter water (*i.e.,* system leachate). Also, off gases above the biotreatment bed were collected and extensive volatilization monitoring was performed. Additionally, Microtox toxicity testing was performed on the biotreatment bed soil and the various media (*i.e.*, groundwater and pipe lysimeter water) associated with the biotreatment plot to evaluate any changes in toxicity resulting from biotreatment. This was also used as an indirect measure of COI catabolites.

2.4.1. Sampling and Analysis Plan: 1986/1987. Sampling of the biotreatment bed soil was performed using a statistical sampling approach where each of the four biotreatment layers was divided into seven subsections. After amendments were applied and treatment beds were tilled, replicate samples were obtained from each subsection and composited to form analytical samples. The soil samples collected were analyzed for a series of parameters including PAHs, PCP, Oil & Grease (O&G), total phenols, and Total Petroleum Hydrocarbons (TPH) (see Table 1 for sampling plan and analytical methods). Control and treated soil samples were also leached using the Toxicity Characteristic Leaching Procedure (TCLP) to determine the potential for constituents to leach in the environment. Several soil core samples were taken from several inches to 5-feet below the biotreatment bed bottom to determine if any COI or toxic metabolites had leached beyond the biotreatment soil bed. The off-gas from the biotreatment bed was sampled during active biotreatment using isolation chambers placed on the surface of the biotreatment bed soil. Samples were collected during the second and third soil loading in 1987. Off-gas samples were collected from the isolation chamber at weekly intervals between 8/4/87 and 12/24/87 for a total of 17 sampling events. The off-gas samples were analyzed for PAH and PCP. Strict QA/QC procedures including the addition of field and trip blanks and duplicate samples were followed during all sampling events.

2.4.2. Sampling and Analysis Plan: 1993. The performance monitoring in 1993 (six to seven years following active biotreatment) comprised of a test pit evaluation, sampling the biotreatment bed soil, bed sub-soils, the storm water retention pond water, ground water, and the pipe lysimeter water. Test pit evaluations were done prior to conducting sampling to identify the total depth of treated soil in the biotreatment cell and to insure collection of the respective biotreatment layer soil samples. A total of nine test pits were excavated: seven test pits equally spaced down the center line of the biotreatment bed and two test pits located on opposite halves in the control plot. Each test pit was approximately 6- to10-feet long and 3- to 6-feet wide with their respective depths ranging from 36-inches to 96-inches. The biotreatment bed bottom and the different soil layers were readily discernible. The depth measurements were used to identify the various biotreat-

Table 1. Analytical methods

	Active Treatment Phase 1986/1987 [1]		Intrinsic Treatment Phase			
			1993 [2]		1997 [3]	
Analytical Parameter	Soil Method	Water Method	Soil Method	Water Method	Soil Method	Water Method
pH	SW846 9045		SW846 9045		SW846 9045C mod	
Total Organic Carbon (TOC)	SW846 9060	SW846 9060	SW846 9060	SW846 9060	EPA 415.1 mod	EPA 415.1
% Moisture	MCA WW 160.3 mod		MCA WW 160.3 mod		EPA 160.3 mod	
Ammonia Nitrogen	MCA WW 3501	MCA WW 350.2	MCA WW 3501	MCA WW 350.2	EPA 350.2 mod	EPA 350.2
Nitrate Nitrogen	MCA WW 353.2 mod	MCA WW 353.1	MCA WW 353.2 mod	MCA WW 353.1	EPA 353.2	EPA 353.2
Ortho-Phosphate as P	MCA WW 365.2	MCA WW 365.2	MCA WW 365.2	MCA WW 365.2	EPA 365.3	EPA 365.3
Total Phenols	SW846 9066	SW846 9066	SW846 9066	SW846 9066		
Pentachlorophenol (PCP)	Kopper's Method	Kopper's Method	Kopper's Method	Kopper's Method	SW846 8270B	SW846 8270B
Polynuclear aromatic hydrocarbons (PAH)	SW846 8310	SW846 8310	SW846 8310	SW846 8310	SW846 8310	SW846 8310
TCLP Non-Volatile Extraction	SW846 1311		SW846 1311			SW846 1311
Grain Size	ASTM D422		ASTM D422			
Total Suspended Solids (TSS)		EPA 160.2		EPA 160.2		EPA 160.2
Microtox Testing		yes		yes		
Microbial Count	yes [4]		yes [4]		yes [4,5]	

1 = all analysis performed by Kopper's
2 = all analysis performed by Enseco-Wadsworth/Alert Laboratories (Wadsworth), Pittsburgh, PA except microbial counts which were perfomed by RETECH, Seattle, WAand Microtox Testing which was performed by ENSR Consulting and Engineering, Fort Collins, CO.
3 = all analysis performed by Lancaster Laboratories, Lancaster, PA except microbial counts which were performed by Carnegie Mellon Research Institue (CMRI), Pittsburgh, PA.
4 = Clark, F., 1985. Methods of Soil Analysis, Vol. 2 pp. 1460-1465. Sahiaris, M.J. Cooney, 1983: and, Applied and Environmental Microbiology, Vol. 25, No. 2, pp. 70-710, February.
4 = Standard Methods for the Examination of Water and Wastewater Sect. 9215B, 1992 ed.
5 = Methods in Applied Soil Microbiology and Biochemistry, p.154

ment layers which were applied during 1986/1987, for visual evaluation if any COI had migrated below the biotreatment cell bottom, and to collect deep soil samples for chemical analysis.

Once the biotreatment depths were measured, the respective depths were divided into a "top half" and a "bottom half" for analytical sampling (*i.e.*, if the total biotreatment depth was 30-inches, the top half represented 0- to 15-inches and the bottom half represented 15- to 30-inches). Each of the seven biotreatment bed test pits were then sampled by withdrawing soil samples from the top half and bottom half of each test pit. For the control plot however, only the "top" 0 to 12-inch portion, which represented the active tilled layer, was sampled. Several soil grab samples at specified sampling intervals were obtained and composited to yield a single analytical sample. Soil samples obtained were analyzed for parameters listed in Table 1. Discrete core soil samples were also collected from below both the biotreatment bed and the control plot bed bottoms at four of the seven test locations to determine leaching/migration of COI from the treated material. The depth samples were collected at depths of 2-, 4-, and 6-feet below the bottom of the biotreatment plot, and below the tilled layer in the control plot. The core samples were analyzed for percent solids, PCP, total phenols and PAHs.

2.4.3. Sampling and Analysis Plan: 1997. Following ten years of passive treatment, a second sampling event was conducted to evaluate further COI reduction due to on-going intrinsic biodegradation. The sampling and analysis activities were nearly identical to the 1993 procedures. However, the 1997 test pits were staggered from the centerline so that previously undisturbed areas were excavated. The biotreatment bed bottom and the different soil layers were again readily discernible. Therefore, the subsequent sampling of the subsurface was easily correlated to the various soil layers. Discrete depth samples were collected at all seven biotreatment bed locations at depths of 2- and 4-feet below the biotreatment bed, but only the 2-foot samples were analyzed (the 4-foot samples were archived). The discrete depth soil samples were analyzed for percent solids, PCP (new method was used), and PAH. No discrete depth samples were collected from below the control plot in 1997.

3. RESULTS

3.1. Active Biotreatment Phase

Results obtained during in 1986/87 revealed that active biotreatment significantly reduced tPAHs, PCP, and O&G concentrations in all four soil loadings after two- to four-months of active biotreatment. The averaged results using combined data from the four loadings in 1986/1987 showed a 68% reduction in tPAH (from 2,976 to 951 mg/kg), 71% reduction in PCP (from 56 to 16 mg/kg soil), and 79% reduction in O&G (from 5,237 to 1,121 mg/kg soil). The reduction in tPAH was primarily due to reduction in the 2- ring (98% reduction), 3-ring (64% reduction) and 4-ring (78% reduction) PAH compounds; only a 5% reduction was observed in the concentration of the 5-ring PAHs, and the 6-ring compounds exhibit no reduction. These results agree with the existing literature in that most of the reduction due to biodegradation is attributable to those PAH compounds that are more desorbable and thus more bioavailable and biodegradable (Connaughton *et al.* 1993; Mihelcic and Luthy, 1991; Pignatello and Xing, 1996; Stucki and Alexander, 1986; Zhang *et al.*, 1998).

Soil TCLP extract testing conducted on several treated and untreated soil samples showed that the leaching potential of PCP and PAHs was reduced by approximately 96% (from 318 to 14 *u*g/L extract) and 58% (from 1,055 to 443 *u*g/L extract), respectively, following 4-months of active biotreatment (biostabilization). The core soil samples taken from beneath the biotreatment bed showed that only one sample contained any PCP and PAHs above the analytical detection limits or background soil concentration. Results of groundwater sampling showed that all samples (both up gradient and down gradient wells) had no detectable PCP or PAH compounds. Moreover, analysis of pipe lysimeter water showed that all PCP and PAH concentrations associated with the control plot and the biotreatment bed were non-detectable. Together, these data indicated that no site chemicals had migrated via leaching from the biotreated bed into the subsurface or the groundwater during this period, and that their removal was attributed to biological processes.

Similar to previous studies (Mueller *et al.*, 1991), off gas monitoring results showed that PCP and carcinogenic PAH emissions were negligible compared to the levels of these compounds in the soil: 0.0004%–0.0012% percent of the total mass of PCP in the soil and 0.0003%–0.0005% percent of total mass of carcinogenic PAHs in the soil. Significant volatilization was measurable only for some of the lower-molecular-weight, noncarcinogenic PAHs; 17.6% of the acenaphthene, 4.5% of the naphthalene, and 1.2% of the acenaphthylene. Approximately 80% of the measurable air emissions occurred during the first two weeks after soil loading and were most pronounced immediately after tilling of the soil.

In an attempt to monitor for the presence of catabolites, Microtox testing of untreated soil, treated soil, groundwater and pipe lysimeter samples determined the relative toxicity of each medium. These data indicated that groundwater and pipe lysimeter water samples were non-toxic. The untreated soil however, was acutely toxic, and after one month of active biotreatment, the toxicity of the soil was significantly reduced. Subsequent soil samples taken and analyzed 2-, 4-, and 6-months after active biotreatment showed further reduction in toxicity. Reduction in soil toxicity corresponded to degradation observed of PCP and PAH components.

3.2. Results of 1993 Performance Monitoring

Sampling performed in 1993 represents the time period of 6–7 years of passive biotreatment following active biotreatment. The test pit evaluations revealed that the "top half" of the biotreatment bed consisted of the second and third loadings applied in 1987, and the "bottom half" consisted of the first loading applied in 1987 and the only loading applied in 1986. The data presented below show that biodegradation of PAHs and PCP continued to occur in soils after the active bioremediation phase, with the rate and extent of biodegradation correlating significantly with the solubility and molecular weight (*e.g.*, bioavailability) of each particular COI.

3.2.1. Soil PCP and tPAH Concentrations. In general, there was little difference between the top half and bottom half soil values for the analytical parameters. For example, similar concentrations were observed for PCP in the top and bottom halves (5.0 mg/kg vs 5.8 mg/kg, respectively) and tPAHs (206 mg/kg vs. 184 mg/kg, respectively). Moreover, PAH concentrations by ring size showed no statistically significant concentration variations between the top and bottom half (data not shown).

3.2.2. Soil Microbiological Indicators. However, approximately 10-times more microbial population counts were measured in the top half of the biotreatment bed ($1.2x10^7$ cfu/g) compared to the bottom half ($1.2x10^6$ cfu/g). The number of bacteria capable of utilizing naphthalene as a sole carbon source was 10-times higher in the top half ($5.6x10^6$cfu/g) than the bottom half ($4.5x10^5$ cfu/g). Other general microbiological indicators suggested that the indigenous, catabolically relevant microflora were active in the soil.

3.2.3. Indicators of COI Biostabilization. For the soil samples that were subjected to the modified TCLP tests, the TCLP extract data for treated soil showed very little leaching of PCP and PAHs. PCP in both the top and bottom half soils extracts were less than 1.0 *ug*/l with the exception of a single sample from the top half that showed PCP at a concentration of 4.6 ug/l in the extract. Total PAH in the extracts of top half samples were reported at <23.1 *ug*/L, and tPAH in the extract of bottom half samples were reported at <16.6 ug/L. Core soil samples collected from the bed bottom showed less than the respective detection limits (<1 *ug*/L) for leachable PCP and PAHs. No PCP or PAHs were detected in any of the water samples.

Microtox testing, which was performed on a composite soil sample obtained from both the top half and bottom half, indicated that the soil was not toxic (data not shown). Together, these data indicated that there was no significant accumulation of any COI leaching from the treated material to the soil directly below the biotreatment bed over a period of approximately seven years (*i.e.,* 1986 to 1993)

3.3. Results of 1997 Performance Monitoring

Performance monitoring data obtained in 1997 (10-years of passive bioremediation) are used herein to summarize the overall performance of the biostabilization strategy. Results presented are averages of replicate samples. It should be noted here that, as with the previous sampling events, PCP and PAH results include data reported below analytical detection limits. In computing average results, these data were recorded as the analytical detection limit.

3.3.1. Soil PCP and tPAH Concentrations. Similar to the results obtained during sampling in 1993, the 1997 analysis indicated little difference (not statistically significant at a 95% confidence limit) between the top half and bottom half soil concentrations for PCP and tPAHs. Average PCP concentrations for top half and bottom half were 6 mg/kg and 13 mg/kg, respectively. For tPAHs, concentrations for top versus bottom halves were 103 mg/kg and 118 mg/kg, respectively. A similar observation was made for PAHs by ring size, with the top and bottom half samples showing similar profiles. For the control plot soil samples, taken between 0- to 12-inches bgs, PCP concentration in the soil was reported at below analytical detection limit of 0.2 mg/kg, and tPAHs in the soil were reported at 3.5 mg/kg.

3.3.2. Soil Microbiological Indicators. As was observed in 1993, there was approximately 10-times greater microbial populations in the top half ($1.8x10^7$cfu/g) of the biotreatment bed compared to the bottom half ($2.2x10^6$cfu/g) which may be attributed to the greater microbial activity around the root zone of the vegetation. However, the proportion of the indigenous microflora capable of utilizing naphthalene as a growth substrate decreased dramatically to $1.4x10^4$cfu/g in the top half and $8.5x10^2$cfu/g in the bottom half.

Table 2. Concentration and percent removal of PCP and PAHs in soil for active and intrinsic bioremediation

	Active treatment phase			Intrinsic treatment phase			
	Untreated soil [1] 1986/1987 Average ± 95% CI (mg/kg)	Treated soil [1] 1986/1987 Average ± 95% CI (mg/kg)	Percent removal 86/87 untreated—86/87 treated (%)	Treated soil [2] 1993 Average ± 95% CI (mg/kg)	Percent removal 86/87 untreated—1993 (%)	Treated soil [3] 1997 Average ± 95% CI (mg/kg)	Percent removal 86/87 untreated—1997 (%)
Pentachlorophenol							
PCP	55.69 ± 9.9	15.92 ± 3.2	71	5.40 ± 1.9	90	9.42 ± 3.3	83
Polynuclear Aromatic Hydrocarbons							
Naphthalene	371.65 ± 232.7	30.73 ± 11.7	92	12.66 ± 2.7	97	1.52 ± 0.9	100
Acenaphthylene	41.91 ± 27.2	4.81 ± 8.1	89	8.16 ± 0.1	81	0.92 ± 0.1	98
Acenaphthene	191.31 ± 46	33.54 ± 21.4	82	9.69 ± 3.2	95	3.34 ± 0.5	98
Fluorene	264.60 ± 75.1	65.36 ± 21.5	75	9.64 ± 3.5	96	0.54 ± 0.7	100
Phenanthrene	551.43 ± 171.2	116.99 ± 41.1	79	16.58 ± 5.2	97	11.16 ± 4.5	98
Anthracene	902.81 ± 232.6	474.37 ± 154.6	47	74.00 ± 24.3	92	41.65 ± 25.5	95
Fluoranthene	319.97 ± 114.7	63.48 ± 29.8	80	10.28 ± 3.1	97	9.62 ± 5.1	97
Pyrene	253.81 ± 99.3	50.53 ± 26.7	80	20.38 ± 7.1	92	6.15 ± 4	98
Benzo (a) Anthracene	52.17 ± 18.9	13.87 ± 5.3	73	3.46 ± 1	93	3.18 ± 1.6	94
Chrysene	50.04 ± 14.3	18.71 ± 4.2	63	5.10 ± 1.2	90	4.68 ± 1.8	91
Benzo (b) Fluoranthene	20.91 ± 4.8	16.93 ± 4	19	6.71 ± 1.2	68	5.16 ± 1.6	75
Benzo (k) Fluoranthene	10.67 ± 3.3	5.27 ± 1.6	51	2.11 ± 0.4	80	2.35 ± 0.7	78
Benzo (a) Pyrene	14.67 ± 4	11.28 ± 3.5	23	7.01 ± 1.1	52	6.84 ± 2.3	53
Dibenzo (a,h) Anthracene	13.08 ± 5.4	20.83 ± 9.6	Increase	0.93 ± 0.2	93	0.48 ± 0.1	96
Benzo (g,h,i) Perylene	10.68 ± 1.1	14.38 ± 5.8	Increase	5.15 ± 1.1	52	7.54 ± 1.5	29
Indeno (1,2,3-cd) Pyrene	9.25 ± 1.5	10.15 ± 5.4	Increase	3.32 ± 0.5	64	5.37 ± 1.1	42
PAH summary							
2-ring	371.65 ± 232.7	30.73 ± 11.7	92	12.66 ± 2.7	97	1.52 ± 0.9	100
3-ring	1,952.05 ± 511.8	695.06 ± 209.4	64	118.06 ± 34.1	94	57.61 ± 29.6	97
4-ring	675.99 ± 245.9	146.59 ± 64.2	78	39.22 ± 11.5	94	23.62 ± 12	97
5-ring	59.33 ± 13.1	54.32 ± 17.7	8	16.76 ± 2.8	72	14.84 ± 4.4	75
6-ring	19.93 ± 2.5	24.53 ± 11.1	Increase	8.47 ± 1.6	57	12.91 ± 2.5	35
Total PAHs	2,975.72 ± 877.1	951.23 ± 230.3	68	195.18 ± 42.7	93	110.49 ± 35.9	96

1. Average of all 1986/1987 untreated/treated soil results excluding the intermediate monthly sample results.
2. Average of all 1993 soil results (i.e., top half/bottom half).
3. Average of all 1997 soil results (i.e., top half/bottom half).
All results reported on a dry weight basis.

This correlated with a significant decrease in the amount of bioavailable PAH. However, the presence of degraders may also reflect those bacteria capable of co-metabolizing other high-molecular-weight PAHs that slowly leach from the soil granules or aggregates.

3.2.3. Indicators of COI Biostabilization. No statistically significant differences were observed in leachable PCP or PAH between the top and bottom halves of the biotreatment bed soil. PCP in the soil extracts from both layers were below analytical detection limit of 1.0 ug/l. Total PAHs in 14 out of 16 soil extracts from both the top and bottom layers were below analytical detection limit of 0.02 ug/l. However, average tPAHs for the top half extract samples was four-times lower than average level measured in the bottom half (4 ug/l vs. 12 ug/l). Individual PAHs reported above detection limits in the extracts were limited to the 2-ring and 3-ring PAHs (data not shown). Thus, there was very little leaching potential of PCP or PAHs from the treated soil. Accordingly, the storm water and groundwater water samples indicated no PCP or PAHs. The pipe lysimeter water also indicated no significant levels of PCP or PAHs.

Data from seven discrete-depth soil core samples collected from 2-feet below the bottom of the biotreatment bed showed an average PCP concentration of 0.20 mg/kg which is very close to the maximum level measured below the treatment bed during active biotreatment in 1986/1987 (0.21 mg/kg) and very close to the level reported for 1997 control plot samples (0.19 mg/kg). These results indicated no migration of PCP beyond the biotreatment bed soil. Average tPAH was reported at 0.43 mg/kg which was a one- to two-orders of magnitude lower than concentrations that were measured during the active bioremediation phase, again indicating no migration of PAHs beyond the biotreatment bed soil.

4. DISCUSSION

Results presented herein indicate no migration of PCP or PAHs from the biotreatment bed surface soils into the subsurface soils or groundwater. The data further suggest that any remaining "leachable" PCP and PAHs are bound to the soil matrix and will not rapidly transfer to the aqueous phase. As discussed below, biostabilization therefore represents a viable approach for treating similarly impacted material, clearly recognizing that the applicability of the approach must be assessed on a site-specific and material-specific basis.

4.1.1. Validation of Biostabilization as an Effective Treatment Strategy. Biostabilization via short-term active bioremediation followed by 10 years of passive bioremediation reduced the soil PCP and tPAH concentrations by 83 and 96%, respectively (Table 2). Residual soil COI concentrations are approximately an order of magnitude greater than natural soil background concentrations of 0.2 mg/kg and 10 mg/kg for PCP and tPAH, respectively. The maximum reduction in PAH was observed for the 2-ring compounds, with the 6-ring PAH compounds showing the least reduction. In fact, no statistically significant reduction in the concentration of the 6-Ring PAH compounds was observed during both active treatment and 10 years of passive biotreatment.

Nevertheless, significant reductions in COI leachability/mobility were observed (Table 3). Here, reduction in constituent leachability is reported as a percent reduction of constituent concentration in the leachate. The reduction in PCP in the soil extract in 1997 exceeded 99%. The reported 1.0 ug/L of PCP in the 1997 treated soil extract is the average of 14 samples, all of which were reported as having PCP concentrations below the detec-

Table 3. Concentration and percent removal of PCP and PAHs in soil extract for active and intrinsic bioremediation

	Active treatment phase			Intrinsic treatment phase			
	Untreated soil [1] 1986/1987 Average ± 95% CI (ug/L)	Treated soil [1] 1986/1987 Average ± 95% CI (ug/L)	Percent removal 86/87 untreated—86/87 treated	Treated soil [2] 1993 Average ± 95% CI (ug/L)	Percent removal 86/87 untreated—1993	Treated soil [3] 1997 Average ± 95% CI (ug/L)	Percent removal 86/87 untreated—1997
PCP	318.15 ± 344.4	14.24 ± 26.5	96	1.26 ± 0.6	100	1.00 ± 0	100
Polynuclear Aromatic Hydrocarbons							
Naphthalene	529.67 ± 523.1	108.99 ± 75	79	8.31 ± 3.5	98	1.48 ± 1.7	100
Acenaphthylene	121.02 ± 106.3	68.67 ± 75.7	43	2.76 ± 0.4	98	0.35 ± 0	100
Acenaphthene	149.66 ± 52.4	89.08 ± 70.7	40	2.24 ± 0.3	99	1.00 ± 0	99
Fluorene	100.02 ± 48.9	14.83 ± 15.4	85	1.25 ± 0.6	99	1.27 ± 1.2	99
Phenanthrene	92.96 ± 51.3	28.90 ± 17.8	69	1.04 ± 0.4	99	0.68 ± 0.6	99
Anthracene	21.42 ± 16.8	13.94 ± 9.2	35	2.38 ± 1.1	89	2.58 ± 2.6	88
Fluoranthene	7.04 ± 6.1	15.06 ± 8.9	Increase	0.33 ± 0.2	95	0.01 ± 0	100
Pyrene	10.22 ± 6.2	51.48 ± 65	Increase	0.64 ± 0.5	94	0.25 ± 0	98
Benzo (a) Anthracene	2.98 ± 3.1	0.93 ± 0.7	69	0.06 ± 0	98	0.02 ± 0	99
Chrysene	4.03 ± 4.5	5.20 ± 5.7	Increase	0.18 ± 0	96	0.10 ± 0	98
Benzo (b) Fluoranthene	2.98 ± 3.1	0.80 ± 0.7	73	0.15 ± 0.2	95	0.02 ± 0	99
Benzo (k) Fluoranthene	0.85 ± 0.7	26.70 ± 50.4	Increase	0.06 ± 0.1	93	0.00 ± 0	99
Benzo (a) Pyrene	2.97 ± 3.1	0.71 ± 0.7	76	0.13 ± 0.1	96	0.01 ± 0	100
Dibenzo (a,h) Anthracene	3.32 ± 3.1	1.03 ± 1.1	69	0.04 ± 0	99	0.02 ± 0	99
Benzo (g,h,i) Perylene	1.88 ± 1.8	16.72 ± 19.2	Increase	0.15 ± 0.1	92	0.10 ± 0	95
Indeno (1,2,3-cd) Pyrene	3.97 ± 3.1	0.06 ± 0	99	0.13 ± 0.1	97	0.05 ± 0	99
PAH summary							
2-ring	529.67 ± 523.1	108.99 ± 75	79	8.31 ± 3.5	98	1.48 ± 1.7	100
3-ring	485.08 ± 216.3	215.42 ± 154.6	56	9.66 ± 2.4	98	5.87 ± 4.4	99
4-ring	24.27 ± 12.9	72.67 ± 73.4	Increase	1.21 ± 0.7	95	0.38 ± 0	98
5-ring	10.11 ± 9.4	29.24 ± 52	Increase	0.38 ± 0.4	96	0.06 ± 0	99
6-ring	5.86 ± 4	16.78 ± 19.2	Increase	0.27 ± 0.3	95	0.15 ± 0	97
Total PAHs	1,054.98 ± 595.4	443.10 ± 332.5	58	19.84 ± 5.4	98	7.95 ± 6.1	99

1. Average of all 1986/1987 untreated/treated soil results excluding the intermediate monthly sample results.
2. Average of all 1993 soil results (i.e., top half/bottom half).
3. Average of all 1997 soil results (i.e., top half/bottom half).
All results reported on a dry weight basis.

tion limit of 1.0 *ug*/L. Although residual PCP in the treated soil is slightly above background soil levels, these data suggest that biodegradation can effectively reduce PCP mobility in the environment such that the residual levels that remain in the soil continue to reduce as they slowly elute and are subsequently biodegraded. Thus the material becomes biostabilized in terms of leachability and mobility.

For tPAHs, overall percent reduction in soil extract in 1997 exceeded 97%. The reported tPAH concentration in treated soil extract of (7.95 *ug*/L) represents the sum of all the ring groups, with approximately 89% of the individual PAHs being reported at detection limit. It is interesting to note that significantly greater reduction in less leachable PAHs occurred during the 6–10 years of intrinsic biotreatment than during the active treatment phase. In general, the 2-ring and 3-ring compounds were most readily reduced during active bioremediation while the 4-, 5-, and 6-ring PAHs were reduced more as a result of passive bioremediation.

Comparing the slope of the decay curve for PCP (and tPAH concentration – data not shown) in soil versus soil extract, it appears that during the intrinsic bioremediation phase, small reductions in soil concentration resulted in even greater reductions in the soil extract concentration (Figure 2). Based on these results, small reductions in soil COI concentration as a result of biodegradation of more leachable and more bioavailable components resulted in orders of magnitude reductions in leachable concentrations. By 1997, the PCP and PAH concentrations in leachate from the treatment bed soil decreased to near background levels (data not shown).

4.1.2 Comparison of Field Results with Results Obtained from Empirical Modeling. To determine chemical loss from the soil, an empirical model modified first order biodegradation equation was used. The equation is given as: $C_t = (C_O - C_R) * e^{-kt} + C_R * e^{-krt}$ where;

C_t = Organic concentration at time t,,

C_O = Initial organic concentration,,

C_R = Organic concentration resistant to biodegradation or not readily bioavailable,

k = First-order decay rate constant for readily bioavailable organics, T^{-1}

k_r = First-order decay rate constant for the recalcitrant organics, T^{-1}

Empirically determined modeling constants $C_{R.}$ k and k_r for the period 1993 to 1997 were obtained by fitting the above equation to data obtained during periods of both active and 1987–1993 passive biodegradation. The parameter estimation exercise utilized a least squares technique to determine these constants (Alcoa, 1997). For the purposes herein, the constant C_R is important to define for the active biotreatment phase since it represents the plateau concentration that marks the end of active biotreatment and also represents the organic concentration that is less bioavailable to biodegradation or not readily bioavailable. The concentration C_O-C_R represents the organic concentration that is readily bioavailable. The first-order release constants k and k_r were also empirically determined to define how fast the readily biodegradable fraction and the resistant fractions decrease, respectively, from the active through the passive bioremediation phases. The resultant values presented in Table 4 (bottom panel) are mostly within the ranges reported in the scientific literature (Howard *et al.*, 1991).

Modeling results are compared with actual 1997 field measurements for PCP, tPAH and PAHs by ring size (Table 4 – top panel). For PCP, there appears to be good compari-

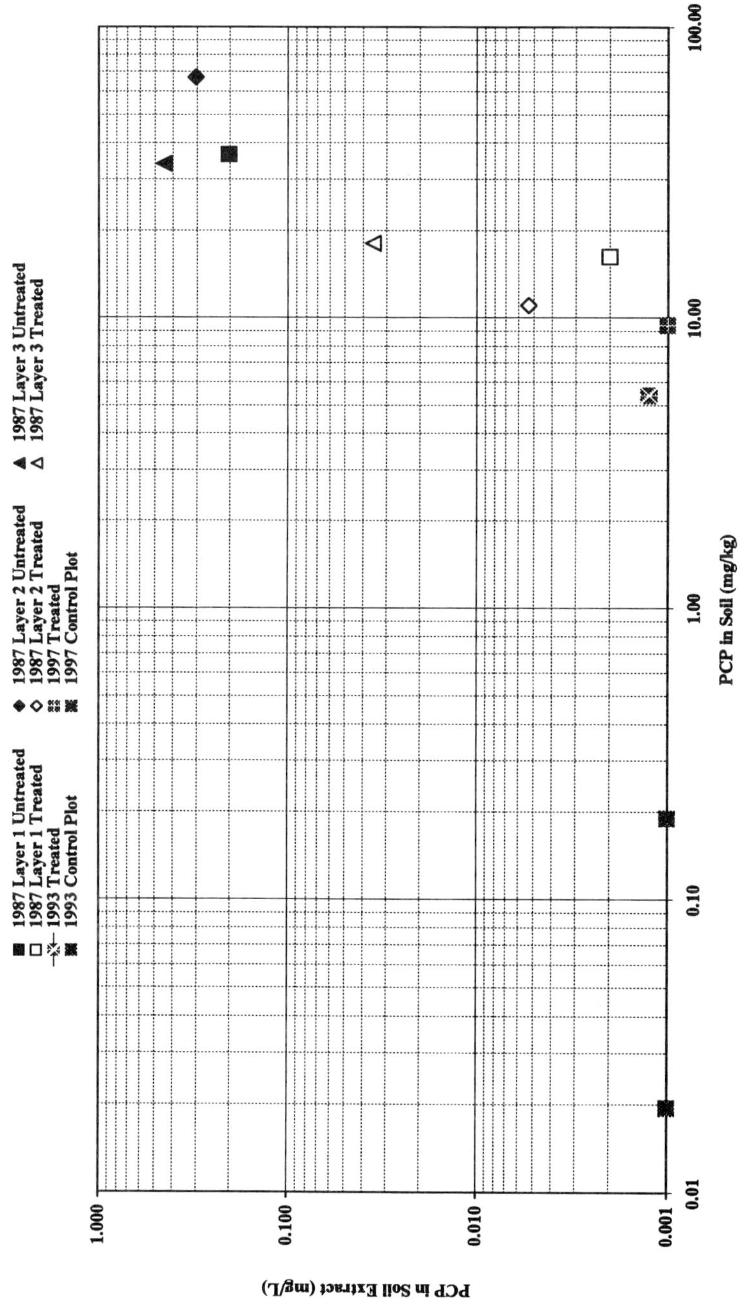

Figure 2. Effect of soil/sludge bioremediation on PCP soil-soil extract concentrations.

Table 4. PCP and PAH concentrations over time empirical modeling results versus field results

Modeling equation:
$C_t = (C_O - C_R) * e^{-kt} + C_R * e^{-krt}$

Constants* :	C_O =	initial concentration
	C_R =	resistant concentration
	k =	first order rate constant for active biodegradation
	k_r =	first order rate constant for intrinsic biodegradation
Variables:	t =	time (days)
	C_t =	concentration at time t

		Concentration at Time t (mg/kg)			
Parameter	Model vs. Field	86/87 untreated = 0 (days)	86/87 treated = 61 (days)	1993 treated = 2281 (days)	1997 treated = 3559 (days)
PCP	Model	56.00	22.07	5.07	3.93
	Field	55.69	15.92	5.40	9.42
2-Ring PAHs	Model	372.00	32.52	12.11	8.25
	Field	371.65	30.73	12.66	1.52
3-Ring PAHs	Model	1952.00	717.36	127.99	59.45
	Field	1952.05	695.06	118.06	57.61
4-Ring PAHs	Model	672.00	169.41	37.83	25.78
	Field	675.99	146.59	39.22	23.62
5-Ring PAHs	Model	59.00	56.88	15.01	6.97
	Field	59.33	54.32	16.76	14.84
6-Ring PAHs	Model	20.00	19.52	8.03	4.82
	Field	19.93	24.53	8.47	12.91
Total PAHs	Model	2976.00	890.02	203.19	121.87
	Field	2975.72	951.23	195.18	110.49

Parameter	CO (mg/kg)	CR (mg/kg)	k (1/day)	k_r (1/day)	$t_{½}$ active (days)	$t_{½}$ intrinsic (days)
PCP	56	8	0.02	0.0002	35	3470
2-Ring PAHs	372	24	0.06	0.0003	12	2310
3-Ring PAHs	1952	503	0.03	0.0006	23	1160
4-Ring PAHs	672	75	0.03	0.0003	23	2310
5-Ring PAHs	59	59	0	0.0006	infinite	1160
6-Ring PAHs	20	20	0	0.0004	infinite	1730
Total PAHs	2976	506	0.03	0.0004	23	1730

son between the field data and the model, with both data sets indicating declining levels of PCP with time. The small variance observed between the field data and the predicted results in 1997 may be attributed to a combination of sample heterogeneity and a change in analytical method in 1997. Nevertheless, this variation does not appear to be significant as there was no significant change in leachable PCP between 1993 and 1997. The model was even more accurate in predicting tPAHs in the soil at various stages of treatment. There was very good correlation between field data and the predicted results for the 2-, 3-, and 4-ring PAH compounds indicating that the model could serve as a good tool to predict soil concentrations for future time periods. For the 5-, and 6-ring PAH compounds, the biostabilization model may still serve as a good predicting tool, but the degree of uncertainty in the predicted values will be greater for this group of compounds.

4.1.3. Alternative End-Point Analysis. Throughout the biostabilization process, residual soil COI concentrations were above background soil levels. However, the absence of chemicals in the sub-soil, storm water, or groundwater indicated that residual PCP and PAHs are strongly bound to the soil and therefore do not pose a threat with regards to leaching to the environment. Thus, acceptance of biostabilization as an effective remedial approach mandates the concomitant acceptance of ARARs or EAE. Such acceptance must be determined on a site-specific basis, but is appropriate for the Nashua site since passive biodegradation effectively removed residual COI they slowly leached from the soil matrix. This supports the concept that active bioremediation followed by proper containment management of the treated material will provide long-term protection of human health and the environment.

5. CONCLUSIONS

The results of a 10-year field study showed that soil PCP and PAH concentrations were significantly reduced after 9–10 years of passive bioremediation following a two- to four-month active biotreatment phase that rapidly stabilized the impacted soil. Since no significant amounts of COI were volatilized or leached to the subsurface or groundwater, it was concluded that the chemicals were released from the soil and subsequently biodegraded. In general, PCP and the 2-ring and 3-ring PAH compounds were most readily reduced during active bioremediation while the 4-, 5-, and 6-ring PAHs were reduced more as a result of passive bioremediation. These results, in conjunction with the soil extract data, indicated that the chemicals were stabilized in terms of constituent leachability and mobility. These data support the theory that a relatively short active bioremediation period usually results in significant constituent reduction to a plateau concentration, and further reduction of soil COI can occur through passive bioremediation over an extended period of time. In other words, soil is effectively biostabilized in that the mobility of the chemical constituents in the treated soil is reduced to such low levels that they no longer serve as a source of constituent release (*i.e.*, no longer act as a leachate source which could potentially impact any associated surface water, groundwater, and underlying soil).

ACKNOWLEDGMENTS

This Project was co-funded by the Aluminum Company of America (Alcoa -Remediation Work Group) and Beazer East, Inc. Denny Fulmer (ES&TD Alcoa) assisted with the 1997 Field work. Data compilation and reporting was done by M. Tabe with the assistance of Janice Fleckenstein.

REFERENCES

Alcoa. 1997. *Active/Intrinsic Bioremediation of Soils Containing PAHs and PCP to Achieve Risk-Based Acceptable Endpoints*. Aluminum Company or America, October, 1997.

Blumer, M. 1976. Polycyclic aromatic compounds in nature. Scientific American 234:35–45.

Connaughton, D., J. Stedinger, L. Lion, and M. Shuler. 1993. Description of time-varying desorption kinetics: release of naphthalene from contaminated soils. Environ. Sci. Technol. 27:2397–2401.

Brown, R., P. Hicks, R. Hicks, and M. Leahy. 1995. Postremediation bioremediation. Bioremediation 3(1):77–84.

EPRI. 1993. *Biodegradability of Pentachlorophenol in the Environment: A Literature Review*. EPRI TR-102172, April 1993.

Heitkamp, M. and C. Cerniglia. 1987. The effects of chemical structure and exposure on the microbial degradation of polycyclic aromatic hydrocarbons in freshwater and estuarine ecosystems. Environ. Toxicol. Chem. 6:535–546.

Howard, P., R. Boethling, W. Jarvis, W. Meylan and E. Michalenko. 1991. *Handbook of Environmental Degradation Rates*, Lewis Publishers, Chelsea, MI USA.

Keystone. 1989.*Pilot-Scale Demonstrations Study Engineered Biodegradation System SM, Appendix to Feasibility Study, Nashua, NH.* Keystone Environmental Resources, Inc., April, 1989.

Koppers. 1987. *Evaluation of an Engineered Biodegradation System at the Nashua, N.H. Site*. Koppers Company, Inc., April, 1987.

Mihelcic, J. and R. Luthy. 1991. Sorption and microbial degradation of naphthalene in soil-water suspensions under denitrification conditions. Environ. Sci. Technol. 25: 169–175.

Mueller, J., S. Lantz, B. Blattmann and P. Chapman. 1991. Bench-scale evaluation of alternative biological treatment processes for the remediation of pentachlorophenol- and creosote-contaminated materials: solid-phase bioremediation. Environ. Sci. Technol. 25:1045–1055.

Mueller, J., J.-E. Lin, S. Lantz, and P. Pritchard. 1993. Recent Developments in Cleanup Technologies: Implementing Innovative Bioremediation Technologies. Remediation/Summer Issue 369–381.

Mueller, J., C. Cerniglia and P. Pritchard. 1996. Bioremediation of Environments Contaminated by Polycyclic Aromatic Hydrocarbons. *In* R. Crawford and D. Crawford (eds), *Bioremediation: Principles and Applications*. Chapter 5, Pages 125–194. Cambridge University Press.

Pignatello, J. and B. Xing. 1996. Mechanisms of slow sorption of organic chemicals to natural particles. Environ. Sci. Technol. 30:1–11.

Pritchard, P.H., J.G. Mueller, J.C. Rogers, F.V. Kremer and J.A. Glaser. 1993. Oil spill bioremediation: Experiences, lessons and results from the *Exxon Valdez* oil spill in Alaska. Biodegradation 3:315–335.

Smith, J.R., R. Tomicek, P. Swallow, R. Weightman, D. Nakles and M. Helbling. 1995. Definition of biodegradation endpoints for PAH contaminated soil using a risk-based approach. In "Hydrocarbon Contaminated Soils – Volume V", P.T. Kostecki *et al* (ed). Amherst Scientific Publishers, Amherst, Mass, USA. Chapter 43, pages 531–572.

Smith, J., M. Tabe, T. Lightfoot, W. Lyman, C. Kaslick, P. Rodrique, M. Khodash, R. Unterman, and R. Rothmel. 1996. Bioremediation of PCB- and PAH-containing sludge/sediment in land treatment units to achieve risk-based endpoints. Proceedings 28th Mid-Atlantic Industrial Waste Conference, Buffalo, NY, USA, July 15–17.

Stucki, G. and M. Alexander. 1986. Role of dissolution rate and solubility in biodegradation of aromatic compounds. Appl. Environ. Microbiol. 53: 292–298.

Zhang, W-E., E. Bouwer and W. Ball. 1998. Bioavailability of hydrophobic organic contaminants: effects and implications of sorption-related mass transfer on bioremediation. Groundwater: Winter, pages 126–138.

26

BIOAUGMENTATION USING REPEATED INOCULATIONS OF XENOBIOTIC-DEGRADING BACTERIA

D. E. Crowley, E. S. Gilbert, A. C. Singer, D. A. Newcombe, and C-H. Yang

Department of Environmental Sciences, University of California
Riverside, California 92521

1. INTRODUCTION

Bioaugmentation is a promising technology for the cleanup of contaminated soils (Vogel, 1996), and has been tested for a variety of soil contaminants. Recent studies, some of which have been conducted in the field, include experiments with alachlor (McFarland *et al.*, 1996), atrazine (Grigg et al., 1997; de Souza *et al.*, 1997), 2,4-D (Digiovanni *et al.*, 1996), carbon tetrachloride (Mayotte *et al.*, 1996), polycyclic aromatic hydrocarbons (Otte *et al.*, 1994), pentachlorophenol (Otte *et al.*, 1994; Lamar *et al.*, 1994), and trichloroethylene (Munakata-Marr *et al.*, 1996; Duba *et al.*, 1996). Viewed optimistically, results from laboratory experiments generally suggest that any compound that can be used as a growth substrate or cometabolized by microorganisms can be remediated through bioaugmentation. Unfortunately, in most field studies soil inoculants have proven to be unreliable, especially when soils are inoculated with only a single application of degrader bacteria. Inconsistencies between the laboratory and the field have been attributed to nutrient limitations, suppression of the introduced bacteria by protozoa, metal or chemical toxicities, limited transport of bacteria through the soil, and low bioavailability of the contaminant, all of which lead to poor survival and activity of the inoculum (Alexander, 1994).

Despite these problems, there is still considerable interest in bioaugmentation, since many highly effective degrader organisms have been identified and cultured in the laboratory, including several that have been patented for use in bioremediation. While biostimulation of the indigenous microflora is viewed as the most practical bioremediation method, soil inoculation remains the only viable alternative in situations where indigenous degraders are not capable of significant degradation, or are not easily enriched by addition of soil amendments (Baud-Grasset *et al.*, 1995). To overcome problems with bioaugmentation, several approaches are being investigated. For example, it may be possible to use plant microbial systems to enhance the survival and activity of degrader bacteria by inoculating

Novel Approaches for Bioremediation of Organic Pollution, edited by
Fass *et al.* Kluwer Academic / Plenum Publishers, New York, 1999.

soils with degrader bacteria that are competent for growth and survival in the rhizosphere (Crowley *et al.*, 1996). There is also the possibility that highly effective bacterial strains may be genetically engineered for better survival, or used with field application vectors to enhance their growth and activity after soil inoculation (Lajoie *et al.*, 1994). Another concept being explored is the use of repeated, inundative applications of degrader bacteria.

Repeated application of soil bacteria using a field fermenter has already been commercialized for delivery of disease suppressive bacteria to control plant diseases, but due to the very recent availability of reliable field fermentation equipment, has not yet been examined for bioremediation. In theory, automated inundative application of degrader organisms should eliminate problems associated with decline in inoculum viability and activity during inoculum storage, and provides a cost effective method for production and delivery of inoculum over large land areas. This chapter reports the results of recent studies examining repeated inoculation for bioremediation of soils contaminated with atrazine and with polychlorinated biphenyls (PCBs). Atrazine is a widely used herbicide that is recalcitrant in most soils due to the absence of microorganisms that are capable of mineralizing this chemical. Bioremediation of PCBs, on the other hand, poses a completely different problem in that biphenyl-degrading microorganisms are ubiquitous in the environment, but must be induced to cometabolize PCBs by an inducing substrate.

2. AUTOMATED FERMENTER SYSTEM FOR REPEATED INUNDATIVE APPLICATIONS OF DEGRADER BACTERIA

To culture degrader bacteria in the field, a commercial fermenter initially developed for application of bacteria for biocontrol of turf grass diseases (BioJect, EcoSoil Systems Inc. San Diego, CA) was used (Figure 1). The automated fermenter system consists of a

Figure 1. Automated fermenter system for on site culture and delivery of xenobiotic degrading bacteria.

temperature controlled stainless steel tank, coupled to a computer controlled system to automatically pump nutrients and inoculum into the fermentation tank where the bacteria are grown to densities up to 10^9 to 10^{10} bacteria per ml of nutrient solution. After the bacteria are produced, the system pumps the bacteria into irrigation water used to treat the site, and the system then undergoes a clean-in-place cycle before the starting the fermentation. When mixed into the irrigation water and used for biocontrol, population densities of 10^5 bacteria per ml are applied over 20 hectares using one automated fermenter. Optimal population densities of degrader bacteria for bioremediation need to be determined and will vary depending on the soil type and pollutant characteristics. The advantage of this technology is that the bacteria added to the soil do not necessarily have to be adapted to the soil conditions, as they must only survive for a period of a few days before the bacteria are added again. Thus even in a soil with mixed contaminants, the bacteria can be applied at intervals that may allow the maintenance of high population densities of degrader organisms.

3. BIOREMEDIATION OF ATRAZINE USING REPEATED APPLICATIONS OF DEGRADER BACTERIA

Atrazine is a model xenobiotic pollutant for bioaugmentation studies because most agricultural soils treated with this herbicide appear to lack an active degrader population that is capable of rapidly mineralizing the residual chemical and its abiotically generated transformation products. Residual atrazine and these metabolites accumulate in the surface soil, which leach into the groundwater on dissolved organic matter. However, while present in the surface soil, residual atrazine may be readily treated by soil inoculation. When applied at normal usage rates, atrazine is present in the soil at concentrations of only 2 to 4 mg kg^{-1}, most of which is sorbed to clay and organic matter, and becomes decreasingly available as a function of aging in soil (Radosevich *et al.* 1997). Thus, bioavailability of atrazine is an important factor that limits the development of a high population density of indigenous degrader organisms. Recently, concern has grown that atrazine and its metabolites are being detected in groundwater above allowable EPA levels (Belluck *et al.*, 1991, and EPA, 1990). Atrazine and other s-triazine herbicides have already been prohibited in Europe because of ground water contamination problems, but are still the most commonly used herbicides in the United States.

Over the past 20 years, many studies have examined the ability of indigenous microbial communities to degrade atrazine in soil (see for example: Kishore *et al.*, 1971; Dousset *et al.* 1997). Most studies show that complete mineralization of atrazine is rare as evidenced by the persistence of atrazine and atrazine metabolites in the soil. For some time, after several decades of research examining atrazine degradation, it appeared that there were no culturable organisms that could mineralize atrazine. Recently, several soil bacteria and a bacterial consortium that can mineralize atrazine in liquid media have been identified (Mandelbaum *et al.*, 1993, Yanze-Kontchou *et al.*, 1994; Alvey *et al.*, 1995; Radosevich *et al.*, 1995). Laboratory studies with these bacteria suggest that bioaugmentation may facilitate the mineralization of atrazine (Alvey and Crowley, 1995; Mandelbaum *et al.*, 1995). However, with a single inoculation, degradation activity slows or ceases before the contaminant has been completely removed from the soil (Radosevich *et al.*, 1997; Alvey and Crowley 1996), which suggest that repeated inoculation or plant-microbial systems may be necessary to sustain an effective population size of atrazine degraders.

3.1. Laboratory Microcosm Studies

Initial experiments examining the potential benefits of repeated inoculation of atrazine contaminated soils with atrazine-degrading bacteria were conducted using microcosms containing soil that had no prior exposure to atrazine, which was artificially contaminated with 100 ppm radiolabeled atrazine-ring-UL-^{14}C. Two different bacterial inocula were used; an atrazine degrading consortium (Alvey *et al.*, 1997) and the atrazine-degrading bacterial isolate *Pseudomonas* sp st. ADP (Mandelbaum *et al.*, 1995). The consortium consisted of a *Clavibacter michiganese*, *Pseudomonas* sp., and an unknown third member, which was later shown to be nonessential (De Souza *et al.*, 1997). At each soil inoculation, late log phase bacteria cultured on atrazine as a sole nitrogen source were stirred into the microcosms to provide a final density of 10^8 cfu g^{-1} soil.

After 35 days, noninoculated soil microcosms mineralized less than 1% of the atrazine. Soils that received a single inoculation of either the *Pseudomonas* sp st. ADP or the consortium mineralized 36 and 17% of the atrazine, respectively. In contrast, soils that were repeated inoculated with *Pseudomonas sp.* st. ADP mineralized 90% of the atrazine by day 35. Repeated application of the consortium resulted in 19.5% degradation of the atrazine by day 6, which continued at a steady rate such that 64% of the atrazine was mineralized by day 35.

Data obtained from this laboratory experiment strongly support the repeated application concept, and showed that repeated applications resulted in a more rapid and complete degradation of atrazine than did a single application of either the consortium or the isolate. The rates of degradation observed after a single inoculation, however, were much lower than those reported by Alvey *et al.*, (1995) Mandelbaum *et al.*, (1995), and Radosevich *et al.*, (1997), suggesting that degradation rates in different soils are highly dependent on soil environmental factors that influence the survival and activity of the inoculum and bioavailability of the contaminant.

3.2. Field Mesocosm Experiments Using an Automated Fermenter

To study the effects of repeated inoculation on degradation of atrazine under field conditions, an experiment was conducted using soil mesocosms, each containing approximately 300 kg of a Dello loamy sand (typic Psammaquent, sandy alluvium; pH 8) contaminated with 100 mg kg^{-1} atrazine in the form of the commercial product, Aatrex. Prior to inoculation, the mesocosms were saturated with water and aged with repeated wet-dry cycles for three weeks. The treatment inoculum was prepared on site in a modified 567 L field fermenter (BioJect, EcoSoil Systems Inc., San Diego, California) batch reactor in 124 L batches. To produce the inoculum, irrigation quality water was passed through tandem, 100 μm and 10 μm filters and an ultraviolet (UV) sterilization unit before entering the fermenter. Float valves located inside the culture vessel maintained the high and low volume settings by turning on and off the water automatically during inoculation cycles. Two air compressor pumps maintained aeration of the inoculum and pressurized 19 L tanks containing concentrated atrazine-glucose-minimal salts media. During each inoculation cycle, timer controlled solenoid valves regulated the amount of concentrated media released to the fermenter vessel.

Atrazine degradation was compared for soils treated with 1, 4, or 8 applications of the atrazine degrading consortium over a 12-week period. To produce the inoculum, atrazine was added to the fermenter as a sole nitrogen source in glycerol media to yield a final concentration of 50 ppm in the medium. Fresh inoculum was used to purge the irrigation lines before each cycle to remove any residual inoculum existing from the previous

Table 1. Degradation of atrazine in soil mesocosms inoculated with single or repeated applications of an atrazine degrading bacterial consortium*

Inoculum Applications	% Atrazine degraded	Atrazine degraders log CU/g soil**
One	0 a	3.8
Four	38 a	3.4
Eight	72 b	5.8

*Inoculum was added at each inoculation to achieve a cell density of approximately 10^6 cfu/g soil. Soils were contaminated with 100 mg kg^{-1} atrazine. The experiment was conducted under ambient conditions at the University of California Agricultural Experiment Station during May through August 1997. Three treatments with four replicate each were organized in a randomized block design.
**CU, Consortium units determined by Most Probable Number assay

application. Applications occurred on average once every 1.5 weeks, with the entire experiment lasting 12 weeks.

Results of this experiment showed that as expected, a single application of the atrazine-degrading consortium did not significantly decrease the concentration of atrazine from 100 mg kg^{-1} over the time course of the experiment (Table 1). Mesocosms receiving 4 applications, alternating with water applications, showed a decrease in atrazine concentration to 62.0 ppm. However, due to high variability between replicates, the concentrations did not differ significantly from the single application treatment. In contrast, treatment of the mesocosms with 8 sequential applications of the degraders resulted in statistically significant removal of 72% of the atrazine after a 12-week period (Table 1). No atrazine metabolites were detected in any of the soils during the experiment.

The relationship between population density and degradation rates was not established in this experiment, but appeared to be dependent on the length of survival in the field between soil inoculation. During the experiment, population numbers of the consortium were monitored to examine population dynamics in relation to inoculation frequency. Most probable number assays of the soil prior to inoculation showed that culturable indigenous atrazine degraders were not present in any of the mesocosms. During the experiment, consortium cells were added to the mesocosms at an average density of 10^5 g^{-1} moist soil at each inoculation. During the 12 week period, population densities of atrazine degraders in the mesocosms that received either a single or four applications of the inoculum maintained approximately $10^{3.8}$ consortium units per gram soil. In comparison, atrazine degraders in the mesocosms that were repeatedly inoculated 8 times maintained cell densities of $10^{5.8}$ consortium units, or almost 3 log units greater than the lower inoculation frequency treatments.

Future experiments include use of the fermenter with the *Pseudomonas* strain ADP degrader to compare its performance with the consortium that was used in the first field experiment. In the microcosm study, the sole degrader performed better than the consortium, which is probably related to elimination of the necessity for diffusion transfer of the intermediate metabolite between the two strains in the bacterial consortium. Ideally, the best system would employ a degrader organism that is constitutive in expressing the atrazine degradation genes, so that atrazine would not have to be used in the growth medium in the field fermenter.

4. BIOREMEDIATION OF PCBs

Biodegradation of PCBs by aerobic bacteria is well-established (Abramowicz, 1990; Focht, 1993), but has not been developed as a bioremediation technology for PCBs because of technical constraints. The initial metabolism of PCBs is cometabolic, meaning that the organisms responsible for PCB transformation gain no energy from the process; hence, the organisms must be sustained by an alternative compound that serves as the growth substrate. Under laboratory conditions, Brunner and Focht (1985) showed that addition of *Acinetobacter* stimulates the rate of PCB mineralization. However, the most detailed field study that has been conducted to date (Harkness *et al.*, 1993), showed that PCB-degrading bacteria inoculated into contaminated river sediments as a single application declined by three orders of magnitude over a 10-day period, and thus may not be practical for bioremediation, depending on the dilution rate of the inoculum in soil. This is similar to the results reported above for soils treated with a single application of atrazine degraders.

One of the main limitations in developing biostimulation or bioaugmentation methods for cleanup of PCBs has been the lack of substrates other than biphenyl that can be used as soil amendments or in a bioreactor to support the growth and induction of PCB-degrading bacteria. Biphenyl, the nonchlorinated analog of polychlorinated biphenyls, is the most widely-studied inducer, and biphenyl-utilizing bacteria have been shown to cometabolize PCBs in both liquid media (Bedard *et al.*, 1986; Kohler *et al.*, 1988) and in soil (Brunner *et al.*, 1985; Barriault and Sylvestre, 1993). Biphenyl is not practical as a soil amendment, since it is classified as a toxic substance, is expensive, and has low water solubility.

Recently, alternative compounds that induce PCB cometabolism have been identified, including ethylbenzene (Seto *et al.*, 1995), naphthalene (Pellizari *et al.*, 1996), and monoterpenes such as carvone (Gilbert and Crowley, 1997). The identification of these new inducing compounds may facilitate bioremediation of PCBs, but has not yet been tested in the field. A structure-activity analysis showed that among the monoterpenes, chemicals with an unsaturated *p*-menthane structural motif cause the strongest cometabolism activity (Gilbert and Crowley, 1997), suggesting that plant-derived isoprenoids may be natural substrates for the enzymes encoded by the *bph* genes in Gram positive bacteria.

The monoterpene, carvone, which is the principal component of spearmint oil, has several properties that make it particularly useful for promoting PCB biodegradation. It is moderately water soluble, inexpensive, and effective at concentrations as low as 50 mg L^{-1}. This substance is a component of spearmint, caraway and cumin, and is registered as a food additive (CAS 2244–16–8). In prior research, induction of resting cells of *Arthrobacter* sp. strain B1B with carvone, resulted in degradation of over 26 major components of Aroclor 1242, including 46 ± 3% and 14 ± 6% of the tetrachloro- and pentachlorobiphenyl congeners, respectively (Gilbert and Crowley, 1997). In comparison to other known inducers with lower water solubility, carvone could be particularly useful for culturing PCB degrader organisms in a fermenter, or for direct application of the chemical to contaminated soils to enhance the degradation activity of indigenous PCB degraders.

4.1. Bioaugmentation of PCB Contaminated Soils Using Repeated Inoculations

Results of laboratory experiments examining repeated applications of *Arthrobacter* B1B are shown in Table 2, which also shows the effects of using carvone-fructose medium for biostimulation of the indigenous degraders in soil contaminated with Aroclor 1242, a commercial PCB mixture (Gilbert and Crowley, 1998). Results of these experiments showed that a single application of carvone induced *Arthrobacter* B1B or of carvone-fruc-

Table 2. Percent degradation of 100 ppm Aroclor 1242 after 9 wks in soil treated with single or repeated applications of carvone-induced *Arthrobacter* B1B or carvone fructose medium

	PCB congener				
Treatment	di-CBs[b]	tri-CBs	tetra-CBs	penta-CBs	total PCBs
Carvone-induced cells, repeat application	88±3 a	41±7 a	11±5 a	3±8 a	27±6 a
Non-induced cells, repeat application	54±9 b	17±7 c	4±5 ab	3±9 a	12±6 b
Fructose & carvone only, repeat application	34±25 c	13±8 cd	6±3 a	5±7 a	10±5 bc
Carvone-induced cells, single application	10±17 d	0±6 e	-1±4 b	2±8 a	0±5 d

[a]Within columns, degradation rates which are marked by the same letter are not significantly different; $p < 0.05$. Statistical analysis by Fischer's least significant difference test.
[b]CB: chlorobiphenyl.

tose medium to soil was ineffective for causing significant PCB degradation. However, repeated applications of the bacteria or carvone both caused the disappearance of PCBs. The most effective treatment was addition of carvone-induced bacteria. After 17 applications, 88% of the dichlorinated congeners were degraded, and 27% of the total PCB was mineralized by the bacteria. These rates of mineralization may be conservative, as they were conducted in soil with no added surfactant to increase PCB bioavailability.

4.2. Biostimulation of Indigenous PCB Degraders Using Repeated Substrate Applications

In addition to using repeated applications of induced bacteria, recent studies suggests that it may be possible to use repeated applications of carvone for biostimulation of indigenous PCB degraders. To identify PCB degrader populations in soils, experiments have since been conducted using carvone or biphenyl soil enrichment. After growth on these substrates, total DNA was isolated from the enrichment culture and subjected to analysis using universal PCR primers to amplify the 16s rDNA, which was then used to identify the predominant culturable PCB degrading bacteria (Figure 2). To identify the

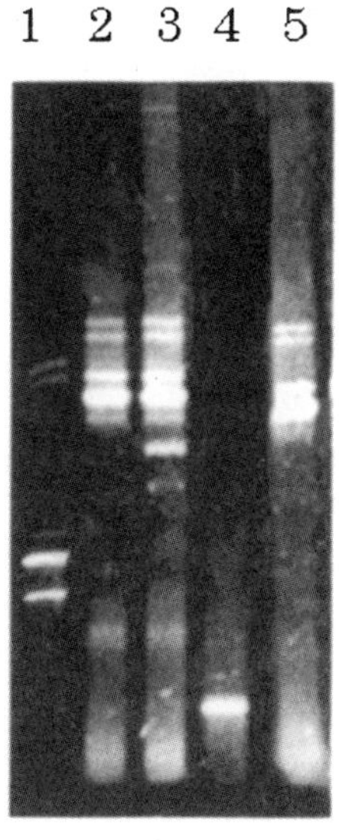

Figure 2. Microbial community DNA analysis using denaturing gradient gel electrophoresis to identify bacteria that grow on carvone and degrade biphenyl. In this experiment, total DNA was isolated from a Dello loamy sand with no known prior exposure to biphenyl, before and after enrichment culture with carvone as a sole carbon source. Lane 1 shows the predominance of two organisms in this soil sample with two other barely visible bands that represent two less predominant bacterial species. The soil was then placed into liquid medium to enrich for carvone- and biphenyl-degrading organisms. The bands in lanes 2 and 3 correspond to individual microbial species that were enriched from the soil after growth in soil suspensions containing biphenyl (lane 2) and carvone (lane 3) as sole carbon sources for bacterial growth. Lane 4 shows the 16s rDNA band position for Arthrobacter B1B. Lane 5 is DNA from bacteria in the carvone enrichment culture that could also grow on biphenyl as a carbon source.

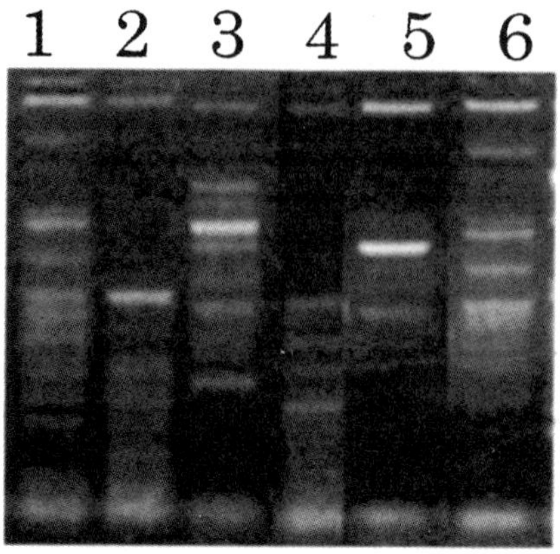

Figure 3. Identification of unique culturable biphenyl-degrading bacteria species having indistinguishable colony morphologies that grow on both carvone and biphenyl. DNA banding patterns generated by PCR amplification of 16S-5S intergenic DNA and enterobacterial repetitive intergenic consensus (ERIC) analysis. Banding patterns were matched to strains cultured on agar media and may be used to develop probes for tracking the population dynamics of these bacteria in the field. Strains shown here were identified by fatty acid methyl ester analysis as indicated by corresponding numbers for each lane.

bacteria, the PCR product DNA, which is a mixture of the same size 16s rDNA from all of the bacteria was separated into distinct bands using denaturing gradient gel electrophoresis (DGGE). The resulting bands represent individual microorganisms, which were matched to band patterns of microorganisms plated on carvone or biphenyl minimal salts agar media. The bands also can be excised from the gel for sequencing and taxonomic identification. Another approach for identification bacterial isolates having indistinguishable colony morphologies on agar media involves the use of enterobacterial repetitive intergenic consensus sequences (ERIC), which generates unique DNA band patterns for each strain (Figure 3).

5. SUMMARY

One of the greatest concerns with environmental contamination has been the lack of methods that can be used to treat low levels of soil contaminants that are dispersed over large land areas. This problem is particularly acute for compounds such as atrazine or polychlorinated biphenyls. When present at high concentrations, these contaminants can be more rapidly treated by excavation and landfilling or by methods such as thermal desorption and vacuum extraction. In contrast, at low concentrations there are no economically viable cleanup methods other than natural attenuation. This has led to the search for low cost in situ treatment methods involving biostimulation and bioaugmentation. Until recently, this latter method has been considered by many to be unreliable due to the problems associated with poor survival and activity of introduced microorganisms as well as their poor transport and distribution in soils. Additionally, the cost of producing, storing, and transporting large quantities of inoculum has precluded bioaugmentation of large land areas.

As suggested by this research, many of the above problems with bioaugmentation may be solved by using repeated inundative applications of bacteria with an automated field fermenter. This approach permits the temporary maintenance of high population densities of active degrader organisms which only need to survive between the application cycles. The use of on site equipment also provides a cost effective method for producing viable cells without the need for off site production, storage, and transport to the field. Yet another application of the automated fermenter would be to produce and apply atrazine degrading bacteria through irrigation water to cause mineralization of residual atrazine after it has been applied for weed control. With field crops that are irrigated, bioaugmentation might permit continued use of this herbicide in fields and orchards that are now subject to possible groundwater contamination.

To monitor the success of biostimulation and bioaugmentation, it is necessary to track the population of degrading bacteria, and their responses to various management treatments. New molecular ecology techniques such as those explored here offer opportunities for monitoring microbial communities, and for evaluating the impact of the introduced microorganisms versus the effects of bioaugmentation on the indigenous community. If repeated applications prove effective for bioaugmentation, there are many other factors that may be further optimized to accelerate the bioremediation process. For example, surfactants may be added along with the bacteria to increase contaminant bioavailability. Combinations of bacteria may need to be used in series to target different organic contaminants and their metabolites. In long term remediation scenarios, it may also be necessary to consider plant microbial systems in conjunction with bioaugmentation, or methods to enhance bacterial mixing in the soil profile. Still another strategy might involve use of repeated applications of bacteria containing degradative plasmids to modify the indigenous soil population. Certainly, in the development of any future strategy involving bioaugmentation, the ability to produce large quantities of inoculum on site and repeatedly inoculate the soil will provide an improved ability to manipulate the soil microbial community and to enhance the bioremediation of soil contaminants.

REFERENCES

Abramowicz, D.A. (1990). Aerobic and anaerobic biodegradation of PCBs: a review. *Crit Rev Biotechnol* **10**:241–251

Alexander, M. (1994). Inoculation. p. 226–245. In Biodegradation and Bioremediation. Academic Press, San Diego, CA.

Alvey, S.A. and Crowley, D.E. (1996). Survival and activity of an atrazine-mineralizing bacterial consortium in rhizosphere soil. *Environ Sci Technol* **30**:1596–1603.

Alvey, S.A. and Crowley, D.E. (1995). Influence of organic amendments on biodegradation of atrazine as a nitrogen-source. *J Environ Qual* **24**:1156–1162

Baud-Grasset, F. and Vogel, T.M. (1995). Bioaugmentation: Biotreatment of contaminated soil by adding adapted bacteria. p. 39–49. In R.E. Hinchee, J. Fredrickson and B.C. Alleman (eds.), *Bioaugmentation for Site Remediation*. Battelle Press, Columbus, OH

Barriault D. and Sylverstre M (1993). Factors affecting PCB degradation by an implanted bacterial strain in soil microcosms. *Can J Microbiol* **39**:594–602

Bedard D.L., Unterman, R.L., Bopp, L.H., Brennan, M.J., Haberl, M.L. and Johnson, C. (1986). Rapid assay for screening and characterizing microorganisms for the ability to degrade polychlorinated biphenyls. *Appl Environ Microbiol* **51**:761–768

Belluck, D.A., Benamin, S.L., and Dawson, T. (1991). Groundwater contamination by atrazine and its metabolites: risk assessment, policy, and legal implications. p. 254–273. In L. Somasundaram and J.R. Coats (eds), Pesticide transromation products: fate and significance in the environment. American Chemical Society, Washington, D.C.

Brunner. W., Sutherland, F.H., and Focht D.D. (1985). Enhanced biodegradation of polychlorinated biphenyls in soil by analog enrichment and bacterial inoculation. *J Environ Qual* **14**: 324–328

Crowley, D., Alvey, S., and Gilbert, E. (1997). Rhizosphere ecology of xenobiotic degrading microorganisms. p 20–38. In E. Kruger, T. Anderson, J. Coats, (eds.) Phytoremediation of Soil and Water Contaminants. American Chemical Society, Washington, D.C.

de Souza, M.L., Wackett, L.P., Boundy-Mills, K.L., Mandelbaum, R.T., and Sadowsky, M.J. 1995. Cloning, characterization, and expression of a gene region from *Pseudomonas* sp. strain ADP involved in the dechlorination of atrazine. *Appl Environ Microbiol* **61**:9:3373–3378.

de Souza, M.L., Newcombe, D., Alvey, S., Crowley, D.E., Hay, A., Sadowsky, M.J., and Wackett, L.P. (1997). Molecular basis of a bacterial consortium: Interspecies catabolism of atrazine. *Appl Environ Microbiol* **64**:178–184.

Digiovani, G.D., Neilson, J.W, Pepper, I.L, and Sinclair, N.A. (1996). Gene transfer of *Alcaligenes eutrophus* JMP 134 plasmid pJP4 to indigenous soil recipients. *Appl Env Microbiol* **62**: 2521–2526

Dousset, S., Mouvet, C. and Schiavon, M. (1997). Degradation of [^{14}C]-terbuthylazine and [^{14}C]-atrazine in laboratory soil mirocosms. *Pest Sci* **49**: 9–16

Duba, A.G. Jackson, K.J., Jovanovich, M.C., Knapp, R.B., and Taylor, R.T. (1996). TCE remediation using in situ, resting-state bioaugmentation. *Env Sci Technol* **30**: 1982–1989

Focht, D.D. (1993). Microbial degradation of chlorinated biphenyls. p 341–407. In Soil Biochemistry. Marcel Dekker, Inc. New York

Forsythe, J.V., Tsao, Y.M., and Bleam, R.D. (1995). Bioremediation: When is augmentation needed? p. 1–15. In R.E Hinchee, J. Fredrickson and B.C. Alleman (eds.), *Bioaugmentation for Site Remediation.* Battelle Press, Columbus OH,.

Gilbert, E.S., and Crowley, D.E. (1997). Plant compounds that induce polychlorinated biphenyl biodegradation by *Arthrobacter* sp strain B1B. *Appl Environ Microbiol* **63**: 1933–1938

Gilbert, E.S., and Crowley, D.E. (1998). Repeated application of carvone induced bacteria to enhance biodegradation of polychlorinated biphenyls in soil. *Appl Microbiol Biotechnol* in press.

Grigg, B.C., assaf, N.A., and Turco, R.F. (1997). Removal of atrazine contamination in soil and liquid systems using bioaugmentation. *Pest Sci* **50**: 211–220

Harkness, M.R., McDermott, J.B., Abramowicz, D.A., Salvo, J.J., Flanagan, W.P., Stephens, J.L., Mondello, F.J., May, R.J., Lobos, J.H., Carroll, K.M., Brennan, M.J., Bracco, A.A., Fish, K.M., Warner, G.L., Wilson, P.R., Dietrich, D.K, Lin, D.T., Morgan, C.B., and Gately W.L. (1993). In situ stimulation of aerobic PCB biodegradation in Hudson River sediments. *Science* **259**: 503–507

Kishore, P.G., Green, R.E. (1971). Microbial degradation of the herbicide atrazine an its 2-hydroxy analog in submerged soils. *Environ Sci Technol* **5**: 426–429

Kohler, H.P.E., Kohler-Staub, D., and Focht, D.D. (1988). Cometabolism of polychlorinated biphenyls: enhanced transformation of Aroclor 1254 by growing bacterial cells. *Appl Environ Microbiol* **54**: 1940–1945

Lajoie, C A; Layton, A C; Sayler, G S. (1994). Cometabolic oxidation of polychlorinated biphenyls in soil with a surfactant-based field application vector. *Appl Environ Microbiol* **60**: 2826–2833

Lamar, R.T., Davis, M.W., Dietrich, D.M., and Glaser, J.A. (1994). Treatment of a pentachlorophenol and creosote-contaminated soil using the lignin-degrading fungus *Phanerochaete sordida*: a field demonstration. *Soil Biol Biochem* **26** 1603–1611

Mandelbaum, R.T., Allan, D.L. and Wackett, L.P. (1995). Isolation and characterization of a *Pseudomonas* sp. that mineralizes the *s*-triazine herbicide atrazine. *Appl Environ Microbiol* **61**:4:1451–1457.

Mandelbaum, R.T., Wackett, L.P., and Allan, D.L. (1993). Mineralization of the s-triazine ring of atrazine by stable bacterial mixed cultures. *Appl Environ Microbiol* **59**: 1695–1701

Mandelbaum, R.T., Wackett, L.P. and Allan, D.L. (1995). Isolation and characterization of a *Pseudomonas* sp that mineralizes the s-trizine herbicide atrazine. *Appl Environ Microbi9ol* **61**: 1451–1457

Mayotte, T.J., Dybas, M.J., and Criddle, C.S. (1996). Bench scale evaluation of bioaugmentation to remediate carbon tetrachloride-contaminated aquifer materials. *Ground Water* **34**:358–367

McFarland, M.J., Salladay, D., Ash, D., and Baiden, E. (1996). Composting treatment of alachlor impacted soil amended with the white rot fungus: *Phanerochaete chrysoporium. Hazardous Waste & Hazardous Materials* **13**: 363–373.

Munakata-Marr, J., McCarty, P.L., Sheilds, M.S., Reagin, M., and Francesconi, S.C. (1996). Enhancement of trichloroethylene degradation in aquifer microcosms bioaugmented with wild type and genetically altered *Burkholderia* (*Pseudomonas*) *cepacia* G4 and PR1. *Env Sci Technol* **30**:2045–2052

Otte, M-P, Gagnon, J., Comeau, Y., Matte, N., Greer, C.W., and Samson, R. (1994). Activiation of an indigenous microbial consortium for bioaugmentation of pentachlorophenol-creosote contaminated soils. *Appl Microbiol Biotechnol* **40**: 926–932

Pellizari, V.H., Bezborodnikov, S., Quensen, J.F. III, and Tiedje, J.M. (1996). Evaluation of strains isolated by growth on napthalene and biphenyl for hybridization of genes to dioxygenase probes and polychlorinated biphenyl-degrading abiity. *Appl Environ Microbiol* **62**: 2053–2058

Radosevich, M., Traina, S.J., and Tuovinen, O.H. (1997). Atrazine mineralization in laboratory-aged soil microcosms inoculated with *s*-triazine-degrading bacteria. *J Environ Qual* **26**:206–214.

Radosevich, M., Traina, S.J., Hao, Y., and Tuovinen, O.H. (1995). Degradation and mineralization of atrazine by a soil bacterial isolate. *Appl Environ Microbiol* **61**: 297–301

Seto, M., Kimbara, K., Shimura, M., Hatta, T., Fukuda, M., and Keiji, Y. (1995). A novel transformation of polychlorinated biphenyls by *Rhodcoccus* sp strain RHA1. *Appl Environ Microbiol* **61**: 3353–3358

U.S. Environmental Protection Agency. (1990). 16 p In Summary Results of EPA's National Survey of Pesticides in Drinking Water Wells. Office of Pesticides and Toxic Substances. Washington, D.C.

Van Veen, J.A., Van Overbeek, L.S., and Van Elsas, J.D. (1997). Fate and activity of microorganisms introduced into soil. *Microbiol Molec Biol Rev* **61**:121–135.

Vogel, T.M. (1996). Bioaugmentation as a soil bioremediation approach. *Current Opinion Biotechnol* **7**:311–316.

Yanze Kontchou, C., and Gschwind., N. (1994). Mineralization of the herbicide atrazine as a carbon source by a *Pseudomonas* strain. *Appl Environ Microbiol* **60**

27

NATURAL ATTENUATION OF AROMATIC COMPOUNDS AND HEAVY METALS IN A LACUSTRINE SAND AQUIFER AT VILLA FARM, UK

Geoff Williams

British Geological Survey
Nottingham, England

ABSTRACT

Excavations in unconsolidated lacustrine sand have been used for over 30 years for the disposal of industrial liquids and sludges containing heavy metals, oils, solvents, acids and alkalis. Wastes have been deposited directly into the saturated zone and a plume extends more than 300 m from the site. Contaminants have stratified at the base of the aquifer approx. 10 m below ground level, where their movement is controlled by the shape of the underlying clay surface. Depth specific sampling at a number of times over a 20 year period has provided data on the movement of the plume and natural attenuation reactions within it. There is a vertical redox zonation between the contaminant plume and the overlying groundwater with methanogenesis at the base, leading upwards through zones of sulphate reduction, iron reduction, manganese reduction and denitrification. Evidence for natural attenuation includes a loss of organic carbon from the plume relative to the conservative species chloride, and the consumption of electron acceptors. Additional evidence includes the presence of a diverse microbial population, intermediate breakdown products, and methane and carbon dioxide. The degradation of dissolved organic carbon can be fitted to a first order rate with a half life of 7.2 years (or decay constant of 0.096 year^{-1}). Attempts to measure degradation by other methods including in situ tracer tests, in situ microcosms, and laboratory batch reactors have been poor because of the slow rate of decay.

A subsidiary active redox zone occurs above the main plume 250 m away from the lagoon due to the upward movement of methane from the polluted groundwater as it rises to the surface and the overall pressure is reduced. Spatial changes in $^{13}C/^{12}C$ and $^{2}H/^{1}H$ in dissolved methane, and carbon dioxide, lend support to anaerobic methane oxidation coupled with sulphate reduction.

Novel Approaches for Bioremediation of Organic Pollution, edited by
Fass *et al.* Kluwer Academic / Plenum Publishers, New York, 1999.

Modelling speciation using the geochemical code EQ3, suggests that Fe and Ni should be precipitated as sulphides and carbonates. The anomalously high aqueous phase concentrations found in the methanogenic (but sulphur depleted zone) are considered to be due to complexation with organics. Complexes of Ni with "fulvic" type material have been identified by size exclusion chromatography and ultra-filtration methods.

1. INTRODUCTION

In order to identify processes controlling contaminant migration in groundwater, the British Geological Survey has been studying the pollution plume developed in a shallow sand aquifer at Villa Farm, 15 km north of Coventry, England. The site consists of a series of disposal lagoons, covering an area of about 2,500 m^2, developed in excavations for sand which intercept the water table. The site fell into disuse in 1981 after receiving a wide variety of industrial wastes for over 30 years. The major waste types include oil/water mixtures, effluent treatment sludges containing heavy metals, acids and alkalis, organic solvents and paint wastes. The distribution of contaminants has been investigated by drilling approximately 100 boreholes to determine chemistry profiles with depth in the aquifer (Williams et al, 1992).

2. SITE DESCRIPTION

2.1. Geology and Hydrogeology

The geology is composed of Quaternary sediments up to 50 m thick which overlie Triassic Mercia Mudstone. Most of the Quaternary deposits consist of relatively impermeable clays overlain by a lacustrine unconsolidated sand (Wolston Sand) which forms a shallow aquifer between 3 and 12 m thick. The Wolston Sand is unconsolidated, relatively well sorted with a median diameter of between 125–350 μm, and consists of clear sub-angular grains of quartz with subordinate feldspars calcite and dolomite, and allocthonous coal fragments. Iron oxy-hydroxides give the sand its characteristic red colour and clay minerals (up to 15%) include illite and kaolinite.

The sand overlies an undulating surface of clay, so that the elevation of the clay determines to a large extent the transmissivity of the sand, thus controlling the groundwater flow direction and migration of contaminants (Mackay et al, 1985). Groundwater flows generally westwards from the site at a velocity of up to 0.3 m.d^{-1}. During disposal operations up to 7,000 m^3.year^{-1} of liquid wastes were discharged producing a recharge mound 3 m above the natural water table, giving rise to radial and downward flow from the lagoons. Very little direct infiltration occurs in the area because of a surface layer of glacial till 1–4 m thick.

Boreholes drilled to define the pollution plume, have been completed in a number of ways to allow sampling at discrete depths. Multi-level samplers allow water to be drawn in by suction from 0.2 m depth intervals throughout the aquifer, and these have formed the principal sampling method. Multi-port screens which have 30 cm length screens at 1m intervals in a 100 mm diameter plain PVC casing have been used to obtain water samples under pressure with a slim submersible pump inserted into a double packer assembly across the screen.

2.2. Inorganic Groundwater Chemistry

The plume of contamination has been detected in boreholes over 300 m to the west of the site (Figure 1). Five vertical profiles of porewater chemistry indicate a gradual tran-

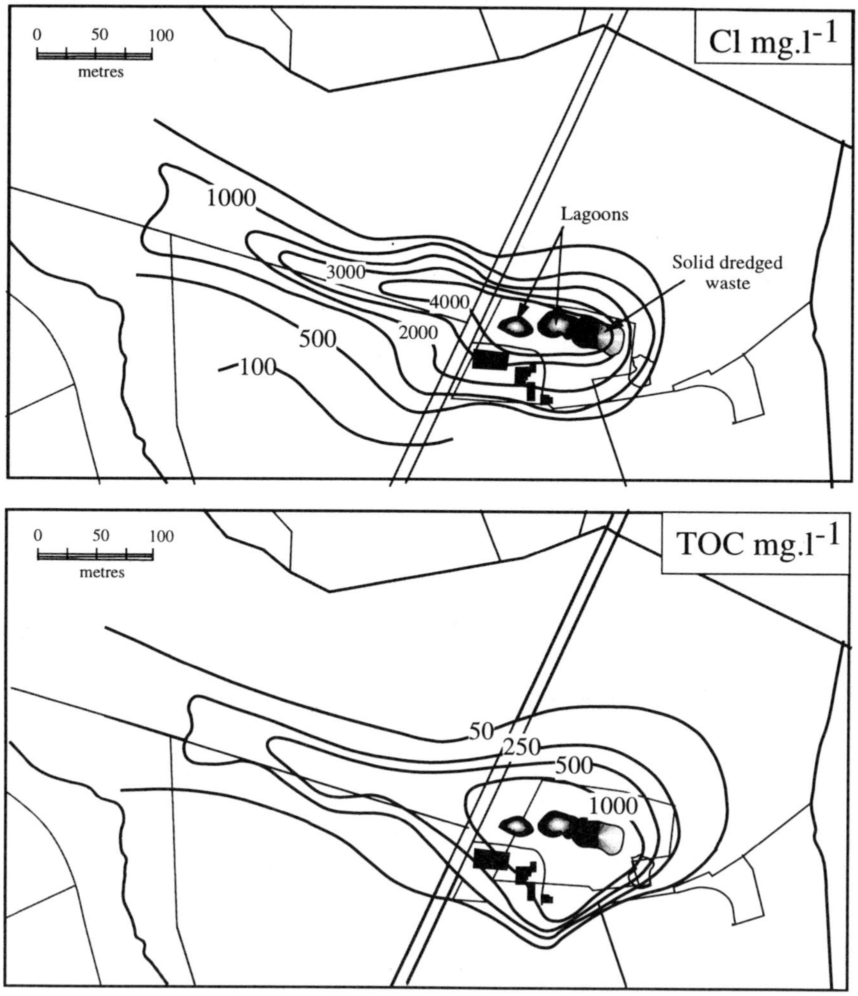

Figure 1. Groundwater plumes of chloride and total organic carbon in 1984.

sition from a heavily polluted and highly reducing zone near the lagoons and at the base of the aquifer, to less reducing and much less contaminated in the natural overlying groundwater. In this transition the predicted order of redox reactions have been identified, namely, aerobic respiration, denitrification, manganese reduction, nitrate reduction, iron reduction, sulphate reduction, and finally methanogenesis (Stumm and Morgan, 1981). Distributions of important and redox sensitive species (NO_3, Mn(II), Fe(III) and SO_4) are displayed in Figure 2.

2.3. Microbiological Characterisation

Total numbers of aerobes and anaerobes have been determined in samples of sediments taken from the aquifer (Dhillon et al, 1989). Both range from 10^4 -10^6 cfu.g^{-1} sediment. Microbial isolates cultured aerobically from sediments in the pollution plume were identified and found to constitute a wide range of organisms (Figure 3, Hitchman and Wil-

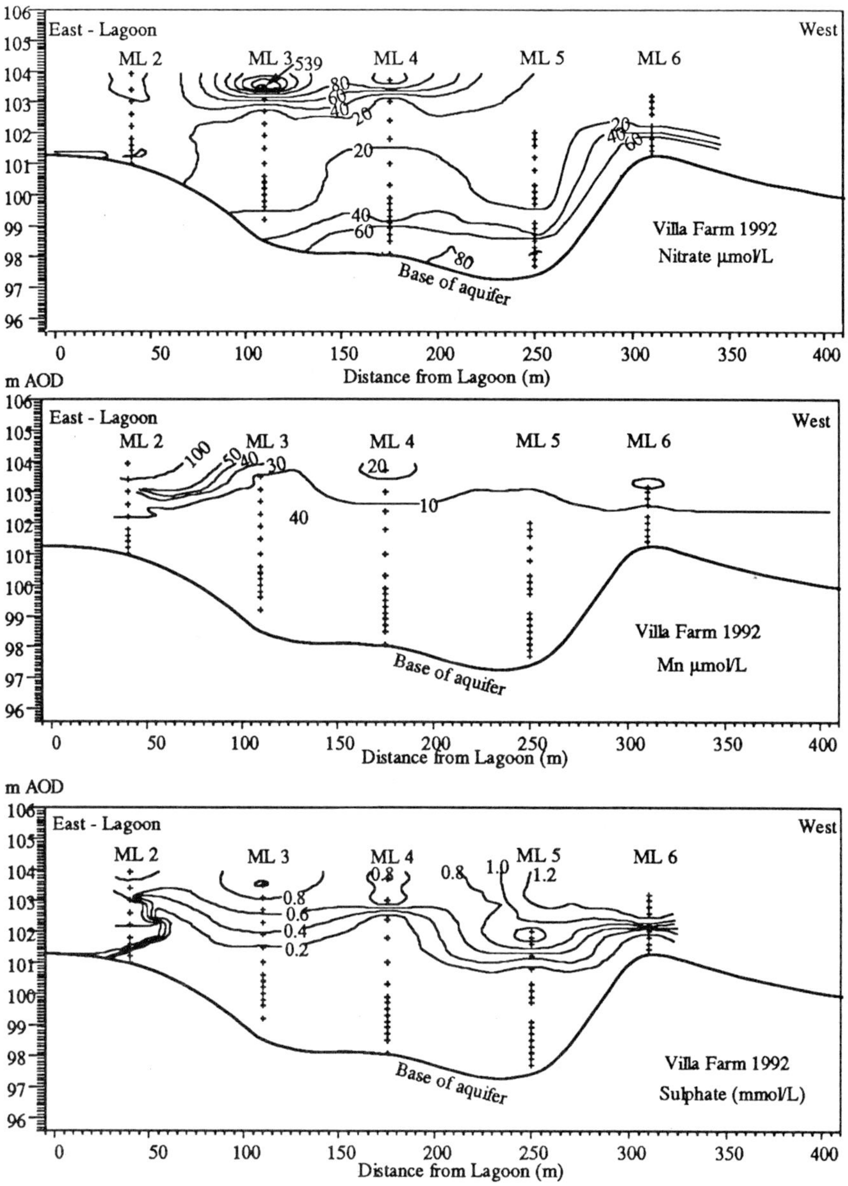

Figure 2. Cross section along plume in 1992 showing redox species.

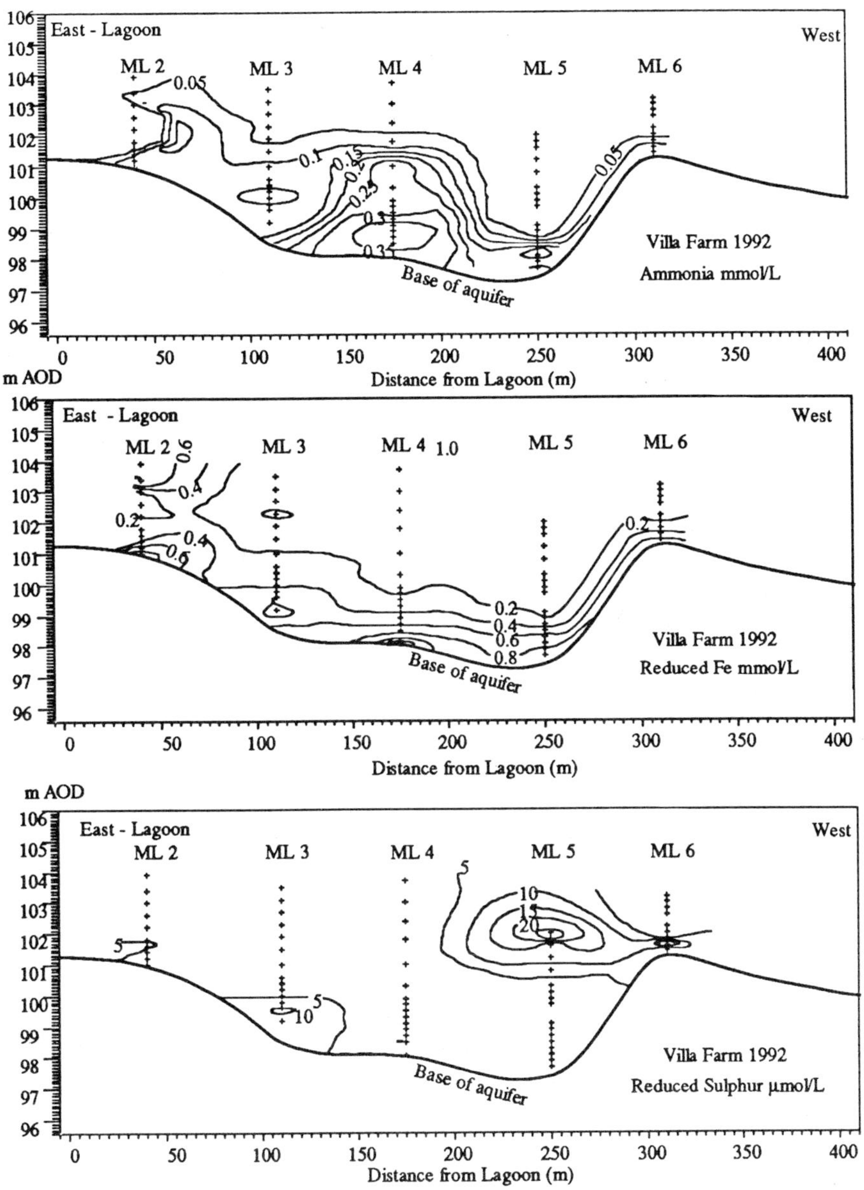

Figure 2. *Continued.*

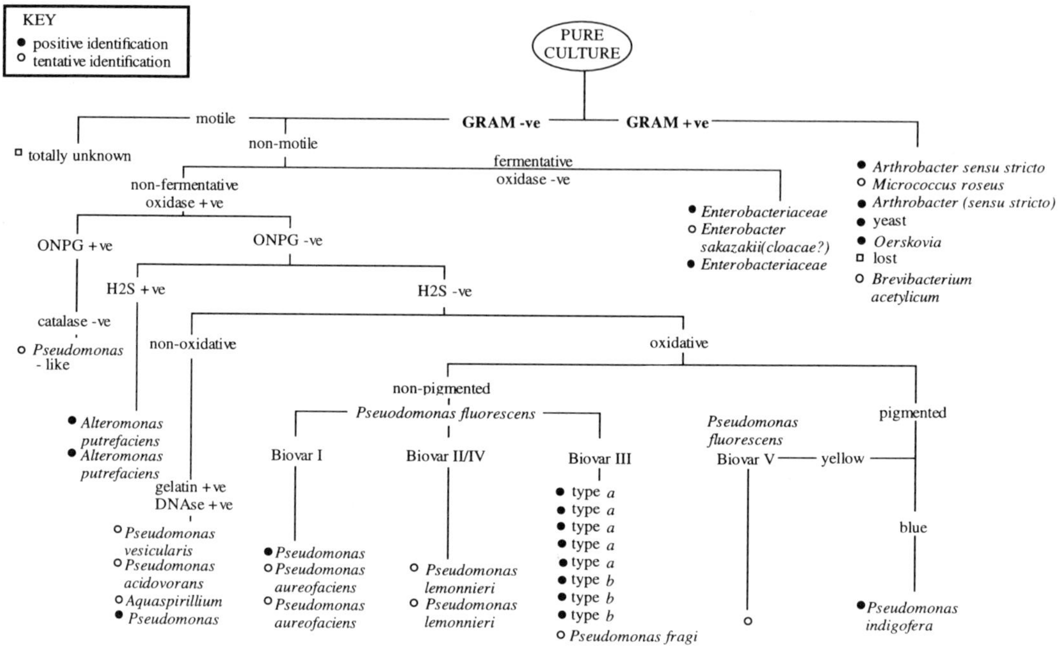

Figure 3. Identification of aerobic isolates (by NCIMB).

liams, 1988). While the performance of a specific microbes has not been studied, there is obviously an indigenous consortium of organisms present in the aquifer.

3. DISCUSSION

3.1. Redox Chemistry

The geochemical profiles augmented by soil gas measurements above the water table have been integrated to provide a conceptual model of the redox zones (Figure 4). In the redox model, absolute concentrations of redox species are not necessarily adequate for allocating a particular water sample to a particular zone. Rather it is the systematic change in redox species with depth or distance from the contaminant source that have been used to define the redox zonation. Methanogenic conditions exist at the base of the aquifer with a transition upwards through sulphate reducing, iron reducing, manganese reducing and denitrifying conditions. A sulphide rich zone exists where there is little dissolved organic carbon in an area of the plume where groundwater flows upwards over a topographic high at the base of the aquifer. Here methane is found in the soil atmosphere above the water table and it is considered that methane is exsolving and moving upwards and initiating secondary redox reactions. This also suggests that methane is acting as a primary substrate initiating reduction of the electron acceptors in the less reduced aquifer above. This is discussed in the next section.

3.2. Evidence for Anaerobic Methane Oxidation

The analysis of methane and carbon dioxide in the soil atmosphere above the water table shows variable methane concentrations not correlated with the plume concentration

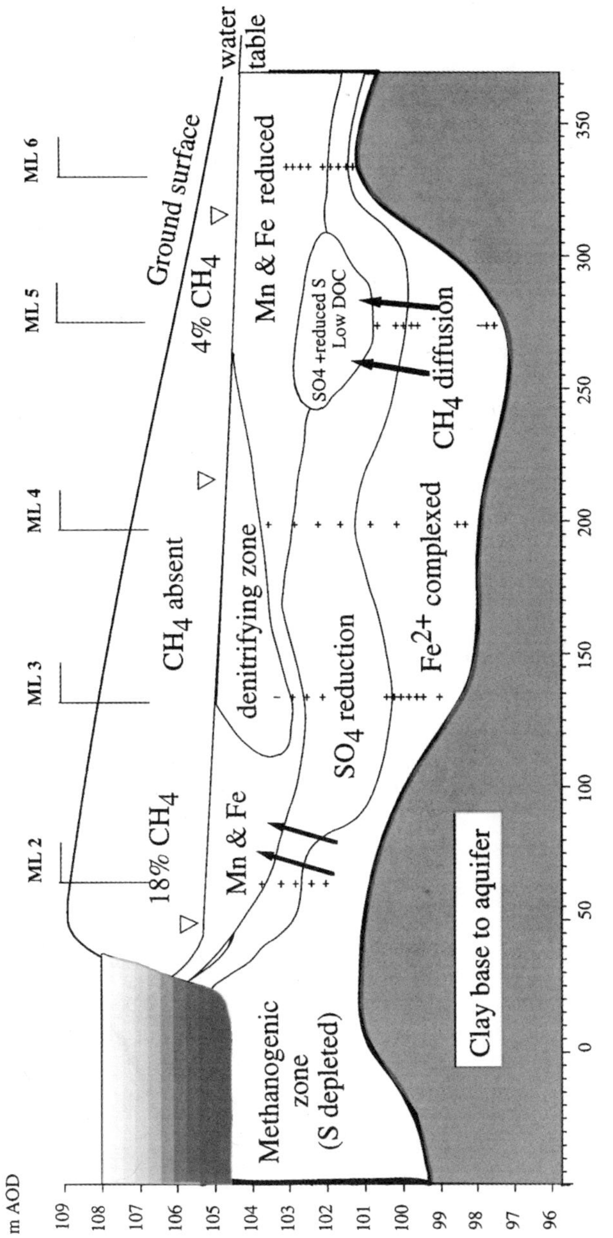

Figure 4. Redox zones in 1992 inferred from geochemical profiles in samplers ML2–ML6.

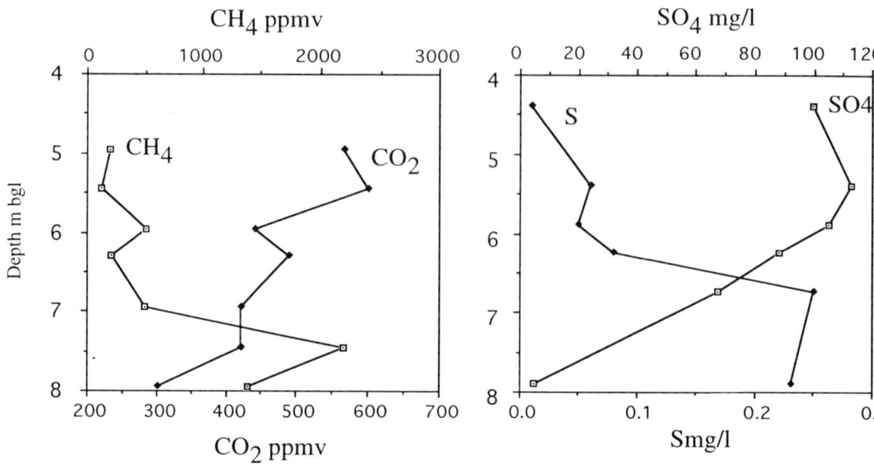

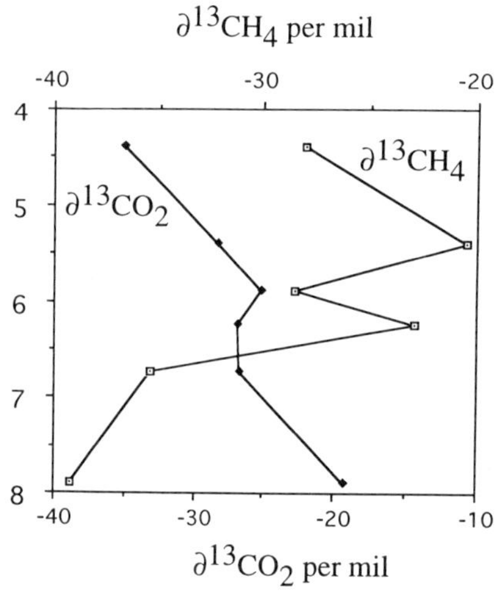

Figure 5. Dissolved gases, carbon isotope ratios and sulphate/sulphide profiles through the plume.

of TOC. The variable distribution of methane in the soil atmosphere may be explained if methane moving upwards through the water column from the base of the aquifer is oxidised to carbon dioxide. The inverse relationship between dissolved methane and dissolved carbon dioxide (Figure 5) suggests that, if this were the case, methane is being oxidised under anaerobic conditions within the zone of sulphate reduction (Hitchman et al, 1990).

The evidence to support anaerobic methane oxidation in anaerobic systems where there is an excess of sulphate, appears to be growing, although microbiologists are still uncertain of the mechanisms involved. The process has been studied mainly in marine, marsh and lake sediments (e.g. Alperin and Reeburgh, 1984; Panganiban et al, 1979), and has been reported in only a few cases in aquifers affected by landfill leachate. Barker and Fritz (1981) found methane only in areas above the groundwater pollution plume from a landfill site where contamination was slight, while Barber and Davis (1986) reported methane only in the most polluted areas. Dalton and Hocknall (1990) reviewed the thermodynamics for anaerobic methane oxidation, and concluded that the process is probably linked to consortium of bacteria which carry out a range of other reduction processes which produce energy. If the energy released is coupled with anaerobic methane oxidation then it could at least give a theoretical basis for the mechanism. It seems that methane itself is not an energy source in the anaerobic oxidation process, since this reaction is thermodynamically unfavourable:

$$CH_4 + 2\,H_2O = CO_2 + 4\,H_2\ \Delta G^{o\prime} + 31.24\ \text{kcal.mol}^{-1}$$

However, it is may be possible for methane to be oxidised if coupled with an energy producing reaction such as sulphate reduction:-

$$CH_4 + SO_4^{2-} + H^+ = HS^- + 2H_2O + CO_2\ \Delta G^{o\prime} - 2.8\ \text{kcal.mol}^{-1}$$

Sulphate reducers are known to utilise hydrogen, so the first reaction becomes potentially more favourable as the hydrogen product is removed. The use of other electron acceptors apart from sulphate such as Fe(III), Mn(IV) and NO_3 is also possible.

Field evidence for anaerobic methane oxidation is provided by the stable carbon isotope data for methane and carbon dioxide (Figure 5). These depth profiles are consistent with the theory that, if methane is metabolised, then residual methane becomes isotopically heavier due to microbial fractionation, and the metabolic product becomes isotopically lighter. The extent of the increase in $\delta^{13}C_{CH_4}$ is around 20‰. Laboratory confirmation of anaerobic methane oxidation has also been provided by Dalton and Hocknall, (1991), who incubated sediment cores from Villa Farm and identified a level of oxidation of between 3 and 40 μmol.g-1 sediment.

3.3. Attenuation of Dissolved Organic Contaminants

The plume has been characterised using GC-MS techniques and found to contain a diverse range of organic compounds (Williams et al, 1984; 1992). Even so only a small proportion of the TOC has been identified and there are significant unknown or unresolved peaks on the chromatograph. The chemical or biological attenuation of TOC has been studied by reference to chloride which is a major input to the site and which forms a useful conservative tracer only being attenuated by dilution and dispersion. The use of laboratory microcosms which simulate physico-chemical conditions in the field are very dependent on the degradation rate. Typically these experiments last several months and changes have only be detected in aerobic systems, while no significant loss of organic occurs anaerobically (Higgo et al. 1996). However, long term monitoring of the plume has shown that there is a systematic change in the TOC/Cl ratio with time (Figure 6). This ratio can be fitted to a first order rate, which suggests that the TOC is attenuated with an effective half life of 7.2 years or decay constant of 0.096 year^{-1}. The importance of this measurement should not be overlooked and provides the basis for predicting future contaminant distributions from the site, and assessing the long term environmental impact.

3.4. Fate of Heavy Metals

Heavy metal sludges have also formed a major input into the lagoon but in general they are strongly attenuated in the plume. Nickel however, is mobile and persists at concentrations of 0.1–0.5 mgL^{-1} within the plume. Geochemical modelling using the speciation code EQ3 (Wolery, 1983) and assuming thermodynamic equilibrium indicates that the

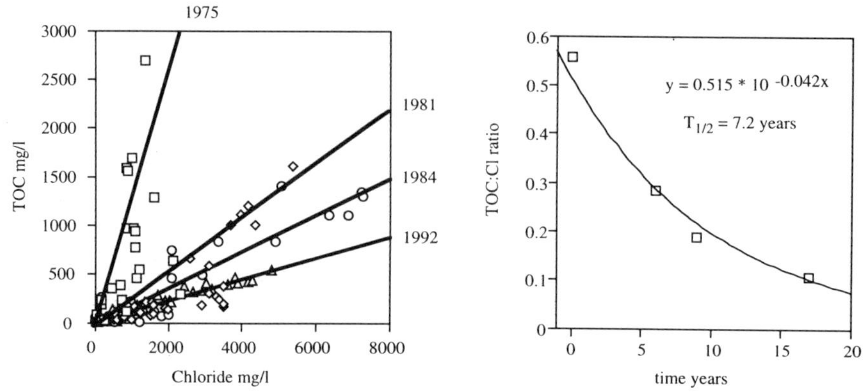

Figure 6. Changes in the ratio of Total Organic Carbon (TOC) to Chloride in groundwater with time.

groundwater is oversaturated with Ni. This may be explained by non-equilibrium, or that other Ni species are present which have not been considered. The analysis of Ni by ICP-AES does not distinguish between soluble species of Ni so it is possible that Ni-organic complexes exist in the heavily contaminated groundwater. In such a complex organic mixture it is not possible to predict Ni- organic complexation because neither the organic composition nor the thermodynamic data for metal complexation are known. Using a direct empirical approach, size exclusion chromatography and ultra-filtration have shown that Ni is in fact associated with relatively high molecular organic compounds (Warwick et al, 1994). Reduced iron is also predicted to be oversaturated in the plume with respect to FeS and $FeCO_3$. By analogy with nickel it is likely that Fe is similarly organically complexed, although his has not been confirmed experimentally.

4. CONCLUSIONS

The long term study of the groundwater pollution plume at Villa Farm has provided evidence for microbially mediated natural attenuation of organic contaminants. Changes in the plume over a period of 25 years has provided an important measurement for TOC attenuation equivalent to a first order rate with a decay constant of -0.096 year^{-1}, or half life of 7.2 years. Detailed sampling through the plume at 20 cm vertical intervals has yielded evidence redox zonation from methanogenic at the base of the aquifer to nitrate reducing conditions at the water table. The redox zonation is complicated by the upward movement of methane which appears to be released from solution as the overall pressure on the aquifer reduces. On moving upwards the methane is oxidised as sulphate is reduced. Support for anaerobic methane oxidation has been provided by fractionation of stable carbon isotopic between methane and carbon dioxide, and the simulation of the process in the laboratory.

The migration of nickel as a soluble organic complex has been determined by size exclusion chromatography and ultrafiltration indicating complexation with fulvic-like material. Geochemical speciation modelling suggests that iron should be precipitated from solution on account of the presence of sulphide and bicarbonate ions. The fact that the plume is oversaturated with respect to pyrite and siderite suggests that iron may also be complexed with organics.

Only by such detailed studies to identify in situ processes and measure reaction rates will it be possible to predict contaminant transport and assess the long term impact of contaminant sources on the environment.

ACKNOWLEDGMENTS

This paper is published by permission of the Director of the British Geological Survey (Natural Environment Research Council).

REFERENCES

Alperin M J and Reeburgh W S, 1984. Geochemical observations supporting anaerobic methane oxidation. In Crawford R L and Hanson R S (eds), Microbial growth on C_1 compounds. American Society for Microbiology. 282–289.

Barber C and Davis G B 1986. The subsurface distribution of methane as an indicator of groundwater polluted by wastes. Proceedings of the Budapest Symposium, July 1986: Monitoring to detect changes in water quality series, IAHS publication no 157.

Barker J F and Fritz P 1981. The occurrence and origin of methane in some groundwater flow systems. Canadian Journal of Earth Science 18, 1802–1816.

Dalton H and Hocknall M D, 1990. Methane oxidation:-is anaerobic methane oxidation possible? Proceedings of Landfill Microbiology R & D Workshop, Harwell. Published by Harwell Laboratories, Oxfordshire, England. ISBN 0-7058-1557-9.

Dalton H and Hocknall M D, 1991. Bacterially-mediated anaerobic methane oxidation—fact or fiction? Landfill Microbiology R & D workshop II. Published by Harwell Laboratories, Oxfordshire, England, ISBN 0-7058-1648-6.

Dhillon H S, Ross C A M, Williams G M, Dart R K, 1989. Aromatic degradation in contaminated cores from the Villa Farm disposal site. U.K. Department of the Environment Report No CWM 005/89.

Higgo, J. J. W., Nielsen, P. H., Harrison, I. Bannon M P and Christensen, T. H, 1996.. The effect of geochemical conditions on the fate of organic compounds in groundwater. Environmental Geology. 27, 335 -346.

Hitchman S. P. and Williams G M, 1988. Characterisation of micro-organisms in a sand aquifer contaminated with industrial wastes. British Geological Survey, Fluid Processes Research Group, Technical Report No. WE/88/15.

Hitchman S P, Williams G M and Ward R S, 1990. Anaerobic methane oxidation in groundwater contaminated with industrial wastes. British Geological Survey, Fluid Processes Research Group, Technical Report No.WE/90/22.

Mackay R, Porter J D, Williams G M, Ross C A M. & Noy D J, 1985. Modelling Mass Transport in the Saturated Zone—A case study. Proc.of the Int. Conference on Water Quality Modelling in the Inland Natural Environment. 10–13 th June, 1985, Bournemouth, England, pp 259–276.

Panganiban A T jnr, Patt T E, Hart W and Hanson R S, 1979. Oxidation of methane in the absence of oxygen in lake water samples. Applied and Environmental Microbiology 37, 303–309.

Stumm W and Morgan J J, 1981. Aquatic chemistry. New York:Wiley-Interscience.

Warwick P, Anderton W, Smith B, and Williams G M, 1994. Nickel speciation in polluted groundwater at Villa Farm. Report of the European Commission EUR 14969 EN.

Williams G M, Ross C A M, Stuart A, Hitchman S P and Alexander L S, 1984. Controls on contaminant migration at the Villa Farm lagoons. Quarterly Journal of Engineering Geology 17 (1) 39–55.

Williams G M, Smith B and Ross C A M, 1992. The migration and degradation of waste organic compounds in groundwater.. Advances in Organic Geochemistry, 1991. Organic Geochemistry, Vol 19 Nos 4–6, pp 531–543.

Wolery, T J, 1983. EQ3NR, a computer program for geochemical aqueous speciation-solubility calculations:users guide and documentation. UCRL- 53414, Lawrence Livermore National Laboratory, Livermore, California.

28

NEW MEMBRANE REACTOR CONCEPT FOR THE BIODEGRADATION OF RECALCITRANT ORGANIC COMPOUNDS

Kurt Peys,[1] Sandra Van Roy,[1] Annemie Ryngaert,[1] Dirk Springael,[1] Carlo Vandecasteele,[2] and Ludo Diels[1,*]

[1]Vlaamse Instelling voor Technologisch Onderzoek (VITO)
Environmental Technology
Boeretang 200, B-2400 Mol, Belgium
[2]University of Louvain
Department of Chemical Engineering
De Croylaan 46, B-3001 Heverlee, Belgium

1. ABSTRACT

Waste water treatments are very often restricted to BOD and nutrient removal. However two problems still arise being heavy metals and toxic organic compounds, resposible for high COD concentrations. The BICMER (Bacteria Immobilized Composite MEmbrane Reactor) stands for the removal and recovery of heavy metals and the complete mineralization of recalcitrant organics. The BICMER functions via specific immolibized bacteria which are responsible for metal removal or biodegradation capacity.

In order to keep cells viable a Membrane Bioreactor was developed. Bacteria are immobilized on a membrane. The membrane separates a nutrients stream from a wastewater stream. The nutrient is necessary to keep the bacteria viable to keep enough biomass (in case of low substrate concentrations), or to add cosubstrate (in case of a need for cometabolism). The bacteria are immobilized in a biofilm on the membrane by a filtration process. Via feeding with nutrients at one membrane side a biofilm is formed at the other side. Nutrient flow to the biofilm is limited by a diffusion process through the membrane and allows the biofilm to be active without excessive growth of the microorganisms. The nutrient depends on the composition of the waste water. Mostly the nutrient will be composed of a nitrogen, a phosphorous and a carbon source.

* Author to which correspondence should be sent. Tel. 32 14 33 51 00; Fax. 32 14 32 03 72; Email dielsl@vito.be

Novel Approaches for Bioremediation of Organic Pollution, edited by Fass *et al.* Kluwer Academic / Plenum Publishers, New York, 1999.

The new reactor concept has the following advantages: low sludge production; specific microorganisms can be used under specific conditions; use of cometabolism; possibility to concentrate the pollutant on the membrane before degradation; low contamination of waste water with bacteria and nutrient; high resistance to shockloadings for toxic compounds; biofilm can survice at very low organic loads; presence of aerobic and anaerobic zones.

The apllicability will be demonstrated by examples on 3-chlorobenzoate, TCE (cometabolism with phenol), PAH and metal cyanides biodegradation. Upscaling will be shown by the treatment of real waste waters as groundwater (polluted by organic solvents) and landfill leachates.

The reactor will offer a possibility to keep specific strains alive by keeping them under optimal and adapted conditions while such strains will not survive in conventional biotreatment systems. Under such conditions it will even be possible to use genetically engineered micro-organisms in this contained system. The reactor concept the possibility to inoculate it with specific bacteria to develop an ideal adapted biofilm. The addition of specific bacteria will improve the biofilm adaptation rate. The concept is also very useful to study biofilm development and events like gene-transfer in biofilms. Results on gene transfer in these biofilms will be shown by the use of REP-PCR, hybridisations and Biolog identifications.

2. INTRODUCTION

Many physico-chemical methods exist for treatment of waste waters contaminated with inorganics (e.g. heavy metals) or organics (e.g. chlorinated aromatics). Most of these methods are based on separation techniques which create smaller streams of highly contaminated pollutants.

During the last twenty years, many bacteria with environmentally beneficial degradation properties have been isolated and investigated. Some bacteria can remove heavy metals from waste water by adsorption (T. Jeffers 1992) or via the induction of metal salt crystallization processes (L. Diels 1989, L. Macaskie 1990). Several bacteria are able to degrade recalcitrant and toxic organic compounds to carbon dioxide and water (P. Chapman 1988). A lot of waste water treatment applications make use of biofilms as they result in several advantages when compared with planctonic cell systems. The successful use of these microorganisms for treatment of waste waters requires the development of safe, well-controlled and economically beneficial reactor configurations. Special attention is paid to hybrid systems wherein microbial degradation is combined with a membrane as a separating device and/or as a contactor for biofilm formation. Therefore a special reactor system was desinged in order to keep high cell densities (immobilized cells) with a high activity (special provision of necessary nutrients). Furtheron problems like low substrate concentrations and the need for cometabolism can be overcome.

The reactor separates the waste water stream to be treated from a small nutrient stream for bacterial maintenance. The nutrient stream can cause and maintain the bacteria induced for bioprecipitation or biodegradation. The two streams are separated by a membrane on which the bacteria are immobilized. As such all bacterial activity is concentrated in the biofilm and the sludge production is limited. This concept is used in Flat Membrane Biofilm Reactors (FMBR) which are used as test systems for bacteria and membranes and in Tubular Membrane Batch Biofilm Reactors (TMBBR) which are used in a continuous treatment and for scale-up (L. Diels 1993a, L. Diels 1993b, L. Diels 1995).

This membrane reactor is an ideal system to study biofilm formation and the genetic events (e.g. gene transfer) occuring in biofilm layers.

3. MATERIALS AND METHODS

3.1. Bacteria

During the last years, bacteria resistant to a variety of heavy metals were isolated and identified (S. Silver 1992). The mechanisms for these resistances are often controlled by plasmid encoded genes or by transposons. A remarkable example of those resistant bacteria is *Ralstonia eutropha* var. *metallotolerans.* The representative strain CH34 was isolated in sediments from a decantation basin of a zinc factory (M. Mergeay 1978). Strain CH34 bears two large plasmids (M. Mergeay 1985, D. Nies 1987) controlling resistance against Cd^{2+}, Co^{2+}, Zn^{2+}, Hg^{2+}, Tl^{+}, Cu^{2+}, Pb^{2+} (pMOL30, 240 kb) and Co^{2+}, Zn^{2+}, Ni^{2+}, Hg^{2+}, CrO_4^{2-}, Tl^{+} (pMOL28, 165 kb). As shown by (A. Nies 1989; A. Nies 1990) resistance to chromate is inducible and based on decreased net accumulation of the metal anion. Resistance to zinc, cadmium, cobalt and nickel are resulting from inducible, energy dependent cation efflux systems (D. Nies 1989a, D. Nies 1989b, R. Siddiqui 1989, C. Sensfuss 1988, J. Collard 1993, H. Liesegang 1993). At increased concentrations of Cd or Zn ions, a removal of these metals from solution is observed during the late log phase and the stationary phase. This accumulation and precipitation is correlated with the concentration and kind of carbon source (lactate or acetate) with the progressive alkalinization of the medium, and also appears to be associated with some outer cell membrane proteins. The alkalinization of the periplasmic space and the bacterial surface is due to the metal efflux system which works together with a proton antiport system continuously generating OH^- ions in the periplasm. The CO_2 produced by the carbon source catabolism is transformed in this alkaline environment into HCO_3^- and CO_3^{2-}. Also metals can bind to functional groups of the periplasmic peptidoglycan and to the outer cell wall polysaccharides and proteins. Special metal induced proteins could be isolated. The metal ions, exported from the cytoplasm, will form metal-bicarbonates and carbonates around the cells. These compounds are in a supersaturated concentration around the cell surface and will crystallize on the cellular bound metal ions, serving as a crystallization focus. In this way a crystallization process is induced on the cells and this will lead to very high metal to biomass ratios (between 0.5 and 5.0 on a weight basis). This trapping could occur at the uptake or the efflux of the metals or more likely during both processes and suggests a kind of biomineralization as reviewed by S. Mann (1992). Once the bioprecipitation process has reached a certain level, nucleation proteins and polysaccharides are released from the cells and bioprecipitation continues on these released foci. At the cell surface new proteins are synthesized and the nucleation can start again. This process is very similar to the $CaCO_3$ formation at the cyanobacteria *Synechococcus* as presented by S. Schultze-Lam (1992).

Metals like Cd, can precipitate at the cell surface and form a kind of crust around the cells. Further on, metals can start to crystallize on this primary metal crust. The bioprecipitation starts with the formation of an amorphous structure followed by a crystallization process.

In order to degrade recalcitrant organics other *A. eutrophus* strains can be used. As a model compound the chloroaromatic 3-chlorobenzoate was used. *A. eutrophus* AE1308 is a transconjugant of the mating between CH34 and the 3-chlorobenzoate degrading *A. eutrophus* JMP134 (E.J. Perkins 1990). This strain (B. Friedrich 1983) bears the same heavy metal resistance genes as CH34 and is able to mineralize 3-chlorobenzoate (CBA) and 2,4-dichlorophenoxy acetic acid (Tfd). In a comparative study with the 3CBA biodegrading *Pseudomonas putida* BN210 it could be concluded that the biodegradation was much faster in the case of the *Pseudomonas* strain.

Ni-lactate is biodegraded by *Alcaligenes xylosoxidans* N31A (Siddiqui et al. 1989) a very highly Ni resistant strain.

TCE (trichloroethylene) biodegradation was done by *Pseudomonas cepacia* G4. The monooxygenase of this strain is induced by toluene or phenol. In the membrane reactor phenol was added at the nutrient side.

The PAH-biodegradation experiments were done with a mixture of strains. The following strains were used. The PAHs on which they can grow are indicated between brackets. *Spinghomonas sp.* LB126 (Fluor), *Mycobacterium gilvum* LB307T (Biph, Phen, Flu), *Mycobacterium sp.* LB501T, *Pseudomonas corrugata* LH117 (Naph), *Pseudomonas putida* LH136 (Naph), *Pseudomonas putida* LH144 (Naph), *Mycobacterium gilvum* LH208 (Phen, Flu, Pyr).

The cyanide biodegradation experiments were done with *Ralstonia eutropha* CH34, *Pseudomonas fluorescens* NCIMB11764 (Harris et al. 1987), and *Pseudomonas fluorescens* B-5040 (Shpak et al. 1995).

3.2. Membrane Reactor

In order to keep cells viable a Membrane Reactor was developed. Bacteria are immobilized in an organo-mineral membrane (Zirfon®), based on polysulfone and ZrO_2 or Al_2O_3. The membrane is casted on a tubular or flat polyester support and separates a nutrient stream from a waste water stream. The nutrient is necessary to keep the bacteria viable to produce continuously metal ligands (in case of heavy metal removal), to keep enough biomass (in case of low substrate concentrations), or to add cosubstrate (in case of a need for cometabolism). The bacteria are immobilized in and on the membrane by a filtration process. Via feeding with nutrients at one membrane side a biofilm is formed at the other side. Nutrient flow to the biofilm is limited by a diffusion process through the membrane and allows the biofilm to be active without excessive growth of the microorganisms (Figures 1 and 2). The nutrient depends on the composition of the waste water. Mostly the nutrient will be composed of a nitrogen, a phosphorous and a carbon source. The general concept is called a BICMER (Bacteria Immobilized Composite MEmbrane Reactor).

3.3. BICMER

The BICMER reactor is based on waste water treatment via a biofilm. The biofilm grows on an organo-mineral membrane, developed by VITO under the name of Zirfon®. The biofilm growth is kept stable by a nutrient stream which flows tangentially at the other side of the membrane and the nutrient compounds are slowly released to the biofilm via diffusion through the membrane.

In that way nutrient and waste water are not mixed, but just separated by the membrane which forms at the same time the support on which the biofilm develops. The concept allows to adapt the nutrient composition to the need of the biofilm bacteria in function of the waste water composition. The biofilm bacteria are chosen in function of the pollutants.

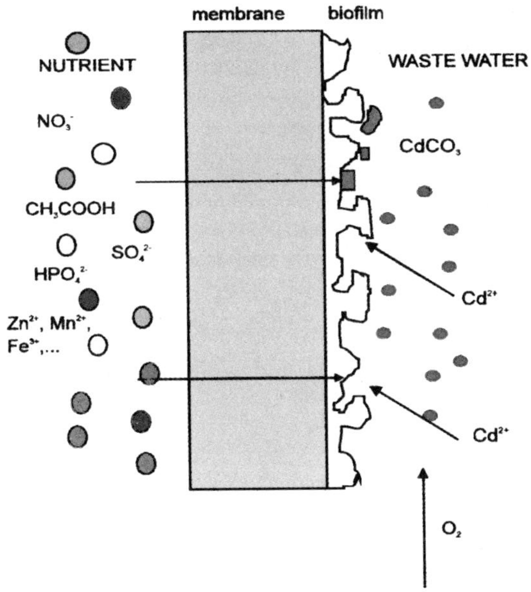

Figure 1. Biofilm reactions with heavy metals.

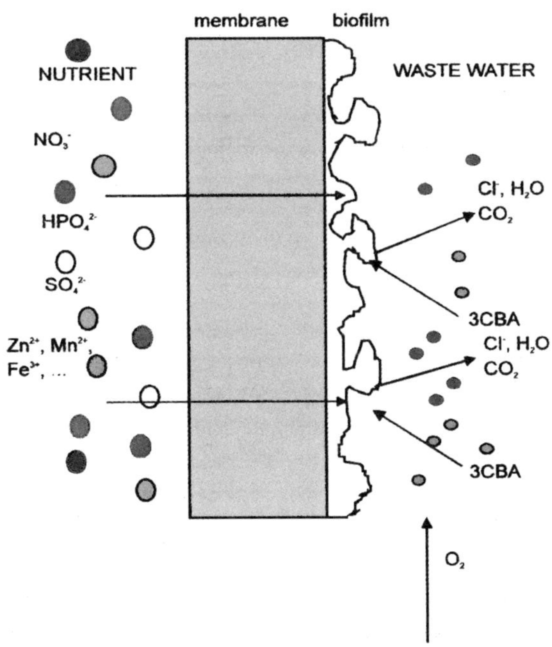

Figure 2. Biofilm reactions with chlorinated aromatics.

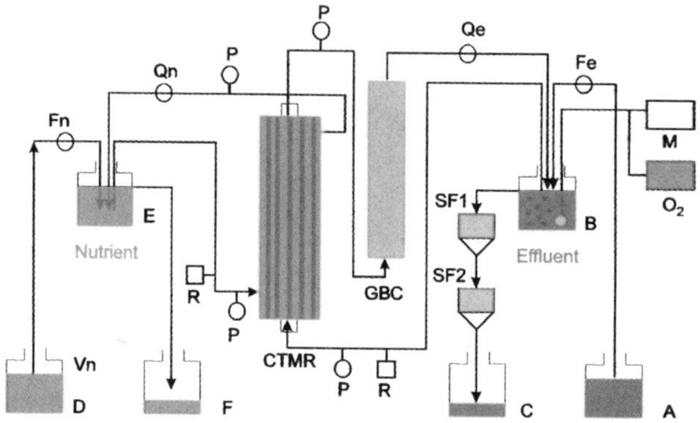

Figure 3. BICMER concept for heavy metal removal. A: effluent stock; B: effluent conditioning; C: effluent output; D: nutrient stock; E: nutrient conditioning; F: nutrient output; TMR: membrane module; M: monitor for pH, O_2 and temperature; O: O_2 provision; P: manometer; R: pressure regulation; Fe and Qe: effluent pumps; Fn and Qn: nutrient pumps; GBC: glass beads column.

4. RESULTS AND DISSCUSSION

4.1. Biofilm

An other important research topic is the study of the biofilm structure. SEM (Figure 4) shows that the biofilm consists of bacterial microcolonies when 3CBA is used as a carbon source. With gluconate a homogenous biofilm develops. This is in agreement with the observations of J. Costerton (Center for Biofilm Engineering, Bozeman). Future research will focus on the distribution of active bacteria in this cluster structure. By means of confocal laser scanning microscopy (CLSM), the origin and structure of the bacterial microcolonies will be studied, as well as the modelling of their growth. Figure 5 shows a CSLM photo of a micro-colony of *P. putida* BN210 grown on 3CBA. A core of bacteria surrounded by a polysaccharide slime layer could be observed.

4.2. Ni-Lactate Removal

The use of the BICMER for removal of metal anions via biosorption and bioprecipitation processes was already presented by Diels et al. (1995). However sometimes metals are complexed with organic molecules. In this configuration it is very difficult to remove

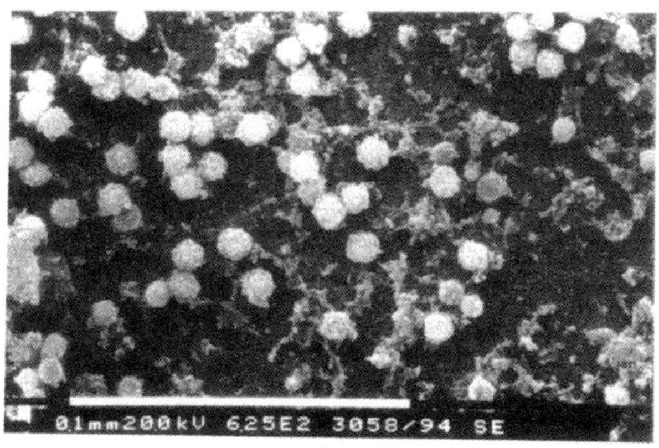

Figure 4. Growth of bacterial microcolonies in a membrane.

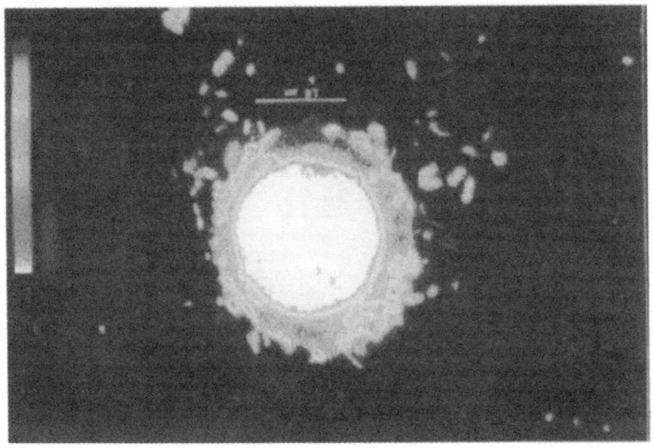

Figure 5. CSLM of a micro-colony from a membrane bioreactor.

them from a solution also by physico-chemical methods. An example is given vfor the removal of Ni-lactate complexes. Ni-lactate is used in the surface treatment industry and indicated as chemical nickel. The removal from waste water is very difficult due to the complexation of Ni to the lactate. Therefore a proces is developed by using a bacteria which can biodegrade lactate and which is resistant against the released Ni. The *Alcaligenes xylosoxidans* N31A degraded the lactate and bioprecipitated the Ni on the biomass. The results of a batch experiment are presented in Table 1.

4.3. Metal Cyanide Complexes

Cyanides and metal-cyanides can be converted to NH_3 and CO_2 by an enzyme system with the overall properties of a dioxygenase:

$$HCN + O_2 + NADH + H^+ \rightarrow NH_3 + CO_2 + NAD^+$$

Cyanide forms complexes with high-stability constants with a wide variety of metals. Many bacteria, including several *Pseudomonas* species, are resistant to metals. Growth of the cyanide-utilising *P. Fluorescens* on metal cyanides as a source of nitrogen was attempted.

Table 1. Ni and lactate concentrations in a membrane bioreactor in function of time

Time (hours)	Ni (mg/l)	Lactate (g/l)
0	14.8	8.4
19	14.3	8.0
46	14.5	8.0
90	12.7	2.0
114	8.3	0
138	3.1	0
162	2.2	0

Table 2. Biodegradation of KCN in a flat sheet reactor

	Input CN^- (mg/l)	Output CN^- (mg/l)
Batch 1		
Control	50	28
CH34	50	4
Batch 2		
Control	50	24
CH34	50	10
Batch 3		
Control	50	18
CH34	50	11

Experiments were also started to study the degradation of (metal)-cyanides in a flat sheet biofilm reactor. For these experiments strain CH34 was chosen, since the behaviour of this strain in the reactor is well known. using a minimal nutrient solution depleted from N as the nutrient wource, and a synthetic waste water containing 50 mg/l KCN, 95 % of the CN^- was degraded after 24 hours (Table 2).

Three batch experiments of 24 hours were done in the reactor. The control reactor was a reactor with no inoculated bacteria. However, the degradation went down after the treatment of succesive batches. This is probably due to a too strict N-limitation, resulting in a bad condition or even death of the biofilm. Therefore, the conditions for CN^- degbra-dation were optimized, not only for strain CH34 but also for the *Pseudomonas* strains B-5040 and NCIMB11764.

Degradation of $Na_3Cu(CN)_2$ and $NaAg(CN)_2$ was the best with the strains B-5040, NCIMB11764. In a treatment of a $NaAg(CN)_2$ containing solution cyanides were degraded and Ag removed by biosorption.

4.4. TCE-Biodegradation

Around 20 ppm of phenol were added in the nutrients solution and a groundwater containing 300 μgTCE/l was treated. The input concentration of TCE was about 300 μg/l. The output concentration was below 2 μgTCE/l. No TCE could be detected in the nutrient solution (could occur by diffusion through the membrane) and no phenol was measured in the nutrient or effluent solution. This indicates that the added phenol did not contaminate the waste water soluition indicating one of the big advantages of the membrane bioreactor. In a non-sterile control reactor some TCE removal was observed but no chloride formation could be measured indicating that no complete biodegradation did occur.

Table 3. Continuous treatment of TCE containing groundwater with a BICMER

Time (hours)	TCE input (μg/l)	TCE output (μg/l)
0	291	3.3
72	333	2.4
168	300	0
554	155	0

The reactor was inoculated with *Pseudomonas cepaciae* G4 and induced with phenol.

Table 4. PAH concentrations at the input and output of the membrane bioreactor

PAH	Input (μg/l)	Output (μg/l)
Naphtalene	2130	12.4
Acenaphthylene	53	1.4
Acenaphthene	93.1	2.7
Fluorene	131	3.5
Phenanthrene	273	8.7
Anthracene	102	3.3
Fluoranthene	109	3.6
Pyrene	82.6	2.7
Benzoanthracene	46.8	1.5
Chrysene	40.9	1.3
Benzo(b)fluoranthene	27.9	0.9
Benzo(k)fluoranthene	24.7	0.7
Benzo(a)pyrene	34.8	1.1
Indeno(1,2,3-c,d)pyrene	21.8	0.6
Dibenzo(a,h)anthracene	5.6	0.2
Benzo(g,h,i)perylene	16.8	0.5

4.5. PAH Biodegradation

As mentioned before several PAH degrading strains were used for the inoculation of a membrane bioreactor. Table 4 indicates the input and output of the PAH molecules at one time.

As can be concluded several of the PAH molecules are cometabolised by the bacteria inoculated on the membrane of the bioreactor. This indicates that a biofilm of specific bacteria is able to degrade a whole series of compounds either by direct biodegradation, either by co-metabolism.

4.6. Gene Transfer

The Flat Membrane Batch Biofilm Reactors are always inoculated with one bacterial species (or in some cases by a mixture of pure cultures). Although the reactors are operated in a semi sterile way, after some time a large heterogenous bacterial population is observed.

Strain *Pseudomonas putida* BN210, a 3-chlorobiphenyl/3-chlorobenzoate mineralizing *in vivo* constructed hybrid strain carrying a natural mobile element involved in degradation of 3-chlorobenzoate, was used as the inoculum in different membrane biofilm reactor configurations in which degradation of 3-chlorobenzate was modelled. Analysis of the bacterial population of the effluent and the formed biofilm lead to the recovery of 3-chlorobenzoate mineralizing bacteria which were clearly different from the initial inoculum. REP-PCR analysis and phenotypic analysis of the isolates showed the occurence of at least 2 different species, i.e., represented by strains KP9 and KP3, which both belonged to the genus *Ralstonia*. The *clc* genes and flanking DNA were cloned in overlapping recombinant cosmids and probes were constructed containing sequences from both inside and outside the *clc* catabolic gene cluster. DNA-DNA hybridizations of Southern blots of restricted total genomic DNA extracts of the isolates using these probes unambigeously demonstrated the presence of the *clc* genes and the flanking sequences in different copies in the chromosome of the isolates. Isolates KP3 and KP9 were able to transfer the *clc* genes to *Ralstonia eutropha* recipients in plate matings. These observations strongly suggest *in situ* horizontal transfer of the *clc*-genes from the initial inoculum into contaminant bacteria which had invaded the reactor. KP9 and KP3 demonstrated a higher maximal growth rate in liquid 3-chlorobenzoate minimal medium batch cultures at both 20°C and 30°C. A competition experiment was performed in which strains BN210, KP3 and KP9 were simultaneously inoculated at equal numbers in a continuous membrane reactor supplied with 3-chlorobenzoate. Analysis of the population in the nutrient and waste stream during the experiment and of the biofilm on the membrane after the experiment demonstrated that KP3 easily outcompeted the 2 other strains. Moreover, a new 3-chlorobenzoate mineralizing strain was recovered. Other reactors were set-up at different time intervals and at different locations using BN210 as the inoculum. In all cases, strain BN210 was lost from the reactor and replaced by other 3-chlorobenzoate mineralizing strains. These isolates were different from strains KP3 and KP9, belonged to the genera *Ralstonia*, *Burkholderia* or *Pseudomonas putida* and contained all the *clc*-element.

In this way it seems that the biodegradation pathway is transferred from BN210 to the contaminating bacteria which are overgrowing BN210 in the biofilm. Probably these bacteria are overgrowing because they adhere or survive better in the biofilm.

These population dynamics are a very important but complex phenomenon occuring all the time in biological waste water treatment systems. In biofilm systems genetic exchange is even improved by the high biomass densities and certain kinds of environmental stress.

In an other experiment we tried to transfer the 3CBA pathway to biphenyl degrading bacteria. A complete biodegradation of 3-chlorobiphenyl could be obtained. And again a transfer of the 3CBA degradation genes to other biofilm bacteria was observed. Until now no transfer of the 3CBA pathway to the biphenyl degrading bacteria could be observed.

5. CONCLUSIONS

The results presented show that the BICMER concept is useful for the removal or recovery of several heavy metals (Cd, Cu, Zn, Co, Ni, Pb, Y, Ge) from waste waters or process streams. The presented system reverts soluble metal ions immediately into insoluble metal carbonate crystals which allows direct revovery of the metals. Such a system can be of interest for the recovery of precious metals. Also other bioprecipitating bacteria can be used to recover specific metals. Also recalcitrant organic compounds can be completely mineralized into CO_2, H_2O and Cl^-. These membrane bioreactors offer several possibilities for treatments of very specific waste streams. It will not be possible to use them as a final end of pipe treatment, but they can be immobilized with specific bacteria especially selected or constructed for the given specific application.

The reactor will offer a possibility to keep such strains alive by keeping them under optimal and adapted conditions while such strains will not survive in conventional biotreatment systems. Under such conditions it will even be possible to use genetically engineered micro-organisms in this contained system. The reactor concept offers the possibility to inoculate it with specific bacteria to develop an ideal adapted biofilm bacteria. In this way a biofilm specially adapted to the waste water will develop. This will go faster than without the addition of specific bacteria. The addition of specific bacteria will improve the biofilm adaptation rate. The concept is also very useful to study biofilm development and events like gene-transfer in biofilms.

The biofilm is protected against toxic compounds and can perform its activity, being bioprecipitation of metals or biodegradation of organics.

This opens also special applications for treatment of waste waters containing heavy metals complexed with the molecules like cyanides, organics, NTA, EDTA, etc.

The new reactor concept has the following advantages:

- low sludge production
- specific microorganisms can be used under specific conditions
- use of cometabolism
- possibility to concentrate the pollutant before degradation
- low contamination of waste water with bacteria and nutrient
- high resistance to shockloadings fo toxic compounds
- biofilm can survive at very low organic loads
- presence of aerobic and anaerobic zones

ACKNOWLEDGMENT

Part of this work was sponsored by the Thüringer Ministerium für Wissenschaft, Forschung und Kultur (409-97016) in Erfurt, Germany.

REFERENCES

Chapman, P.J. 1988. Constructing microbial strains for degradation of halogenated aromatic hydrocarbons. In G.S. Omenn (ed.), Environmental biotechnology: reducing risks from environmental chemicals through biotechnology, Basic Life Sciences, vol. 45, New York Plenum Press, pp. 81–95.

Collard, J.M., Provoost, A., Taghavi, S., and Mergeay, M. (1993). A new type of *Alcaligenes eutrophus* CH34 zinc resistance generated by mutations affecting regulation of the *cnr* cobalt-nickel resistance sysstem. *J. Bacteriol.* **175**:779–784.

Diels, L., Qinghan, D., van der Lelie, D., Baeyens, W., and Mergeay, M. (1995). The *czc* operon of *Alcaligenes eutrophus* CH34: from resistance mechanism to the removal of heavy metals. *J. industrial Microbiology* **14**:142–153.

Diels, L., Sadouk, A., and Mergeay, M. (1989). Large plasmids governing multiple resistances to heavy metals: a genetic approach. *Toxic.Environm. Chem.* **23**:79.

Diels, L. 1990. Accumulation and precipitation of Cd and Zn ions by *Alcaligenes eutrophus* strains. Biohydrometallurgy, Canmet SP89–10. J. Salley, R.G.L. McCready and P.L. Wallace (eds.), 369–377.

Diels, L., Van Roy, S., Mergeay, M., Doyen, W., Taghavi, S., and Leysen, R. 1993a. Immobilization of bacteria in composite membranes and development of tubular membrane reactors for heavy metal recuperation. In R. Paterson (ed.), Effective Membrane Processes: New Perspectives, Kluwer, Dordrecht 275–293.

Diels, L., Van Roy, S., Taghavi, S., Doyen, W., Leysen, R., and Mergeay, M. 1993b. The use of *Alcaligenes eutrophus* immobilized in a tubular membrane reactor for heavy metal recuperation. In A.Z. Torma, M.L. Apel and C.L. Brierley (eds.), Biohydrometallurgical Technologies. Jackson Hole, The Minerals, Metals and Materials Society, 133–144

Diels, L., Van Roy, S., Somers, K., Willems, I., Doyen, W., Mergeay, M., Springael, D., and Leysen, R. (1995). The use of bacteria immobilized in tubular membrane reactors for heavy metal recovery and degradation of chlorinated aromatics. *J. Membrane Science* **100**:249–258.

Friedrich, B., Meyer, M., and Schlegel, H.G. (1983). Transfer and expression of the herbicide degrading plasmid pJP4 in aerobic autotrophic bacteria. *Arch. Microbiol.* **134**:92.

Harris, R.E., Busch, A.W., and Knowles, C.J. (1987). Microbial cyanide and nitrile metabolism. Sci. ProJ. Oxf. **71**:293–304.

Jeffers, T.H., Ferguson, C.R., and Bennett, P.G. (1992). Biosorption of metal contaminants using immobilized biomass. A laboratory study, Bureau of Mines, RI9340

Liesegang, H., Lemke, K., Siddiqui, R.A., and Schlegel, H.G. (1993). Characterization of the inducible nickel and cobalt resistance determinant *cnr* from pMOL28 of *Alcaligenes eutrophus* CH34. *J. Bacteriol.* **175**:767–778.

Mann, J.S. (1988). Molecular recognition in biomineralization. *Nature* **332**:119.

Mergeay, M., Houba, C., Gerits, J. (1978). Extrachromosomal inheritance controlling resistance to cadmium, cobalt and zinc ions: evidence from curing in a Pseudomonas. *Arch. Int. Physiol. Biochim.* **86**:440–441.

Mergeay, M., Nies, D., Schlegel, H.G., Gerits, J., Charles, P., Van Gijsegem, F. (1985). *Alcaligenes eutrophus* CH34 is a facultative chemolithotroph with plasmid-bound resistance to heavy metals. *J. Bacteriol.* **162**:328–334.

Nies, D., Mergeay, M., Friedrich, B., and Schlegel, H.G. (1987). Cloning of plasmid genes encoding resistance to cadmium, zinc and cobalt in *Alcaligenes eutrophus* CH34. *J. Bacteriol.* **169**:4865–4868.

Nies, A., Nies, D., and Silver, S. (1989a). Cloning and expression of plasmid genes encoding resistance to chromate and cobalt in *Alcaligenes eutrophus*. *J. Bacteriol.* **171**:5065–5070.

Nies, D.H., Nies, A., Chu, L., Silver, S. (1989). Expression and nucleotide sequence of a plasmid-determined divalent cation efflux system from *Alcaligenes eutrophus*. *Proc. Natl. Acad. Sci. USA* **86**:7351–7355.

Nies, D., and Silver, S. (1989b). Plasmid-determined inducible efflux is responsible for resistance to cadmium, zinc and cobalt in *Alcaligenes eutrophus*. *J. Bacteriol.* **171**:4073.

Nies, A., Nies, D., and Silver, S. (1990). Nucleotide sequence and expression of a plasmid-encoded chromate resistance determinant from *Alcaligenes eutrophus*. *J. Biol. Chem.* **265**:5648.

Perkins, E.J., Gordon, M.P., Caceres, O., and Lurquin, P. (1990). Organisation and sequence analysis of the 2,4-dichlorophenol and dichlorocatechol oxidative operons of plasmid JP4. *J. Bacteriol.* **172**:2351.

Schultze-Lam, S., Harauz, G., and Beveridge, T.S. (1992). Participation of cyanobacterial S layer in fine grain mineral formation. *J. Bacteriol.* **174**:7971–7981.

Senfuss, C., and Schlegel, H.G. (1988). Plasmid pMOL28-encoded resistance to nickel is due to specific efflux. *FEMS Microbiol. Lett.* **55**:295.

Siddiqui, R.A., Benthin, K., Schlegel, H.G. (1989). Cloning of pMOL28 encoded nickel resistance genes and expression of the genes in *Alcaligenes eutrophus* and *Pseudomonas* spp. *J. Bacteriol.* **171**:5071–5078.

Silver, S., and Walderhaug, M. (1992). Gene regulation of plasmid- and chromosome-determined inorganic ion transport in bacteria. *Micriobiol. Rev.* **56**:195–228.

Shpak, V.E., Podolskaya, V.I., Ulberg, Z.R., and Shpak, E.A. (1995). Degradation of metal cyanide complexes in microbe dispersions. *Colloid Journal* **57**:102–105.

Zeyer, J., Wasserfallen, A., and Timmis, K.N. (1985). Microbial mineralization of ring-substituted anilines through an ortho-cleavage pathway. *Applied and Environmental Microbiology* **50**:447–453.

INDEX